Earth and Environmental Sciences Library

The *Earth and Environmental Sciences Library* (EESL) is a multidisciplinary book series that showcases innovative approaches and in-depth reviews, aimed at enhancing the role of the Earth and Environmental Sciences communities. The series also offers valuable guidance for stakeholders, decision-makers, policymakers, international organizations, and NGOs.

Topics of interest include oceanography, the marine environment, atmospheric sciences, hydrology, soil sciences, geophysics, geology, agriculture, environmental pollution, remote sensing, climate change, water resources, and natural resource management.

To further advance these critical areas, the *Earth and Environmental Sciences Library* invites contributions from the Earth and Environmental Sciences communities in the form of edited volumes, monographs, textbooks, conference proceedings, and more. The series fosters knowledge exchange and collaboration across disciplines, encouraging comprehensive discussions and solutions to pressing environmental challenges.

Manoj Kumar Jindal • Kiran Patni
Editors

Water Quality-Uranium Solubility Nexus: Unlocking the Secrets of Uranium Solubility

Uranium Chemistry in Water

Springer

Editors
Manoj Kumar Jindal
Indian Institute of Technology Jodhpur
Jodhpur, Rajasthan, India

Kiran Patni
Bhimtal Campus
Graphic Era Hill University
Nainital, Uttarakhand, India

ISSN 2730-6674 ISSN 2730-6682 (electronic)
Earth and Environmental Sciences Library
ISBN 978-3-031-98908-7 ISBN 978-3-031-98909-4 (eBook)
https://doi.org/10.1007/978-3-031-98909-4

This Springer imprint is published by the registered company Springer Nature Switzerland AG
The registered company address is: Gewerbestrasse 11, 6330 Cham, Switzerland

Preface

The *water quality-uranium solubility nexus* is a complex issue that has recently garnered significant attention within both the scientific and governmental communities, particularly among those focused on the drinking water supply and the commitment to providing safe water for populations. Uranium solubility, however, remains an enigma that is intricately linked to the complex interplay between water chemistry and the multifaceted interactions of uranium with water. This nexus has become a growing concern for several reasons, all of which are explored in this book. This book provides a holistic understanding of the correlation between soluble uranium and water quality parameters. It provides detailed information on the various factors influencing uranium solubility in water, integrating both natural and anthropogenic influences. This book systematically addresses the properties and influences shaping uranium solubility while also outlining future research directions to better understand and manage this issue.

Serving as an essential resource, this book is invaluable for education, research, and industrial applications. This work consolidates critical information about the impact of uranium solubility, making it a key reference for government initiatives, research groups, policymakers, and stakeholders involved in water quality management and environmental safety. This book also provides valuable insights into assessing uranium solubility while exploring a platform for future research paths in this crucial area. The intricate relationship between uranium solubility and water quality has been thoroughly explored, shedding light on the complex factors governing uranium behavior across diverse water systems. It emphasizes how specific conditions facilitate uranium solubility in water and paint a vivid picture of the global uranium landscape, highlighting the diverse patterns of uranium occurrence in aquatic environments. This book serves as a comprehensive treatise on uranium solubility and its distribution, providing a critical foundation for addressing the challenges uranium poses to water safety and environmental health.

This timely work is not only a valuable resource for the scientific community but also for policymakers and industries grappling with the implications of uranium in water. It enhances our understanding of the challenges and solutions related to uranium solubility and serves as an indispensable reference for anyone working in environmental science, water quality management, and radioactive contamination. This book is designed for researchers, environmental professionals, policymakers, and students and provides a well-rounded perspective on the critical issue of uranium contamination in water. Equipped with knowledge, insights, and innovative approaches, readers will gain a deeper understanding of these challenges. We hope that this book inspires action toward safeguarding water resources for future generations. The increasing levels of uranium in various water systems, particularly in groundwater, highlight the urgency of understanding, monitoring, and addressing uranium contamination. This book highlights potential pathways to address this critical issue.

Jodhpur, Rajasthan, India Manoj Kumar Jindal
Nainital, Uttarakhand, India Kiran Patni

About This Book

This book provides a holistic understanding of the correlation between soluble uranium and water quality parameters, presenting the intricate dynamics that govern uranium solubility in diverse water systems. Uranium, one of the most prevalent and significant radioactive isotopes, has gained considerable attention due to its complex behavior in aquatic environments. Despite its importance, the solubility of uranium remains a critical enigma, intricately tied to the dynamic interplay between water chemistry and uranium interactions with water. The *Water Quality–Uranium Solubility Nexus* addresses this concern, providing a comprehensive exploration of the factors influencing uranium solubility in water, integrating both natural and anthropogenic influences. It consolidates essential information in one place, making it an invaluable resource for researchers, policymakers, education, and industries involved in environmental science and water management. Through a systematic and forward-looking approach, this book highlights key research directions and provides a critical foundation for future studies. This book serves as a crucial reference for understanding and mitigating the challenges associated with uranium contamination in water, benefiting scientists, environmental experts, and anyone concerned with water quality and safety.

Contents

Background

Understanding Uranium and Its Ionic Forms 3
Xiaonan Ruan, Longfang Ren, Bing Shi, Huaqing Liu, and Taotao Qiang

Water Quality and Its Diverse Source 11
Suvarshitha Pusuluru, Sai Kumar Punna, Karrun Velmurugan, Selvarajan Ethiraj, and Melvin S. Samuel

The Crucial Link: Why Water Quality and Uranium Solubility Matter 21
Kamyar Yaghmaeian, Mohammad Rezvani Ghalhari, Niloufar Borhani Yazdi, Farshad Hamidi, Aysan Morovvati Zarajabad, Mahla Mazloumian, Narges Ebrahimi, and Hossein Arfaeinia

Water Quality Parameters Influencing Uranium Solubility

Impact of pH on Uranium Solubility 43
Manoj Kumar Jindal and Kiran Patni

Temperature Influence on Uranium Solubility 55
Benjamin Uchechukwu Nwaka, Chukwuemeka Ehirim, I. Fabian Ezema, Emmanuel Obinna Ogueri, and K. Sudeep Kumara

Dissolved Oxygen and Its Role in Uranium Solubility 75
Brian Youden, Andrew Carrier, Ken Oakes, Mark Servos, Runqing Jiang, and Xu Zhang

Redox Potential: A Key Player in Uranium Fate 91
Yvon Ukwishaka, Kashyap Komal, Ayush Kumar Singh, Manish Kumar, Biswa Mohan Sahoo, Pardeep Singh, and Preetismita Borah

Effect of Salinity on Uranium Solubility 107
Divya Patel

Organic Matter: Friend or Foe to Uranium Solubility 115
Rashmi Raghav, Kamal Kant Tiwari, and Pankaj Kandwal

Impact of Physicochemical Parameters on Uranium Mobility and Solubility in Water 127
Vikram Singh, Manoj Kumar Jindal, and R. S. Aswal

An Overview on the Effect of Heavy Metals, Microorganisms, and Radioactive Elements on Uranium Solubility 141
Mohammad Hassan Ehrampoush, Hossein Arfaeinia, Farshad Hamidi, Danial Nayeri, Fatemeh Mortezazadeh, Mahdi Ahmadi Nasab, and Niloufar Borhani Yazdi

Unveiling the Chemistry Behind Uranium Solubility and Interfering Agents 161
C. G. Poojitha, P. Manohar, B. S. Rohini, Bhupender Singh, and A. Pyngrope

Uranium Solubility in Various Water Systems

Exploring Uranium Solubility in Freshwaters (Groundwater & Surface Water) 187
Sai Kumar Punna, Suvarshitha Pusuluru, Karrun Velmurugan, Selvarajan Ethiraj, and Melvin Samuel

Unveiling Uranium Behavior in Polluted Water: Seawater, Wastewater, and Industrial Water 199
Bhupender Singh, Krishan Kant, Nandita Kanwar, Naresh Tanwer, Rajesh Kumar, Sakshi Yadav, and Kamlesh Kumar Sharma

Water Chemistry Influence on Uranium Solubility Across Diverse Water Systems: A Comparative Study 219
Benjamin Uchechukwu Nwaka, Emmanuel Obinna Ogueri, I. Fabian Ezema, and Mamta Bisht

Current Scenario of Soluble Uranium in Water

Understanding the Combined Effects of Multiple Water Quality Parameters 237
Taufiq Ahamad, Rohit Singh Sajwan, Manoj Kumar Jindal, and Shubham Sharma

Natural and Anthropogenic Factors: Shapers of Soluble Uranium Levels 271
Kashyap Komal, Ayush Kumar, Yvon Ukwishaka, Sarthak Ranjan, Manish Kumar, Biswa Mohan Sahoo, Pardeep Singh, and Preetismita Borah

Implication for Uranium Transport and Remediation Strategies 291
Manjulata Yadav and Vijita Diwan

Conclusion: Summary and Recommendations for Future Research

Emphasis on Real-World Application: Uranium Solubility in Real-World Samples: A Global Report........................... 311
Abhishek Joshi, Krishna Pal Singh, R. S. Aswal, and R. C. Ramola

Water Quality–Uranium Solubility Nexus: Integrated Approaches, Challenges, and Future Directions 325
Manoj Kumar Jindal

About the Editors

Manoj Kumar Jindal Dr. Manoj Kumar Jindal is an active researcher who has completed a PhD in Applied Chemistry and is currently focusing on environmental chemistry/science research, particularly in the fields of water and radiation, including water quality, quantity, sustainability, and water chemistry. He has worked on the uranium-water nexus. He has over 10 years of experience in research and teaching. Currently, he is a researcher at the Center for Sustainable Drinking Water, IITJ-Jal Jeevan Mission at the Indian Institute of Technology Jodhpur (IITJ), Rajasthan, India. Previously, he worked at the Divecha Centre for Climate Change, Indian Institute of Science (IISc), Bangalore, India, for 2 years and at IIT Bhilai for 3.5 years. He has authored over 40 research publications in indexed journals. He is also the author/editor of five books and numerous book chapters. He serves as a reviewer for many journals and book publishers, including Elsevier, Springer Nature, Taylor & Francis, Wiley, and Oxford University Press. He is an Editorial Board Member of *Discover Sustainability*, published by Springer Nature.

Kiran Patni Dr. Kiran Patni is an Assistant Professor in the Department of Chemistry at Graphic Era Hill University, Bhimtal Campus, Uttarakhand, India. She received her PhD in Chemistry from Kumaun University, Nainital, Uttarakhand, India, where her research focused on the spatial distribution of uranium and other water quality parameters that affect the suitability of water for drinking purposes. Dr. Patni's research interests include water quality assessment, water pollution control, and water resource management. She has published several research papers and book chapters in peer-reviewed journals and has presented her work at national and international conferences. Dr. Patni is a dedicated and passionate teacher, committed to providing her students with high-quality education and preparing them for careers in environmental science and water resource management. She is also actively involved in outreach activities and public education programs to raise awareness of water quality issues and promote sustainable water management practices.

Background

Understanding Uranium and Its Ionic Forms

Xiaonan Ruan, Longfang Ren, Bing Shi, Huaqing Liu, and Taotao Qiang

Abstract Uranium, as a radioactive element and strategic resource, is of great significance for environmental protection, nuclear energy utilization, and health risk assessment. With the exploration of uranium in recent years, the study of soluble uranium has become important for understanding and separating uranium from different places. Therefore, the physical and chemical properties, geological background, applications, and environmental impact are elaborated in detail by exploring the form of uranium present in solution, oxidation state, solubility, complexation behavior, acid-alkalinity, and redox properties. The intrinsic relationship between uranium and uranium ions was clarified, as well as the characteristics, distribution, transformation and migration behavior of the dominant species of uranium in diverse water environment were further elaborated. The research significance of the dissolution and migration behavior of uranium in natural ecosystems, public health risk, uranium resource management and sustainable use of energy was introduced. These findings can provide scientific guidance for accurately predicting uranium behavior and risks in water quality, provide a theoretical basis for chemical research on water uranium, and promote research progress in effectively separating and utilizing uranium in the future. This will not only improve environmental monitoring and pollution control strategies but also promote innovative applications in the fields of nuclear energy and materials science.

Keywords Properties of uranium · Ionic forms of uranium in water · Uranium chemistry

X. Ruan · L. Ren · B. Shi · H. Liu · T. Qiang (✉)
College of Bioresources and Materials Engineering, National Demonstration Center for Experimental Light Chemistry Engineering Education, Shaanxi University of Science & Technology, Xi'an, Shaanxi, People's Republic of China

M. K. Jindal, K. Patni (eds.), *Water Quality-Uranium Solubility Nexus: Unlocking the Secrets of Uranium Solubility*, Earth and Environmental Sciences Library, https://doi.org/10.1007/978-3-031-98909-4_1

1 Properties of Uranium

1.1 Historical and Geological Background

The history of uranium dates back to 1789, when Martin Heinrich Klaproth, a chemist from the Prussian Kingdom, discovered uranium in pitchblende ore and named it after Uranus, the planet, with the element symbol U. Later, Eugene-Melchior Peligot first separated the metal uranium in 1841, which was known as uranium glass due to its yellow–green color used for glass and ceramic dyeing until Henri Becquerel discovered its radioactivity in 1896 and was verified by Marie Curie. From 1934 onward, further research by Enrico Fermi and others made uranium the fuel for the nuclear energy industry. Since 1965, the Nuclear Energy Agency (NEA) and the International Atomic Energy Agency (IAEA) have jointly compiled and regularly updated the Red Book, documenting and collecting historical overviews of the uranium industry in terms of uranium resources, production, demand, and environmental activities worldwide, with cooperation from their member countries.

As a strategic resource, uranium remains the main fuel source for the nuclear energy industry today, providing reliable low-carbon energy for countries around the world and reducing greenhouse gas emissions. In addition, uranium and its isotopes play unique roles in medical diagnosis and treatment and are also important in scientific research, such as for determining geological samples, tracking environmental and biological processes, and studying nuclear reactions and properties. As a naturally radioactive element, uranium releases alpha particles during decay, with the longest-lived isotopes U-238 (half-life of 4.47 billion years) and U-235 (half-life of 704 million years) being the most common. They are often used to estimate the age of the Earth [1, 2]. Additionally, uranium is a metal element with variable oxidation states that is controlled by oxygen fugacity and oxidation–reduction sensitivity; it exhibits geochemical behaviors such as weathering, migration, and enrichment in ore formation during surface circulation processes [3]; and penetrates the evolution of natural environments and human production and life.

1.2 Physical Properties

Uranium is the heaviest element in nature, the actinide with an atomic number of 92 [4] and an atomic mass of 238.02891 u, and it exists in nature as only three isotopes (U-234, U-235, U-238), in addition to 24 artificial isotopes. In particular, uranium-235 is the only naturally fissile isotope of uranium, which is cleaved into fragments by thermal neutrons, triggering chain fission that releases a large amount of energy (195 MeV) and plays a very important role in nuclear reactions. The purified uranium metal is silver-white with a slight blue hue and has the characteristics of high hardness, high density, malleability and radioactivity. The occurrence states of various geological bodies can be roughly divided into three types: uranium minerals (such as uranyl mineral precipitation and crystalline uranium ore), isomorphic mixed crystals (zircon and apatite, etc.) and dispersed adsorption states (adsorbed on the surface of minerals and organic matter by electrostatic and hydrogen bonding). The average content of uranium in the Earth's crust is 2–4 ppm; in sea water, it is 3 ppm; in natural soil, rock and water, it ranges from 1 to 10 ppm [5, 6]. Although uranium is widely distributed, its content is very low; under specific geological conditions, it is able to undergo mineralization and is often stored in pitchblende, mica uranium ore, vanadium potassium uranite, monazite and other 170 kinds of uranium-bearing geological bodies.

1.3 Chemical Properties

Uranium metal powder reacts with cold water. In air, the chemical properties of the surface formed by a dark oxide layer are more active, and a free state of uranium metal is difficult, as almost all nonmetallic elements and their compounds can react with uranium. The valence electron shell structure of uranium is $5f^3\ 6d^1\ 7s^2$, in which the *5f/6d* orbital gives uranium unique properties that are different from those of transition metals and lanthanides. Based on the bonding effect of the unsaturated electron layer, uranium has variable valence characteristics and can present +3 ~ +6 valence states, but the +4 and +6 valence states are the main ones [5, 7]. U and U^{3+} have strong reducibility, and U and U^{3+} can reduce H^+ in water to H_2 in solution and in the absence of oxidants, whereas U^{3+} is often used for the catalytic conversion of small molecules because of its high chemical activity [8–10]. U^{4+} easily forms crystalline uranium deposits (pitchblende and uranite), which exhibit superconducting properties when stimulated by external conditions [11]. Stable minerals formed by U^{5+} have also been found to exhibit interesting magnetic properties [12–17]. U^{6+} forms complexes with a variety of anions in the form of UO_2^{2+}, and its inherent luminous properties are used in the design of sensors and optical materials; U^{6+} has high solubility and is the main form of uranium migration [18–20]. On the basis of the above chemical properties and the stability of different valence states of uranium, the common complex behavior of uranium includes anion and cation complexes with uranium. Cationic complexes are most common with lithophile elements (such as potassium, rubidium, and strontium), followed by iron and sulfur. O^{2-} is the main anion that binds uranium, and the binding of OH^- and F^- with uranium occasionally occurs in individual uranium minerals. Secondly, acid ions form complexes with uranium and other metal elements together to form uranium-rich derivatives with different properties. For the oxide of uranium, the phase state in the uranium oxide system is very complex, and the common oxide states of uranium(IV) and uranium(VI), respectively, correspond to uranium dioxide (UO_2) and uranium trioxide (UO_3); other existing oxides include uranium monoxide (UO), uranium pentoxide (U_2O_5), and uranium peroxide ($UO_4{\cdot}2H_2O$). U_3O_8 is the most stable uranium oxide and one of the most common in nature. At ambient temperatures, UO_2 is gradually converted to U_3O_8 and is the most commonly used uranium fuel in nuclear reactors. Both uranium oxides are not easily soluble in water and are relatively stable in many chemical environments, so they are often stored and disposed of as uranium oxides.

1.4 Biochemical Properties

The biochemical toxicity of uranium mainly comes from chemical toxicity and radioactive toxicity. The chemical toxicity of uranium comes from the heavy metal properties of uranium. More than 60% of the internal environment of animals and the human body is liquid. After removing the intolerant uranium oxides, uranium oxides and uranium dust particles easily form aerosols, which can slowly dissolve in body fluids to form uranyl ions, which participate in biochemical effects, damage tissue functions and organs, and eventually deposit in the kidney, liver, bone and skin. This leads to respiratory and skin system damage, chromosome damage and genotoxicity. Radioactive toxicity of uranium refers to the radiation damage caused by its radioactivity to tissues and organs. The decay of uranium is accompanied by α, β, and γ rays. Organisms absorb the energy of rays, resulting in molecular and atomic-level ionization and broken bonds in the cell material, resulting in nuclear mutagenesis, cell degeneration and other irreversible damage, resulting in cancer, tumors and other radiological biological effects.

2 Ionized Uranium

2.1 Species and Distribution

Uranium can be quickly dissolved in hydrochloric acid at room temperature or nitric acid under heating conditions, and the ionic forms mainly include U(IV) and U(VI). U^{3+} is unstable, has strong reducibility, and is easily oxidized to U^{4+} by air at room temperature. With a potential of 3.8, U^{4+} can exist only in strongly acidic solutions. Like Ti^{4+}, U^{4+} has a weak base in aqueous solution. When the U^{4+} potential is close to the amphoteric oxide region, it will hydrolyze to form $U(OH)^{3+}$ when the pH increases. U^{5+} can exist only in solutions with a pH of 2–4, and disproportionation reactions occur outside this range to form U^{4+} and UO_2^{2+}. The ionic potential of U^{6+} is 7.4, and it has amphoteric characteristics in the amphoteric oxide region. It is weakly alkaline in acidic and neutral media, resulting in the formation of acidic salts (UO_2SO_4, $UO_2(CO_3)_3^{4-}$, etc.). In alkaline media, it is weakly acidic, resulting in the formation of insoluble alkaline salts ($K_2U_2O_7$, $Na_2U_2O_7$). Since U^{6+} can be rapidly hydrolyzed in solution to produce stable uranyl ions or other salts of different oxidation states, especially those containing oxygen ions, the chemical formula

UO_2^{2+} has a linear structure consisting of U–O bonds, and the ligands can be bonded to uranyl ions in the horizontal plane centered on the uranium atom. In addition to cooperating with many inorganic anions to form corresponding complexes, the order of coordination ability is as follows: CO_3^{2}- ≥ O_2^{2}- > OH- > F- > SO_4^{2}- > Cl- > NO^{3}- > Br- > I^{-}, common uranyl nitrate hexahydrate, uranyl oxalate dihydrate, uranyl nitrate and alkali metal M nitrate to form $MNO_3{\cdot}UO_2(NO_3)_2$ uranyl complex salts. Although there are more Cl^- complexes in seawater, most ore-forming fluids, and water around underground waste storage, the complex $UO_2(CO_3)_n^{2(n-1)-}$ formed by UO_2^{2+} and CO_3^{2-} greatly enhances the mobility of uranium ions in water systems [21–24].

2.2 Dominant Species in the Water Media

Under different water quality conditions, the ion form of uranium will differ; generally, when the pH is less than 2.5, uranium is a single uranyl ion that forms in the solution. When the pH is greater than 2.5, uranyl ions begin to hydrolyze, and a polymerization reaction multinucleate or even cages, resulting in hydroxide precipitation. The order of hydrolysis of various uranium ions in aqueous solution is U^{4+} > UO_2^{2+} > U^{3+} > UO^{2+} [25]. In acidic environment, U^{6+} and uranate ions (UO_2^{2+}) are the main dominant species, with high solubility and migration. In the neutral environment, U(VI) remains the dominant species, which is stable and exists in hydrated form, forming $[U(H_2O)_6]^{2+}$. In alkaline environments, U(VI) ions may combine with hydroxide ions to form uranates (such as $UO_2(OH)_2$) or to form other U(VI) complexes, causing the uranium ions to precipitate.

2.3 Conversion Between Ionized Uranium Forms

The transformation of uranium ions in aqueous solutions is affected by many factors, including acid–base, redox and ion concentrations. In terms of acidity and alkalinity, changes in pH directly affect the morphology of uranium ions. Under acidic conditions, uranium (VI) ions easily form and enhance the stability and migration of uranium (VI) in water. Under alkaline conditions, uranium (IV) may be converted into uranium (IV), which easily precipitates and further reduces the mobility of uranium. In reducing or anoxic environment, uranium exists in the form of U^{4+}, and under appropriate conditions, appropriate oxidants or reducing agents can achieve the mutual conversion of U(IV) and U(VI). For example, oxidants ($K_2Cr_2O_7$, $Fe_2(SO_4)_3$, and $KMnO_4$) can promote the formation of U^{4+} and enhance its migration in water. Reducing agents (Al, Cd, and Fe^{2+}) can promote the reduction of U^{6+}, resulting in reduced mobility of uranium. In addition to reducing UO_2^{2+} ions to U^{4+} and partial reduction to U^{3+}, U^{3+} can be rapidly oxidized to U^{4+} in the air. In addition, anions (CO_3^{2-}, SO_4^{2-}, Cl^-) react with uranium, affecting its solubility and migration characteristics, and cations (Mg^{2+}, Na^{+}, Ca^{2+}) can form insoluble salts with uranium ions, reducing the migration of uranium.

2.4 *Migration of Ionized Uranium*

The migration of uranium ions is affected by many factors, including their chemical form, environmental conditions, and characteristics of the soil and water. In water, uranium ions can migrate through dissolution, diffusion and adsorption. Due to its strong water solubility and mobility, U^{6+} has a stronger migration ability in water bodies than uranium U^{4+}; although U^{4+} cannot exist in flowing water systems, but migration behavior at high temperatures has been reported [26–28]. In addition, the migration of uranium is also affected by geological conditions, with a high adsorption capacity for clay and slow-moving water bodies, which may limit the migration of uranium. Soil rich in organic matter can increase the solubility of uranium through complexation reactions, thus promoting uranium migration. The migration of uranium ions has an important impact on ecology and the environment, which may lead to the pollution of water and soil and affect the survival and health of organisms. Therefore, understanding the migration characteristics of uranium ions has far-reaching significance for environmental monitoring, nuclear waste management and ecological protection.

3 Significance of Study on the Behavior of Uranium and Uranium Ions

The presence of ionic uranium greatly promotes spatial mobility and the diffusion of potentially toxic elements from minerals to sediments and soils, as well as surface runoff pollution, and increases the risk of uranium readsorption in the environment. This phenomenon affects the composition of the soil microbial community and interferes with the development of biological diversity through chemical and radiation pathways [29].Although the migration of uranium is usually guided by U(VI), this paradigm meets the observation results at medium and low temperatures. Some studies have shown that U^{4+} migrates under high-temperature conditions [30, 31]. In addition, uranyl complex ions in the form of anions also have strong mobility and do not have blocking ability in geological media. Therefore, for nuclear materials and waste management and construction of engineering barriers, it is common to reduce conditions and build buffer-backfill materials to prevent uranium migration leaks and accidents [32–34]. As terrestrial uranium resources have gradually ceased to be the optimal solution for sustainable energy development, it is imperative to explore unconventional uranium resources. Ionic uranium is widely distributed and has a very large volume, and its ionic state transformation and existence are the main modes of uranium migration in water and the core of effective separation and purification of uranium from water. The enrichment and capture of uranium ions can be achieved by combining advanced techniques such as electrochemistry, photochemistry and biochemistry. According to the survey, the sea contains a large amount of soluble uranium (up to 45 billion tons), so the extraction of uranyl ions from seawater is becoming a viable strategy to supplement and replace land resources.

Given that the radioactive activity of uranium and its potential threat to human beings and life on earth will last for hundreds of thousands of years, on the basis of research on uranium for nearly three centuries and the utilization and development of nuclear energy, research on the behavior of uranium and ionized uranium is related to global issues such as environmental protection and governance, resource utilization and management, and public health risk assessment; therefore, its advantages and disadvantages in economic construction and peaceful development should receive more attention to promote the sustainable use of energy and resources in the future and the innovative development of science and technology.

References

1. Gaschnig RM et al (2014) Onset of oxidative weathering of continents recorded in the geochemistry of ancient glacial diamictites. Earth Planet Sci Lett 408:87–99
2. Hofmann S, Münzenberg G (2000) The discovery of the heaviest elements. Rev Mod Phys 72:733–767
3. Sun S-S, McDonough WF (1989) Chemical and isotopic systematics of oceanic basalts: implications for mantle composition and processes. Geol Soc Lond, Spec Publ 42:313–345
4. Prohaska T et al (2022) Standard atomic weights of the elements 2021 (IUPAC technical report). Pure Appl Chem 94:573–600
5. Grenthe I et al (2006) Uranium. The chemistry of the actinide transactinide elements, pp 253–698
6. Abney CW, Mayes RT, Saito T, Dai S (2017) Materials for the recovery of uranium from seawater. Chem Rev 117:13935–14013. https://doi.org/10.1021/acs.chemrev.7b00355
7. Langmuir D (1978) Uranium solution-mineral equilibria at low temperatures with applications to sedimentary ore deposits. Geochim Cosmochim Acta 42:547–569
8. Chen W et al (2003) Synthesis, structure, and photoelectronic effects of a uranium– zinc–organic coordination polymer containing infinite metal oxide sheets. J Am Chem Soc 125:9266–9267
9. Halter DP, Heinemann FW, Bachmann J, Meyer K (2016) Uranium-mediated electrocatalytic dihydrogen production from water. Nature 530:317–321
10. Falcone M, Chatelain L, Scopelliti R, Živković I, Mazzanti M (2017) Nitrogen reduction and functionalization by a multimetallic uranium nitride complex. Nature 547:332–335
11. Jiao L et al (2020) Chiral superconductivity in heavy-fermion metal UTe2. Nature 579:523–527
12. Burns PC, Finch RJ (1999) Wyartite: crystallographic evidence for the first pentavalent-uranium mineral. Am Mineral 84:1456–1460
13. Wander MC, Shuford KL (2012) A theoretical study of the qualitative reaction mechanism for the homogeneous disproportionation of pentavalent uranyl ions. Geochim Cosmochim Acta 84:177–185

14. Skomurski FN, Ilton ES, Engelhard MH, Arey BW, Rosso KM (2011) Heterogeneous reduction of U6+ by structural Fe2+ from theory and experiment. Geochim Cosmochim Acta 75:7277–7290
15. Mougel V et al (2012) Uranium and manganese assembled in a wheel-shaped nanoscale single-molecule magnet with high spin-reversal barrier. Nat Chem 4:1011–1017
16. Chatelain L, Walsh JP, Pécaut J, Tuna F, Mazzanti M (2014) Self-assembly of a 3d–5f Trinuclear single-molecule magnet from a pentavalent uranyl complex. Angew Chem 126:13652–13656
17. Nocton G, Horeglad P, Pécaut J, Mazzanti M (2008) Polynuclear cation– cation complexes of pentavalent uranyl: relating stability and magnetic properties to structure. J Am Chem Soc 130:16633–16645
18. Liu W et al (2019) Introducing uranium as the activator toward highly stable narrow-band green emitters with near-unity quantum efficiency. Chem Mater 31:9684–9690
19. Xie J et al (2017) Highly sensitive detection of ionizing radiations by a photoluminescent uranyl organic framework. Angew Chem 129:7608–7612
20. Wang Y et al (2018) Emergence of uranium as a distinct metal center for building intrinsic X-ray scintillators. Angew Chem 130:8009–8013
21. Millero FJ, Feistel R, Wright DG, McDougall TJ (2008) The composition of standard seawater and the definition of the reference-composition salinity scale. Deep-Sea Res I Oceanogr Res Pap 55:50–72
22. Rudnick R, Gao S, Holland H, Turekian K (2003) Treatise on geochemistry. Crust 3:1–64
23. Dargent M et al (2013) Experimental study of uranyl (VI) chloride complex formation in acidic LiCl aqueous solutions under hydrothermal conditions (T= 21 C–350 C, Psat) using Raman spectroscopy. Eur J Mineral 25:765–775
24. Bargar JR, Reitmeyer R, Davis JA (1999) Spectroscopic confirmation of uranium (VI)– carbonato adsorption complexes on hematite. Environ Sci Technol 33:2481–2484
25. Infante I et al (2007) A Fock space coupled cluster study on the electronic structure of the UO2, UO2+, U4+, and U5+ species. J Chem Phys 127
26. Paquette J, Lemire RJ (1981) A description of the chemistry of aqueous solutions of uranium and plutonium to 200°C using potential-pH diagrams. Nucl Sci Eng 79:26–48
27. Lemire RJ, Tremaine PR (1980) Uranium and plutonium equilibriums in aqueous solutions to 200 °C. J Chem Eng Data 25:361–370
28. Parks GA, Pohl DC (1988) Hydrothermal solubility of uraninite. Geochim Cosmochim Acta 52:863–875
29. Antunes M, Santos A, Valente T, Albuquerque T (2020) Spatial mobility of U and Th in a U-enriched area (Central Portugal). Appl Sci 10. https://doi.org/10.3390/app10217866
30. Montreuil J-F, Corriveau L, Potter EG (2015) Formation of albitite-hosted uranium within IOCG systems: the Southern Breccia, Great Bear magmatic zone, Northwest Territories, Canada. Mineral Deposita 50:293–325
31. Timofeev A et al (2018) Uranium transport in acidic brines under reducing conditions. Nat Commun 9:1469
32. Swift PN, Bonano EJ (2016) Geological disposal of nuclear waste in tuff: Yucca Mountain (USA). Elements 12:263–268
33. Hedin A, Olsson O (2016) Crystalline rock as a repository for Swedish spent nuclear fuel. Elements 12:247–252
34. Grambow B (2016) Geological disposal of radioactive waste in clay. Elements 12:239–245

Water Quality and Its Diverse Source

Suvarshitha Pusuluru, Sai Kumar Punna, Karrun Velmurugan, Selvarajan Ethiraj, and Melvin S. Samuel

Abstract The rapid proliferation of uranium production, combined with escalating anthropogenic activities, has led to pervasive contamination of drinking water sources by uranium, posing significant environmental and public health concerns. In addition, intensive riverbed mining practices have severely altered aquatic ecosystems, impairing water quality, habitat integrity, and biodiversity. Uranium further impairs risks by persisting in aquatic environments and resisting conventional water treatment technologies. This chapter critically reviews the occurrence, sources, pathways, and ecological impacts of uranium in drinking water systems. This study highlights the influence of riverbed mining on water clarity, sediment transport, and aquatic life and discusses microbial risks in graywater reuse systems through quantitative microbial risk assessment (QMRA) frameworks. Synthesizing recent findings, this chapter identifies pressing knowledge gaps, advocates for improved monitoring strategies, and outlines the need for integrative management approaches to safeguard aquatic ecosystems and human health.

S. Pusuluru
Department of Engineering Data Science, University of Houston, Houston, TX, USA

S. K. Punna
Department of Materials Science and Engineering, University of Houston, Houston, TX, USA

K. Velmurugan
Department of Mechanical Engineering, Indian Institute of Technology – Kharagpur, Kharagpur, West Bengal, India

S. Ethiraj
Department of Genetic Engineering, College of Engineering and Technology, School of Bioengineering, SRM Institute of Science and Technology, Kattankulathur, Tamil Nadu, India

M. S. Samuel (✉)
Department of Bioengineering, Saveetha School of Engineering, Saveetha Institute of Medical and Technical, Chennai, India

Department of Civil, Construction and Environmental Engineering, Marquette University, Milwaukee, WI, USA

M. K. Jindal, K. Patni (eds.), *Water Quality-Uranium Solubility Nexus: Unlocking the Secrets of Uranium Solubility*, Earth and Environmental Sciences Library, https://doi.org/10.1007/978-3-031-98909-4_2

Keywords Riverbed mining · uranium contamination · Water quality degradation · Quantitative microbial risk assessment (QMRA)

1 Introduction

Microorganisms and aquatic vegetation are crucial for maintaining the ecological balance and health of riverine environments [1]. Sediment transport from rivers to oceans represents a fundamental component of global denudation and geochemical cycles and acts as a key indicator of soil erosion and land degradation [2]. These natural processes are influenced by seasonal fluctuations in water flow, river network characteristics, and sediment deposition rates, all of which vary with geography and time [3]. However, the growing demand for riverbed materials has led to widespread, often illegal, mining activities, including in agricultural and floodplain areas, which significantly disrupt riverine processes. Such activities affect the river's flow regime, structural integrity, and ecological functions while also disturbing riparian zones and reducing biodiversity in both terrestrial and aquatic ecosystems [4, 5]. Heavy metals have increasingly come to symbolize the Anthropocene epoch, reflecting humanity's profound impact on the environment [6]. Since the conclusion of World War II, the global production of heavy metals has grown exponentially, closely paralleling the surge in the human population [6]. Innovations in metal production technologies have further expanded their applications across diverse sectors [7]. However, at the end of their useful life, a significant portion of these plastics are disposed of in landfills [8]. Uranium is a naturally increasing radionuclide in granite and other mineral deposits. It has the potential to enter local water, air and food commodities in various amounts through leachate from soil deposits, release in mill tailings, emit from the nuclear industry, dissolve in phosphate fertilizers and incinerate coal and other energy. Canadian drinking water has ranged from nondetectable uranium concentrations to high uranium concentrations in some areas (>700 μg/L), especially in some private wells and ground water sources. Natural uranium consists of the 238 isotopes of uranium in its hexavalent state, which is associated with oxygen as the uranyl ion UO_2^{2+}. Uranium consists of U-235 (0.72%) and U-234 (0.01%) isotopes. It is classified as a lithophilic element and is usually found in large masses of minerals. Uranium deteriorates because of both alpha and gamma emissions. Compared with other radioactive nucleotides, natural uranium has a very low level of radioactivity because of its extremely long half-life. When uranium, a naturally occurring radioactive heavy metal, is present in groundwater, rivers, lakes, or other aquatic systems at concentrations that could endanger human health and the environment, it is referred to as uranium contamination in water bodies. This problem is especially problematic in areas where there is nuclear activity, uranium mining, or certain geological formations. Water resources include uranium (U), a naturally occurring radioactive element, as a result of natural meteorological processes. As a result, screening drinking water for this purpose is now unavoidable. It is found in practically every area of the environment and is the heaviest naturally occurring element. Because of its widespread distribution, U is thought to be a major contributor to the radiation dosage that the general public receives [9]. Certain anthropogenic operations, such as uranium mining and processing and the disposal of spent nuclear fuel, can partially pollute drinking water sources with uranium. However, local geological deposition is the primary cause of uranium pollution [10, 11]. Uranium can enter aquifers through the leaching of uranium-containing minerals such as coffinite, pitchblende, and uraninite. UO^{2++}, which is relatively more stable and highly mobile, is created when uranium host rock oxidizes to the +6 state and joins forces with O_2. When water is in contact with water, it dissolves and seeps into adjacent water sources, such as groundwater reservoirs.

2 Effects of Riverbed Mining on River Morphology, Hydrology, and Biodiversity

This disrupts the natural sediment transport regime, which is critical for maintaining riverbed structure and habitat continuity. Lateral Channel Instability: Riverbed mining contributes to channel widening, bank erosion, and instability, which are key indicators of lateral morphological disturbance [4]. The effects of mining are not limited to physical changes; they also significantly degrade water quality. Mining operations increase suspended sediment loads, increase turbidity levels and reduce water clarity. Furthermore, the lowering of riverbeds due to excavation disrupts the groundwater table and inhibits natural groundwater recharge, contributing to aquifer depletion [12]. Aquatic biodiversity is particularly sensitive to these disturbances. Elevated turbidity negatively affects primary producers such as phytoplankton and disrupts the habitat conditions required by zooplankton, macrobenthic organisms, and fish species. Notably, fish species such as salmon and trout, which rely on clean, gravel-rich substrates for spawning, are adversely affected by increased sedimentation and habitat fragmentation resulting from riverbed mining [13]. These alterations can directly harm fish populations or indirectly affect them by changing habitat availability and structural complexity [14]. Moreover, differences in hydraulic conditions between untouched, multichannel river stretches and narrowed, incised mining zones are associated with reduced fish diversity and abundance. Disruption of migratory pathways and spawning grounds further exacerbates the decline in fish populations and overall riverine health [13].

3 Effects on Water Quality

Intensive sand mining operations significantly deteriorate the quality, clarity, and ecological value of river water. The physical removal of sedimentary material from riverbeds adversely affects surface water characteristics, contributing to increased turbidity, electrical conductivity, and total dissolved solids (TDS) concentrations [4]. Uranium can enter the food chain through water and soil, possibly harming the health of the community. Uranium processing and mining can pollute aquatic ecosystems for more than 100 years, threatening downstream communities and wildlife. Although a small quantity of uranium can affect fish, it can accumulate in the food chain and cause deformities and propagative difficulties for aquatic species. Uranium contamination in water can lead to metabolic dysfunction and neurotoxicity. It has been reported that it can lead to chemical toxicity and ultimately cause human carcinogen. The highest uranium amount was detected in the thermal groundwater from Mszczonow and Cieplice, while the lowest was detected in some places in Uniejow and Poddeice. In some places, the uranium concentration ranges are higher than 25 μg/L in the Riihimaki waterworks. In Tanzania, Singida Urban District, the uranium concentrations in soil, rock, and water sources are high. This resulted in higher uranium concentrations in surface and underground water sources and drinking water in Singida. The uranium concentrations in these areas clearly exceeded the drinking water limit of uranium. The uranium concentration in Myingyan Township, Myanman, exceeded the safe health-based reference values in most wells [15].

Techniques such as laser fluorimetry (LF) and fission track analysis (FTA) were used to measure the uranium concentration in the drinking water. Using the above two techniques, drinking water samples collected from Punjab were assessed. The results obtained from the LF and FTA techniques are comparable, and the drinking water concentration ranges from approximately 3.2–60.5 ppb [16]. Compared with FTA, LF analysis has advantages such as simplicity, sensitivity, sample preparation and rapid analysis. Similarly, the uranium and other heavy metal contents were assessed in Sirhind canal water and groundwater in the Malwa Belt Punjab, and the uranium content in the surface water of the Sirhind canal downstream was below the detection limit. However, in some areas, the uranium concentrations were 30–366 ppb [17]. In India, various studies have examined the distribution of uranium in drinking water from different states. The study revealed that 54% of the collected water samples presented concentrations less than 1 mg/L, and only 4 samples presented higher uranium concentrations (>10 mg/L). Similarly, a study conducted in the Bathinda district of Punjab state, India, reported that the uranium concentration exceeded the permissible level of the WHO standard for drinking water. Another study was conducted in Kerala and analyzed 830 samples from different sources during the pre- and postmansoon seasons. The concentration of uranium ranged from <0.5 to 12.54 μg/L in the premansoon season and <0.5–5.93 μg/L in the postmansoon season, which is well within the drinking water standard limit [18]. The studies conducted in Tarragona, Spain and Hungary as well as the assessment of Kyrgyzstan collectively highlighted the presence of natural radioactivity in drinking water and its potential implications. In Tarragona, approximately 23% of the drinking water samples exceeded the alpha gross index parameter, indicating the presence of natural radionucleotides. A study analyzed 951 samples with uranium concentrations above 0,01 μg/L, with concentrations of up to 166 μg/L. The mean and median concentrations were 5 μg/L and 1.6 μg/L, respectively. The study revealed that 2.6% of the analyzed samples exceeded the WHO-acceptable concentrations of 30 μg/L and 44.2 μg/L, which resulted in the exceedance of the German maximum acceptable concentrations recommended for infant food preparations. In Brazil, the level of alpha gross and beta natural radioactivity in surface, underground, and drinking water exceeds the recommended level [19, 20].

Additionally, the movement of heavy trucks can endanger road safety and cause structural damage to transportation infrastructure [21]. Studies highlight that illegal mining activities have led to the degradation of agricultural lands and turned excavation pits into mosquito breeding grounds during the rainy season, increasing the spread of vector-borne diseases [22, 23]. In rural Bangladesh, riverbank erosion attributed to sand mining has forced entire communities to relocate, with at least 29 households having lost their homes and farmland. Excessive extraction of river sand has also rendered many rural water systems unusable, contributing to seasonal water shortages during dry periods [24]. Furthermore, the deterioration of water quality from mining activities compromises water used for drinking, bathing, and domestic purposes. This degradation increases the likelihood of infectious disease transmission and poses long-term health risks to communities. In developing regions, environmental damage to hydrological systems often correlates directly with negative public health outcomes [25]. Workers engaged in sand mining are particularly at risk due to direct exposure to polluted river water, including contaminants such as diesel and suspended particles. Rivers altered by pit excavation and redirection during flood events may become hazardous, and even life-threatening [22]. These environments also disrupt fish breeding and spawning habitats, resulting in a sharp decline in economically important species [26]. The collapse of benthic communities affects not only aquatic food chains but also the diets of terrestrial and amphibian species. The construction of physical barriers and extraction from key habitats can drastically reduce reproductive success in fish populations. Therefore, systematic field studies are urgently needed to identify critical breeding and migratory pathways. The management of human interventions in freshwater ecosystems must become a priority to preserve ecological integrity and ensure long-term sustainability [27]. Exposure to uranium primarily occurs through the ingestion of contaminated water. Biomonitoring studies have confirmed the widespread presence of uranium in human blood samples, indicating ongoing exposure among the general population. The biological resilience of these substances poses significant metabolic challenges, as they tend to accumulate in tissues over time. Uranium, a well-documented radioactive material, has a half-life of approximately 4–5 years in the human body and can remain in environmental reservoirs for several decades.

- **Rivers and Surface Waters**: In these systems, industrial wastewater is the primary contributor, accounting for 80–90% of direct uranium discharges. Effluents from fluoride-related industries are especially prevalent, discharging uranium-rich wastewater that travels via tributaries and interconnects waterways, disseminating pollutants across wide regions [28].
- **Lakes and Reservoirs**: Landfill leachate represents the predominant source of uranium contamination in these settings. Research conducted at Baoshan Reservoir in Taiwan revealed that leachate accounted for contamination in more than 61% of the sampled sites, whereas urban runoff contributed approximately

39% [29]. Furthermore, the application of microplastic-containing agricultural materials, such as plastic films used for moisture retention and certain pesticides, can lead to surface runoff and rainwater-mediated transport of microplastics into nearby water bodies [30].
- **Groundwater Systems**: Uranium infiltration into groundwater typically occurs through the percolation of contaminated surface water. Although uranium concentrations in Chinese groundwater remain relatively low at present, the isolated nature of aquifers presents a challenge for remediation if contamination occurs. This is of particular concern given that approximately 70% of the population in China relies on groundwater for drinking water, and nearly 40% of agricultural irrigation also depends on these reserves [31, 32].

Although indirect sources contribute less than 5% of total uranium emissions, their importance lies in the complexity of precursor degradation and the challenges in accurately quantifying their impact.

4 Health Hazards Linked to Uranium in Drinking Water

Long-term exposure to high quantities of uranium in drinking water can be harmful to one's health, even though low levels are usually regarded as harmless. As a proven carcinogen and heavy metal, uranium has the capacity to cause cancer in people. It can also harm the kidneys and cause other health problems, particularly in susceptible groups such as young children and pregnant women. The maximum contamination limit (MCL) of uranium in drinking water is set at 30 micrograms per liter (μg/L) by the Environmental Protection Agency (EPA). Accordingly, uranium levels in public drinking water systems should not exceed 30 μg per liter. Additionally, the World Health Organization (WHO) has established a 15 μg/L uranium guideline limit for drinking water. The maximum permissible limit (MPL) for uranium in drinking water in India is 60 micrograms per liter (μg/L), as determined by the Atomic Energy Regulatory Board (AERB). This implies that public drinking water systems should be limited to 60 μg of uranium per liter of water.

Comparison of Various Uranium Determination Techniques in Water The various uranium determination techniques discussed in the literature can be separated into two categories: laboratory techniques and "in situ" techniques [33]. Considering their accuracy, the following fundamental procedures are part of the laboratory techniques used: vaporization or preconcentration; acidification; coprecipitation; calcification; separation of acid or carbonate; and extraction and setup for calculation. Initially, the total alpha and beta activities of each sample were assessed. The gathered data provide guidance for selecting the approach that will be utilized. Additional standards for selecting the best approach are based on the geological and hydrochemical characteristics of uranium. The starting pH was always regulated.

The Use of Photometry In 6 N hydrochloric acid, U(IV) and arsenazo(III) combine to generate a red complex acid. Under these circumstances, U(VI) is reduced, with Zn present, to U(IV) of ascorbic acid (water quality). Uranium content measurement in water. This compound absorbs the most light at a wavelength of 655 nm. By applying a set of criteria, a calibration, a curve is created, and the uranium content is calculated.

The Method of Liquid Scintillation (LSC) Liquid–liquid extraction is a chemical process [34] in which a liquid scintillation cocktail of di-isopropil naphthalene (200–300 millilitres of toluene and 60 millilitres of HDEHP were combined with millilitres of OptiPhase HiSafe 2 as a complexing agent. At the sampling location, the samples were acidified to a pH of 2–3 via concentrated nitric acid.

Fluorimetry with Lasers This technique relies on the luminescence of uranyl ions in solution. To improve the yield, we employed Fluran I, a product of the Scintrex Company, to obtain the initial yield. The sample signal was read both before and after the inclusion of the uranium analysis (UA-3 User Guide and Analytical Procedures, Scintrex Ltd.) reagent. The signal of a sample with an unknown uranium concentration is contrasted with that of a standard.

Gamma Spectrometry Gamma spectrometry was used to verify a few samples. Evaporation is used to preconcentrate the samples to 0.45 l; under certain situations, a low concentration is anticipated to be 50 ml. Merck produces uranyl nitrate ore [35], which was dissolved and utilized as a reference solution to determine the uranium isotope concentration.

Application of Quantitative Microbial Risk Assessment (QMRA) Future research must prioritize the development of standardized protocols for monitoring uranium across diverse freshwater environments, coupled with advanced analytical techniques capable of detecting low-concentration contaminants. Longitudinal studies are needed to elucidate the fate, transport, and bioaccumulation of uranium within aquatic food webs, especially under varying climatic and hydrological conditions. In the context of riverbed mining, future efforts should focus on ecosystem-scale assessments to quantify the cumulative impacts on biodiversity, geomorphology, and groundwater recharge dynamics. Moreover, expanding the scope of quantitative microbial risk assessment (QMRA) to encompass a broader range of pathogens, including emerging viral and fungal species, will increase the predictive capacity of public health models in water reuse systems. Integrating machine learning and predictive analytics could further refine risk assessments and support the design of resilient, adaptive water management strategies. Cross-sector collaboration among scientists, policymakers, and industry will be crucial to drive innovation and ensure the sustainable protection of global water resources.

References

1. Padmalal D, Maya K (2014) Sand mining: environmental impacts and selected case studies. Springer
2. Walling DE, Fang D (2003) Recent trends in the suspended sediment loads of the world's rivers. Glob Planet Chang 39:111–126
3. Rodriguez LG, McCallum A, Kent D, Rathnayaka CM, Fairweather H (2023) A review of sedimentation rates in freshwater reservoirs: recent changes and causative factors. Aquat Sci 85(2)
4. Kamboj V, Kamboj N, Sharma S (2017) Environmental impact of riverbed mining: a review. Int J Sci Res Rev 7:504–520
5. Kamboj V, Kamboj N, Sharma A., Bisht A (2022) Fish diversity associated with environmental parameters in impacted area of Ganga River, India. In Proceedings of the national academy of sciences, India section B: biological sciences, 93, pp 79–90
6. Brandon JA, Jones W, Ohman MD (2019) Multidecadal increase in plastic particles in coastal ocean sediments. Sci Adv 5(9):eaax0587
7. Geyer R, Jambeck JR, Law KL (2017) Production, use, and fate of all plastics ever made. Sci Adv 3(7):e1700782
8. Rukmana D, Salman D, Alimuddin I, Arsyad (2020) Economic and environmental impacts of sand mining activities at Sadang River, Pinrang Regency, South Sulawesi. IOP Conf Ser Earth Environ Sci 575:012043
9. McCall W, Christy TM, Christopherson T, Issacs H (2009) Application of direct push methods to investigate uranium distribution in an alluvial aquifer. Ground Water Monit Remidiat 29:65–76
10. Babu MNS, Somashekar RK, Kumar SA, Shivanna K, Krishnamurthy V, Eappen KP (2008) Concentration of uranium levels in groundwater. Int J Environ Sci Technol 5:263–266
11. USEPA (2009) National primary drinking water regulations. United States Environmental Protection Agency, Washington, DC
12. Malik DS, Jaiswal D, Bharti R, et al. (2022) Anthropogenic impact on fish faunal diversity and their habitat ecology in the Ganga River and its tributaries, Uttarakhand. In Springer proceedings in earth and environmental sciences, pp 53–71
13. Kondolf GM (1997) Hungry water: effects of dams and gravel mining on river channels. Environ Manag 21:533–551
14. Kanehl PD, Lyons JD (1992) Impacts of in-stream sand and gravel mining on stream habitat and fish communities, including a survey on the Big Rib River. Wisconsin Department of Natural Resources, Bureau of Research, Marathon County, p 155
15. Bacquart T et al (2015) Multiple inorganic toxic substances contaminating the groundwater of Myingyan Township, Myanmar: arsenic, manganese, fluoride, iron, and uranium. Sci Total Environ 517:232–245
16. Prabhu S et al (2012) Application of fission track technique for estimation of uranium concentration in drinking waters of Punjab. J Radioanal Nucl Chem 294(3):443–446
17. Kailley SK, Virk HS (2020) Uranium and heavy metal contamination of Sirhind canal water and groundwater in the Malwa belt of Punjab and its mitigation strategies. Res Rev J Toxicol 10(1):34–41
18. Shalumon C et al (2021) Analysis of uranium and other water quality parameters in drinking water sources of five districts of Kerala in southern India and potability estimation using water quality indexing method. HydroResearch 4:38–46
19. Barescut J et al (2011) Gross alpha and beta activities in surface, underground and drinking waters of a high natural radioactivity region of central South Bahia state, Brazil. Radioprotection 46(6):S63–S67
20. Csondor K et al (2020) Transient nature of riverbank filtered drinking water supply systems—a new challenge of natural radioactivity assessment. J Environ Radioact 211:106072
21. Pitchaiah PS (2017) Impacts of sand mining on environment: a review. SSRG Int J Geoinform Geol Sci 4:1–5

22. Mingist M, Gebremedhin S (2016) Could sand mining be a major threat for the declining endemic Labeo barbus species of Lake Tana, Ethiopia? Singap J Trop Geogr 37:195–208
23. Rastogi R, Kumar V (2017) The environmental impact of river sand mining. Int J Tech Non-Tech Res 8:329–334
24. Chu, E. and Karr, J.R., 2016. Environmental impact: concept, consequences, measurement. Elsevier eBooks
25. Parker D (1996) Environmental assessment and auditing of mining operations—an international perspective. In Proceedings of IBC UK Conf. on the environmental management of mining operations, London
26. Choudhary M, Nautiyal P, Nautiyal R et al (2023) Status of fish assemblage structure in the Ganga and Indus riverine systems of the Western Himalaya. WWP 9:613–638
27. Thakur K, Kumar R, Brar B (2021) A review on freshwater fish diversity of India and concept of DNA barcoding in fish identification. Egypt J Aquat Biol Fish 25:667–693
28. Wang R, Wang Y, Gao Y et al (2022) Emerging and legacy per- and polyfluoroalkyl substances in the rivers of a typical industrialized province of China: spatiotemporal variations, mass discharges and ecological risks. Front Environ Sci 10:986719
29. Jiang JJ, Okvitasari AR, Huang FY, Tsai CS (2021) Characteristics, pollution patterns and risks of perfluoroalkyl substances in drinking water sources of Taiwan. Chemosphere 264:128579
30. Opperman JJ, Luster RA, McKenney BA, Roberts MS, Meadows AW (2010) Ecologically functional floodplains: connectivity, flow regime, and scale. J Am Water Resour Assoc 46:211–226
31. Zhang R (2007) Investigation of photolysis and photooxidation reaction mechanism of fluorinated organic compounds (Ph.D. dissertation, Fudan University)
32. Zhang G et al (2018) Occurrence, fate and risk assessment of antibiotics in water environment of China (in Chinese). Environ Chem 37:1491–1500
33. Environmental Protection Agency (EPA) (1997) National primary drinking water regulations; analytical methods for radionuclides. 40 CFR Part 141. DOE, USA
34. McDowell W, Coleman CF (1974) Combined solvent extraction–liquid scintillation method for radioassay of alpha emitters. In Proceedings of international solvent extraction conference
35. Richter S, Alonso A (1999) The isotopic composition of commercially and available uranium chemical reagent. J Anal At Spectrom

The Crucial Link: Why Water Quality and Uranium Solubility Matter

Kamyar Yaghmaeian, Mohammad Rezvani Ghalhari, Niloufar Borhani Yazdi, Farshad Hamidi, Aysan Morovvati Zarajabad, Mahla Mazloumian, Narges Ebrahimi, and Hossein Arfaeinia

Abstract This chapter examines the critical nexus between water quality parameters and uranium solubility, highlighting the intricate interplay that governs uranium mobility in aqueous environments. Uranium, predominantly present in the hexavalent state as a uranyl ion (UO_2^{2+}), exhibits varying solubilities influenced by factors such as pH, redox potential, temperature, dissolved oxygen, salinity, and organic matter content. This chapter delves into the chemical and physical properties of uranium, elucidating the mechanisms by which these parameters affect its solubil-

K. Yaghmaeian
Center for Water Quality Research (CWQR), Institute for Environmental Research (IER), Tehran University of Medical Sciences, Tehran, Iran

Department of Environmental Health Engineering, School of Public Health, Tehran University of Medical Sciences, Tehran, Iran

M. R. Ghalhari · N. B. Yazdi · F. Hamidi · A. M. Zarajabad
Department of Environmental Health Engineering, School of Public Health, Tehran University of Medical Sciences, Tehran, Iran

M. Mazloumian
Department of Environmental Health Engineering, School of Health, Shahid Sadoughi University of Medical Sciences, Yazd, Iran

N. Ebrahimi
Department of Health Education and Promotion, School of Public Health, Tehran University of Medical Sciences, Tehran, Iran

Student's Scientific Research Center, Tehran University of Medical Sciences, Tehran, Iran

H. Arfaeinia (✉)
Addiction and Lifestyle Research Center, Bushehr University of Medical Sciences, Bushehr, Iran

Department of Environmental Health Engineering, Faculty of Health and Nutrition, Bushehr University of Medical Sciences, Bushehr, Iran

M. K. Jindal, K. Patni (eds.), *Water Quality-Uranium Solubility Nexus: Unlocking the Secrets of Uranium Solubility*, Earth and Environmental Sciences Library, https://doi.org/10.1007/978-3-031-98909-4_3

ity. Case studies are presented to illustrate uranium behavior in different water systems, including groundwater, surface water, and polluted water, emphasizing the role of anthropogenic activities. The health implications of uranium solubility are critically analyzed, considering its nephrotoxic and radiotoxic effects. The environmental consequences, including bioaccumulation and ecosystem disruption, are also discussed. Current research findings are synthesized to identify existing knowledge gaps and suggest future research directions. By providing a comprehensive understanding of the factors influencing uranium solubility, this chapter underscores the importance of monitoring and managing water quality to mitigate the risks associated with uranium contamination.

Keywords Water quality · Uranium solubility · Sources of uranium in water

1 Introduction

1.1 Importance of Water Quality

Water quality plays a fundamental role in maintaining environmental and public health, influencing ecosystems, human well-being, and industrial processes. The physicochemical properties of water, including pH, temperature, dissolved oxygen, and the presence of organic and inorganic constituents, govern its suitability for consumption, agricultural use, and industrial applications [1]. Pollutants in water bodies, such as heavy metals, radionuclides, and organic pollutants, pose significant risks to both ecological systems and human health, necessitating continuous monitoring and management strategies [2].

Among the various pollutants, uranium presents a unique challenge because of its dual nature as a chemically toxic heavy metal and a radioactive element. The solubility and mobility of uranium in aquatic environments are heavily influenced by water quality parameters, such as pH, redox potential, and the presence of complexing agents such as carbonate and phosphate ions [3]. Groundwater pollution with uranium is a critical concern, as it can lead to prolonged exposure through drinking water, contributing to nephrotoxicity and other adverse health effects [4]. Studies have shown that uranium solubility is significantly enhanced under alkaline conditions because of the formation of uranyl–carbonate complexes, whereas reducing environments tend to immobilize uranium in its less soluble U(IV) state [2, 4].

Given the increasing global concerns over water scarcity and pollution, understanding the intricate relationship between water quality and uranium solubility is essential for developing effective remediation strategies and ensuring safe water resources. This book explores the fundamental interactions governing uranium behavior in aqueous environments, emphasizing the importance of water chemistry in mitigating uranium pollution and protecting both human and environmental health.

1.2 Overview of Uranium Solubility

The solubility of uranium is a complex process influenced by various environmental parameters, including its oxidation state, mineral composition, and geochemical interactions. A comprehensive understanding of these factors is essential for predicting uranium mobility in contaminated environments and nuclear waste disposal sites. The following sections discuss key aspects influencing uranium solubility. Uranium predominantly occurs in two oxidation states: U(IV) under reducing conditions and U(VI) in oxidizing environments. In metamorphic fluids, U(IV) is the dominant species, exhibiting low solubility (~10^{-12} molal) [2]. In contrast, U(VI) is significantly more soluble (1.3×10^{-7} molal) in oxidized fluids, particularly at lower temperatures [2]. Uranyl minerals, particularly uranyl-carbonate complexes, play crucial roles in controlling uranium solubility in contaminated soils [5]. Geochemical modeling suggests that these complexes represent the most stable uranium species, significantly influencing its transport and leaching behavior [5]. Research indicates that uranium solubility progressively increases in carbonate-rich environments, with a documented solubilization rate of 0.16 $\mu g\ m^{-2}\ h^{-1}$ under oxic conditions [6]. The presence of carbonate minerals enhances uranium dissolution, highlighting the critical role of environmental chemistry in uranium mobility [6]. Although uranium solubility is governed primarily by oxidation states and mineral interactions, understanding these processes is vital for informing strategies for environmental remediation and nuclear waste management.

1.3 The Connection Between Water Quality and Uranium Solubility

Uranium solubility in aqueous systems is strongly affected by several water quality parameters, such as pH, temperature, and the presence of dissolved ions. A thorough understanding of these interactions is essential for evaluating the environmental risks posed by uranium contamination in water bodies. Uranium is more soluble under slightly basic conditions (pH $\approx$ 7.5) than under acidic conditions (pH $\approx$ 4.5) because of the increased formation of uranyl carbonate complexes in alkaline environments [3]. Uranium leaching reaches its optimum at about 50 °C, a temperature at which carbon species become most soluble, thereby promoting the mobilization of uranium [3]. The presence of calcium and bicarbonate ions can increase uranium solubility by forming stable complexes. Moreover, elevated levels of total dissolved solids can further influence uranium mobility [7]. Uranium pollution is associated with both chemical and radiological hazards. In groundwater systems, for example, high uranium concentrations can lead to adverse health effects such as kidney damage [8]. Given the complex interplay between uranium and various water quality parameters, remediation efforts must account for these factors to effectively control uranium solubility and mobility in polluted environments.

2 Chemical and Physical Properties of Uranium

Uranium ranks 49th in abundance in the Earth's crust, with levels comparable to those of beryllium, tin, and arsenic. It is quite common in seawater and is the 26th most abundant element, along with arsenic and vanadium. With a density of 19.1 kg/m^3, uranium is a highly dense metal similar to gold and has the highest atomic number (92) among all naturally occurring elements on the periodic table. Uranium is lithophile because it tends to stay near the surface of the earth and tightly binds with oxygen. It is naturally present in all soils, minerals, rocks, and water sources. It can also originate from a diverse range of human-made sources such as phosphate fertilizers, mine waste, fly ash from power plants, and military activities [9]. Uranium is the most abundant element found in nature and has both chemical and radiological characteristics. A total of 27 uranium isotopes linked to nuclear reactions or high-energy experiments have been identified, with the most prevalent being ^{234}U, ^{235}U, and ^{238}U [10]. In different radioactive tests, these isotopes transform into stable lead isotopes and emit alpha, beta, and gamma rays. Uranium exists in four primary isotopic compositions, namely, natural uranium (99.27% ^{238}U, 0.72% ^{235}U, and 0.005% ^{234}U), enriched (higher amounts of ^{234}U and ^{235}U), depleted (lower levels of $U^{235}U$ postprocessing, and ^{236}U), plutonium, americium, and other transuranium elements formed in nuclear reactions. Uranium isotopes possess a lengthy half-life, aiding in diminishing their radiological activity in their natural surroundings [10]. Uranium is typically found in varying concentrations in the Earth's crust, with a total abundance of 1.7 ppm. In the upper crust, the concentration is 2.7 ppm, whereas in the oceanic crust, it is 0.5 ppm [11]. Despite these quantities, uranium may still be discharged into the environment through natural processes and human activities, resulting in possible health hazards, particularly in industrial regions [11, 12]. Other factors influencing the movement and dispersal of uranium include pH, redox potential, complex formation, and adsorption onto organic substances [13]. Table 1 shows the physical and chemical properties of uranium [10].

2.1 *Ionic Forms of Uranium in Water*

Uranium is a naturally present element found in underground water sources. When uranium-containing minerals dissolve in groundwater, they seep into drinking water. The quantity of uranium present in well water can differ and will change on the basis of how much uranium is present in nearby rocks [14]. Uranium enters surface water from the air through particle deposition or from the erosion of rock and soil. Uranium found in surface water can spread over vast distances to ponds, rivers, and oceans. The presence of uranium in water is altered through the creation of complexes and redox reactions. The oxidized versions of uranium are quite soluble and can be removed from rocks to move through the environment. Soluble uranium precipitates under highly reducing conditions [14].

Table 1 The physical and chemical characteristics of various uranium compounds

Property	Description/Value
Weight of individual atoms or molecules	238.0289
Formula of a chemical	U
Synonyms	Uranium I
Rewordings	
CAS Registry No.	7440-61-1
Color	Silvery-white
State of matter	Solid
Smell	No data
Melting point, °C	1135
Boiling point, °C	4131
Autoignition temperature	20 °C (cloud), 100 °C (layer)
Solubility	
Water	Insoluble
Other solvents	Soluble in concentrated acids; insoluble in alkalis, alcohols
Density, g/cm^3	18.06–19.1
Partition coefficients	Not relevant
Vapor pressure	0 mmHg at 20 °C; 1 mmHg at 2450 °C
Henry's law constant	Not relevant
Flashpoint	No data
Flammability limits	Not relevant
Conversion factor	1 µg = 0.69 pCi

Uranium is found in states III, IV, V, and VI, but the main states found in nature are IV and VI [9]. Various factors, such as the formation of complexes, hydrolytic processes, redox characteristics [15], and adsorption in natural waters, can affect the chemical behavior of uranium species in aqueous solutions. The outcome is determined by the concentrations of minerals and/or colloids, among other factors [15–17].

The most common type of uranium found in water is the uranyl ion (UO_2^{2+}), which can be influenced by factors such as pH and temperature in the environment. Strongly reducing U(VI) yields the U^{4+} cation; however, when exposed to air, it quickly reverts to UO_2^{2+} through oxidation. The pentavalent UO^+_2 ion can be reversibly generated by reducing UO_2^{2+}, but it quickly divides into U(IV) and U(VI). The reduction of U(IV) results in the formation of unstable trivalent $U^{3+,}$ which oxidizes in aqueous solution [17]. The uranyl ion (UO_2^{2+}) is frequently found in balance with complexes such as $[UO_2(H_2O)_5]^{2+}$ and $[UO_2Cl(H_2O)_4]^+$, with stability constants revealing their varying levels of abundance [18]. Different coordination geometries of UO_2Cl^+, $UO_2SO_4^-$, and $UO_2OH_4^{2-}$ complexes have been identified on the basis of temperature and ionic strength [19]. In neutral to alkaline environments, calcium uranyl carbonate complexes such as $[(UO_2)_{24}O_4(OH)_{12}(CO_3)_{36}]^{4-}$ are predominant, emphasizing the importance of carbonate interactions in uranium speciation [20].

Molecular dynamics simulations have shown that uranyl ions can establish durable hydration shells, impacting their actions in water habitats [21]. Although the uranyl ion is important in uranium chemistry, its behavior in natural waters can be complex due to its interactions with other ions and environmental factors, resulting in a variety of complexes. Comprehending these dynamics is crucial for successful environmental control and remediation plans.

2.2 *Solubility Mechanisms*

Several factors, such as the oxidation state, pH, water chemistry, and presence of agents, affect the solubility of uranium in water [22]. Uranium is found mainly in four oxidation states in water solutions, and the uranyl cation (UO_2^{2+}) is the most soluble [17]. In the presence of oxygen, uranium moves more easily, especially as U(VI), whereas U(IV) is less soluble and typically forms UO_2 precipitates [23]. Uranium (VI) solubility in brines without carbonate is influenced mainly by hydrolysis at elevated pH values, whereas borate complexation is a factor at pH values of 8–9 [24]. The solubility of uranyl minerals, such as those in metacompilers, exhibits a U-shaped relationship with pH, reaching its lowest solubility at pH 8.5 [25]. In silicate liquids, the solubility of uranium increases with hyperalkalinity and is greater in oxidizing environments and compositions rich in sodium [26]. Higher levels of bicarbonate increase solubility through the creation of durable uranyl-carbonate complexes [25]. The levels of uranium in natural waters can change greatly due to geological factors, redox conditions, and the presence of different ions [9]. Uranium is mobile under acidic conditions but becomes immobilized in strongly reducing environments [27]. The release of uranium into groundwater from natural sediments is affected by the breakdown of minerals containing uranium, removal from mineral surfaces, the creation of aqueous complexes, and a reduction in precipitation. Iron-manganese (oxy)hydroxides and silicate minerals play a key role in holding uranium in sediments, including a substantial portion as absorbed ions [28]. Although the solubility mechanisms of uranium are well researched, additional studies are needed to comprehensively grasp the effects on environmental health and safety due to the intricate interactions in natural waters.

3 Water Quality Parameters that Influence Uranium Solubility

The solubility of uranium in groundwater is governed by several water quality parameters, including pH, ionic strength, and the presence of complexing agents. A thorough understanding of these factors is essential for evaluating uranium contamination in aquifers. Uranium is more soluble in mildly alkaline environments (pH ~7.5) than in acidic environments (pH 4.5), as higher pH promotes the formation of uranyl carbonate complexes, thereby increasing uranium mobility [3]. In Indian groundwater, a pH of approximately 7.5 has been linked to competitive dissolution–precipitation processes that influence uranium speciation [7]. Elevated ionic strength, particularly due to high concentrations of calcium and total dissolved solids (TDS), can reduce uranium solubility by facilitating the formation of insoluble mineral phases such as calcite [29, 30]. The presence of major anions and cations, including sulfate and bicarbonate, plays a significant role in determining uranium distribution and mobility in groundwater [30]. The oxidation–reduction potential of groundwater is a critical factor in uranium speciation. Under reducing conditions, uranium predominantly exists as U(IV), which is significantly less soluble than U(VI) [30].

A mixed redox environment can enhance uranium mobilization by promoting the dissolution of iron and manganese oxides, releasing uranium into solution [7]. While these parameters generally govern uranium solubility, variations in local geological and hydrological conditions can lead to deviations in uranium behavior, making it challenging to predict uranium mobility across different aquifer systems.

4 Case Studies

Various case studies have shown that uranium contamination in water can lead to serious health hazards [31]. These studies investigated the presence of uranium in drinking water and how it may affect human health [32]. Groundwater uranium levels can range significantly, from 0.10 μg/L to 571.7 μg/L [32, 33]. In a study near the Manchanabele Reservoir in Bengaluru, India, uranium concentrations in groundwater varied from 0.88 to 581.47 ppb, with approximately 66% of samples falling within the WHO's recommended limit of 30 ppb [31]. High concentrations of uranium in water have been associated with increased chances of developing cancer and other health issues [34]. Sheep tissue can also contain concentrations of uranium from nearby water, as uranium builds up in the food chain [34]. In Kamianske, Ukraine, a case study noted how uranium was released from tailings into nearby water sources, leading to a notable rise in contamination levels [35]. Efficient ways to remedy this situation include membrane filtration and ion exchange, resulting in more than 97% uranium removal efficiency [36]. Household filters are advised as a temporary measure to lower uranium levels in well water and minimize risks. Furthermore, monitoring indoor radon levels is recommended because of its connection to both uranium and leukemia [34]. Although these studies emphasize the risks of uranium found in water, they also highlight the challenges of addressing this issue and stress the importance of continuous monitoring and advancements in technology [37].

4.1 Uranium Solubility in Various Water Systems

Various forms of uranium are found in water sources due to runoff from naturally occurring deposits, its discharge from mining byproducts, or nuclear activities, as well as the burning of coal and other fossil fuels. Naturally, formed phosphate deposits used for fertilization can also contain high concentrations of uranium (up to 150 mg $kg^{-1)}$, leading to contamination of groundwater [38]. The chemical components present in groundwater are not regularly tested as indicators of water quality, even though they have toxic effects. Uranium and its offspring can be deadly when found in elevated concentrations in groundwater. Natural U is the heaviest element found in its original state, without any added or removed components.

^{238}U is the most common isotope in water by mass, with a long half-life of 4.5×10^9 years. The original process of the geochemical cycle of U commences with rock undergoing chemical weathering in the oxidized zone on land, leading to its release into water with consistent isotopic ratios of ^{238}U and ^{235}U at 137.88:1 [39]. The quantity of U found in groundwater is influenced by its presence in bedrock, the nearness of rocks rich in U, soil, the pH and oxidation state of U, and the levels of other substances in the water that can create U complexes (e.g., $CO_3.^{2-}$, F^-, $PO_4\ ^{3-}$) [38]. Dissolved uranium is typically present in natural waters in small amounts, with elevated levels usually found in mineralized, thermal, and saline waters [40]. U can be found in surface waters in three different oxidation states: U^{4+} (U[IV]), UO_2^+ (U[V]), and UO_2^{2+} (U[VI] or uranyl ions). In low redox potential waters (anoxic waters), U is present as U^{4+} and/or $^{UO2+}$. Uranium in the IV oxidation state has a significant tendency to form solid precipitates such as uraninite (UO_2[s]) and remain in place, whereas UO_2^+ creates soluble complexes. Alternatively, in oxygenated waters, uranium appears as UO_2^{2+} and creates durable, easily soluble ionic and/or neutral complexes that are very movable and significantly contribute to uranium transportation in the process of weathering [38]. The prevailing species of U depend on the pH-Eh conditions and the concentration and accessibility of complexing ions. Uranium shows increased mobility in slightly alkaline environments (pH 7.5) as opposed to acidic conditions (pH 4.5), leading to notable leaching at moderate temperatures (approximately 50 °C) because of increased carbonate species [3]. The redox and complexation reactions of U are greatly impacted by hydrolysis, as hydrolytic reactions can restrict solubility or affect sorption to particles [9].

In the drinking water sector, the US Environmental Protection Agency has suggested that the acceptable level of ^{238}U in the drinking water standard should be 20 μg l^{-1}. Additionally, the levels of U in plants increased in a linear fashion as the concentration of U in irrigation water increased, particularly in lettuce and radish [41]. The categorization of uranium as a chemo-toxicological and radiological substance is recognized. It is considered the sole radioactive element where chemical toxicity is the limiting factor in risk evaluations [42]. The increase in the use of water with concentrations of U higher than those recommended by the EPA (Environmental Protection Agency) has raised concerns about potential toxicological and radiological risks for human consumers. The main way people are exposed to uranium is by drinking well water, which accounts for approximately 99% of the toxicological and radiological doses received by human consumers. This finding indicates that relying on alternate water sources or treating high U levels in well water could significantly reduce exposure, especially for Nambe residents, who rely on well water [41]. The estimated safe level of uranium in water is approximately 2.5–38 times higher than the current EPA limit of 30 μg U l^{-1} [43]. The EPA's current restriction on kidney hazards from prolonged exposure is cautious and aimed at protection, assuming the accuracy of a safety factor and toxicity threshold of 1 μg U g^{-1} kidney [44]. In regard to the chemical quality of drinking water and health issues, it is crucial to emphasize the dangers of high levels of uranium. Importantly, health problems related to water can occur because of the toxicity of this element. This type of toxicity is commonly found in industrial zones in developing nations, where people likely consume locally sourced food and water [45]. Understanding how the solubility of uranium changes under different conditions is essential for evaluating the safety of nuclear waste storage facilities over time. Although uranium is not very soluble in natural waters overall, specific areas with high concentrations can be a threat to the environment, so continuous monitoring and research are needed to comprehend its intricate interactions in various water systems [46]. Different origins of uranium in underground water sources are shown in Fig. 1.

4.2 *Impact of Anthropogenic Activities*

Although uranium originates primarily from natural sources, human activities such as mining for uranium, coal, phosphate, oil, and gas; the use of phosphate fertilizers in agriculture; industrial activities; and military operations can also increase uranium levels in nearby water sources. The amount of uranium present in surface and subsoil water varies greatly and is influenced primarily by the local geology

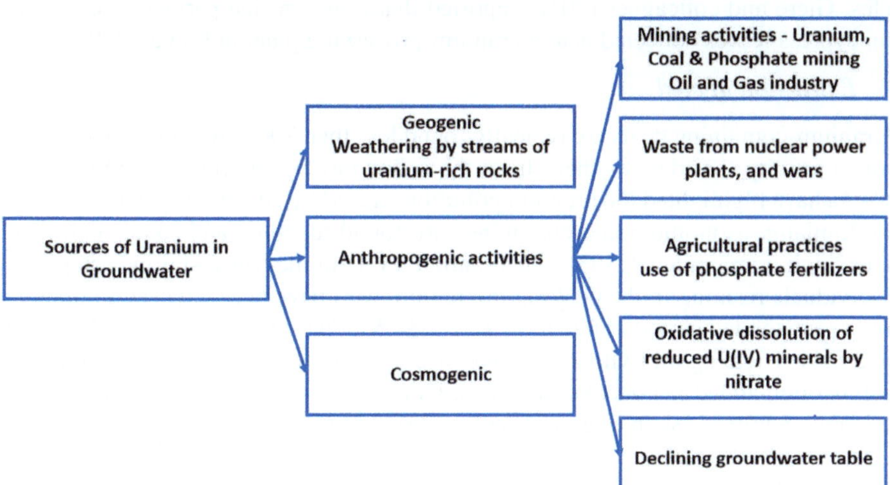

Fig. 1 Different origins of uranium in underground water sources [47]

[48]. Mining and extraction operations such as exploratory drilling, cutting, blasting, transportation, stockpiling, and processing/leaching can emit dust with uranium and other harmful substances into the air and nearby water sources, affecting local soil, plants, and animals [47].

1. ***Uranium mining***

In 2019, Kazakhstan led in terms of uranium production from mines, with 42% of the world's supply, whereas Canada followed with 13%, Australia with 12%, and India holds the 11th position in terms of uranium production globally. Water in uranium mining regions can become polluted because of mining and milling activities. Furthermore, poor disposal of uraniferous tailings and waste within a uranium mine site can lead to uranium pollution of nearby surface water, groundwater, soil, and air [47]. Both open-pit and underground mining activities necessitate the removal of water from the mining site, resulting in water pollution. Furthermore, surface water runoff and seepage from waste rock piles introduce uranium into nearby water sources. During the settling process, tailings and ore stockpiles also play a role. In certain regions of the US, mine water containing uranium is frequently released into nearby areas without adequate removal processes, leading to possible uranium exposure from crops growing in contaminated soils [49]. Consequently, local populations are experiencing an increased rate of long-term illnesses [50].

2. ***Debris from nuclear power facilities and global conflicts***

In the nuclear industry, activities such as nuclear fuel production, processing, and disposal can lead to the release of uranium, potentially posing risks of exposure in the environment or workplace. The military arsenals of three NATO (North Atlantic Treaty Organization) countries—the USA, Great Britain, and France—contain depleted uranium (DU) munitions [51]. The high density of DUs makes them ideal for tank armor and certain bullets, allowing them to penetrate enemy armor vehicles. Giere and colleagues (2012) reported that uranium-rich particles were found on cypress leaves collected near a uranium-processing plant in France [52].

3. ***Extraction of coal***

Uranium-containing coals are typically found less than 200 vertical feet beneath the unconformity that divides them from the sediments below. Recent findings from China have highlighted the uranium pollution caused by coal mining in the Yili area of Xinjiang, as uranium and coal mines are found together, and some coal mines have high uranium levels [53]. The uranium levels in the hair and urine samples of individuals residing in the coal mining region, who have been exposed to uranium over an extended period, range from 22.2–634.5 ng/g (with an average of 156 ng/g) and 8.44–761.6 ng/ml (with an average of 202.6 ng/ml), respectively. Although fly ash is often added to concrete materials, its radioactivity is similar to that of other common additives, such as granite and red brick [47].

4. ***Mining for phosphate rock***

Water quality is impacted by phosphate rock extraction and processing through the discharge of industrial effluents containing toxic metals, heavy metals, and radioactive elements such as uranium into water sources. The typical amount of uranium found in phosphate rocks is approximately 75 micrograms per gram, causing a substantial release of uranium when mine water is discharged. Furthermore, the large quantities of dust released also lead to poisonous air in mining regions and nearby areas. Significant levels of fluoride and the release of radon gas are additional issues that have direct impacts on human health [54].

5. ***Industry of oil and natural gas***

The oil and gas sector heavily relies on radiation generators containing sealed and unsealed radioactive sources, some of which may pose a threat to the environment and human health if not handled and controlled correctly. Furthermore, substantial quantities of naturally existing radioactive material can be found in reservoir rock during production, maintenance, and operations [55]. Radioactive elements such as uranium are naturally present in different soil and rock types, as well as in the water flowing through them, due to oil and gas extraction processes. It includes high levels of heavy metals and may also contain substantial quantities of another cancer-causing radioactive element, radium. Drilling and cutting up dirt and rock in fracking creates drilled cuttings with high levels of uranium and thorium, potentially increasing the cancer risk for workers in the oil and gas industry and nearby residents [47].

6. ***Agricultural practices***

The extensive use of fertilizers, a major source of uranium in groundwater, is leading to a notable amount of uranium entering agricultural soils [56]. Researchers have analyzed uranium levels in commonly used phosphate fertilizers in Brazilian farmlands as well as fertilizers from other nations. In Brazilian phosphate fertilizers, the amount of uranium is between 5.17 and 54.3 μg/g, whereas in other countries, it varies from 3.2 to 221 μg/g. Furthermore, many commercial fertilizers not only increase the leaching of uranium into groundwater but also lower the soil pH. Depending on the pH and atmospheric CO_2 pressure, transportable uranium species can move from soil to aquatic systems [45, 56]. Hence, prolonged application of fertilizers containing uranium can greatly increase the levels of uranium in fertilized farmlands. Furthermore, if soluble uranium is transported to soils and groundwater, it becomes accessible to plants for absorption through their roots [57].

7. ***Nitrate pollution***

Nitrates are supplied to the soil in the form of fertilizers to ensure that plants receive the necessary nitrogen they need for growth. Excessive nitrate under acidic conditions can aid in the oxidative breakdown of U(IV) minerals with reduced properties. Following nitrate oxidation, uranium is able to create chemical compounds with carbonate and calcium ions in groundwater, facilitating the movement of uranium through the soil column and into the groundwater. Approximately 78% of the groundwater geochemical data gathered from the High Plains (HP) and Central Valley (CV) aquifers in the United States show levels that surpass the US EPA's maximum allowable limits, largely due to the presence of nitrate [58].

8. ***Decreasing depth of groundwater***

Human actions, specifically the excessive use of groundwater for farming, industry, and homes, can lead to falling groundwater levels that create conditions that support the development of soluble uranyl carbonate complexes. Contaminants such as uranium and arsenic in the soil can be released and contaminate water bodies as the water level decreases [59].

5 Health and Environmental Implications

5.1 Uranium Toxicity

Soil and surface water systems contain many uranium ions (UO_2^{2+}), which are the principal contributors to the chemical toxicity of uranium [60]. Owing to its radioactive characteristics, researchers have thoroughly investigated the effects of uranium on human health. Both chemical and radiological toxicity can result from uranium exposure. While exposure to natural and depleted uranium usually results in chemical harm, enriched uranium mainly poses radiological risks. The only exception, which also includes a small radiological component, is prolonged exposure to insoluble uranium compounds [61]. The toxicity of U is influenced mostly by its chemical formula and route of exposure [62]. The most pronounced systemic effects are from highly water-soluble uranium compounds: uranyl nitrate, fluoride, uranium hexafluoride, and tetrachloride [63]. Poorly water-soluble chemicals, such as uranium tetrafluoride and ammonium diuranate, are limited by low levels of systemic toxicity [45]. Insoluble uranium compounds, such as uranium dioxide, trioxide, peroxide, and triuranium octaoxide, exhibit comparatively lower systemic toxicity [64].

Recent studies have investigated the toxic effects of uranium across various species [64, 65]. This study identified airborne particles, food and water consumption, and skin absorption as the three main uranium exposure pathways [66]. Endocytosis is the probable mechanism through which U enters cells. Radiotoxicity is not the principal contributor to uranium toxicity; instead, chemical toxicity is the primary factor [67].

5.2 *Impact on Human Health*

Uranium may enter water supplies via natural deposits or industrial activities [9]. While it occurs naturally in small amounts, it becomes harmful at increased concentrations [68]. The principal health consequence of exposure to high uranium levels is renal impairment [62]. Nonetheless, the majority of observed effects in humans have resulted from elevated short-term exposure, and several occupational studies have indicated a reversal of renal effects following the cessation of exposure [69]. Prolonged exposure may lead to renal consequences [70]. The health impacts of exposure to naturally occurring uranium in drinking water are linked to the chemical toxicity of uranium [62]. The USEPA established a drinking water standard for uranium, primarily on the basis of its chemical toxicity, while also considering the potential carcinogenic effects of prolonged exposure to uranium in drinking water under the premise that any radioactive substance may induce cancer [71]. The possible effects of uranium on several body systems, particularly the reproductive system and developmental processes, continue to be a focus of continuing research [45, 72].

5.3 *Environmental Consequences*

Uranium, a naturally occurring radioactive element, is recognized for its applications in nuclear energy and weaponry [72]. Nevertheless, the presence of uranium in aquatic habitats has substantial environmental implications [45]. The presence of uranium in water can lead to various substantial environmental consequences: uranium contamination can impair aquatic ecosystems [62]. Uranium can accumulate in aquatic organisms, leading to bioaccumulation and biomagnification within the food chain [73]. Fish and other creatures exposed to high uranium levels may experience developmental and reproductive difficulties, affecting biodiversity [74].

Uranium can substantially degrade the quality of groundwater. Contaminated water sources present hazards for consumption, agriculture, and recreational activities [75]. The World Health Organization (WHO) has set a recommended threshold of 30 μg/L for uranium in potable water owing to its potential to induce renal impairment and increase cancer risk [76]. Uranium that permeates aquatic systems can ultimately accumulate in sediments, leading to soil pollution. This impacts the surrounding ecology and may also reduce agricultural yield if crops are grown in contaminated soils [77].

The extraction and processing of uranium frequently lead to habitat loss. The extraction of uranium often entails considerable land disturbance, potentially affecting local ecosystems and displacing wildlife populations [78]. Uranium pollution can disrupt food networks within ecosystems. The consumption of uranium-contaminated vegetation by herbivores may have harmful effects on their predators throughout the food chain [79, 80].

6 Current Research and Future Directions

6.1 Recent Findings

The solubility of uranium in water enables groundwater containing it to move through sandstone aquifers, where it engages with strata rich in organic carbon. Uranium strongly attracts organic carbon, which causes an amorphous organouranium complex to form near these high-carbon deposits. When the pH is less than 5, the main species of U(VI) in water is the soluble uranyl ion UO_2^{2+}, which dissolves quickly in oxygenated water. When the pH is between 5 and 9, $UO_2(OH)_2$ is the main species, and when the pH is greater than 9, $UO_2(OH)_0$ becomes the main species. U(IV) can be found as $U(CO_1)_4^{+4}$, U_4O_9, and UO_2 in complex systems such as natural waters that are anoxic and have pH values of 6, 7, and 8. In oxic water at pH 9.2, the predominant species is $UO_2(CO_3)_3^{-4}$ [81, 82].

In the Deoghar region of India, 150 groundwater samples were analyzed for uranium, and additional water quality indicators were assessed in relation to these samples. Research has indicated that the oxidation–reduction potential (ORP) and the contents of calcium, phosphate, nitrate, and magnesium are positively correlated with the uranium content. This finding indicates that uranium may exist in groundwater as a dissolved salt with these characteristics. Uranium levels were negatively correlated with temperature, electrical conductivity, salinity, dissolved oxygen (DO), and total dissolved solids (TDS). These investigations indicate that the solubility of uranium in groundwater can be influenced by specific water quality parameters [82]. Research in geochemical modeling has examined uranium speciation and its correlation with water quality parameters. The study revealed that higher uranium levels in the baseline environment were caused by alkaline groundwater and surface water, which are high in calcium. These waters help U (6+) complexes move around and become free. The uranium enrichment in gneiss and granite host rocks is also significant. The solubility of uranium can be influenced by alkalinity, calcium concentration, pH, and various other water quality characteristics [83]. To predict groundwater quality, researchers in California's Central Valley developed a regional model that incorporates aquifer data and soil properties. Research has identified soil alkalinity and calcium concentrations as the primary factors influencing uranium contamination in groundwater. This indicates that uranium solubility in groundwater is significantly affected by soil conditions, which subsequently influence water quality [84].

6.2 Future Research Needs

Future research should focus on uranium solubility and other water quality indices, such as microbial activity, organic matter, and trace elements. Changes in water quality over time, such as seasonal changes in temperature, rainfall, or pollution levels, help us learn more about how uranium dissolves over time [82, 85]. Using better geochemical models that include the complex interactions between uranium and different chemicals in water, such as calcium, phosphate, and carbonate, may help us obtain a better understanding of how uranium dissolves in different situations [72]. Using both natural and human-made factors, advanced computer programs with artificial intelligence can be used to predict how uranium will behave in different types of water and under different conditions [86, 87]. Because the solubility of uranium may change when it interacts with soil and sediments, more research should look into how different types of soil (such as clay and sand) and sediments affect uranium's ability to move through groundwater [88]. The surface chemistry of soils and sediments must be examined for its impact on uranium retention and release across different water quality conditions [89]. An examination of uranium solubility in various aqueous sources, including irrigation water, urban runoff, and industrial effluents, is essential. These sources often include compounds that may either increase or decrease uranium solubility [47]. Comparing how uranium dissolves in freshwater and saltwater helps us understand how salinity and the minerals that are present affect how uranium behaves. In places where uranium levels in water are higher than what is considered safe, more research needs to be done on how aquatic organisms can use dissolved uranium and how it might affect the environment [75, 90]. Future research should focus on determining the long-term health risks of drinking water, especially in regard to changing uranium levels and the effects that occur when uranium is mixed with other water pollutants [91].

7 Conclusion

The pollution of the environment with uranium is of particular concern since it has dual toxicological impacts—both chemical and radiological. Although uranium occurs naturally in the Earth's crust, human endeavors such as mining, industrial processes, phosphate fertilizer use, and military actions can substantially increase uranium concentrations in water, soil, and air. Uranium is added to surface water through atmospheric deposition, rock and soil erosion, and industrial waste, whereas in groundwater, mineral dissolution, desorption, and redox reactions control its mobility.

The two dominant oxidation states of uranium, U(IV) and U(VI), are mainly associated with its solubility and transport. U VI) is highly soluble and mobile under oxidizing conditions, whereas U(IV) is more stable and tends to precipitate, restricting its dispersal in groundwater. In addition, the mobility of uranium in aquatic systems is also controlled by environmental factors such as pH, redox potential, and complexing agents such as carbonate and phosphate, which influence its potential for contamination.

The health effects of exposure to uranium—particularly its nephrotoxicity—and its potential for bioaccumulation and biomagnification in the environment highlight the need for strict environmental control and monitoring. A detailed understanding of the geochemical processes of uranium and their interactions with natural and anthropogenic processes is essential for devising effective mitigation strategies. With sustainable management practices, reduced industrial emissions, and improved water treatment technologies, the risks of uranium contamination can be minimized to safeguard human health and ecological systems.

Water chemistry is a primary regulator of the solubility, mobility, and bioavailability of uranium in aquatic systems and has significant implications for environmental and public health. The geochemical conditions present in terms of pH, redox potential, and complexing ligands such as carbonate and phosphate dictate different chemical forms and transport routes of uranium.

References

1. Borhani Yazdi N et al (2024) Degradation of piroxicam and celecoxib from aqueous solution by high-energy electron beam as a sustainable method. Heliyon 10(21):e39839
2. Zhang M et al (2023) The immobility of uranium (U) in metamorphic fluids explained by the predominance of aqueous U (IV). Minerals 13(3):427
3. Briganti A et al (2017) Understanding uranium behaviour in a natural rock–water system: leaching and adsorption tests on the Tufo Rosso a Scorie Nere ignimbrite (Viterbo area, Central Italy). Environ Earth Sci 76:1–14
4. Abolli S et al (2024) Understanding uranium distribution: a systematic review and meta-analysis in the context of drinking water resources. Res Eng 22:102152
5. Zhao D et al (2012) Impact of water quality parameters on the sorption of U(VI) onto hematite. J Environ Radioact 103(1):20–29
6. Noubactep C et al (2006) Characterizing uranium solubilization under natural near oxic conditions. In: Uranium in the Environment: mining Impact and Consequences. Springer, Berlin, Heidelberg, pp 425–435
7. Sujathan S, Singh A (2023) Investigation of potential drivers of elevated uranium prevalence in Indian Groundwaters with a unified speciation model. Environ Sci Technol 57(5):1970–1986
8. Meher P et al (2015) Post monsoon spatial distribution of Uranium in water of Alaknanda and Ganges River. Iran J Radiat Res (IJRR) 13(1):95–99
9. Smedley P, Kinniburgh D (2023) Uranium in natural waters and the environment: distribution, speciation and impact. Appl Geochem 148:105534
10. Keith S et al (2013) Toxicological profile for uranium. Agency for Toxic Substances and Disease Registry (US), Atlanta, GA
11. Chi G et al (2019) Uranium-rich diagenetic fluids provide the key to unconformity-related uranium mineralization in the Athabasca Basin. Sci Rep 9(1):5530
12. Bone SE et al (2019) Complexation by organic matter controls uranium mobility in anoxic sediments. Environ Sci Technol 54(3):1493–1502

13. Rathod AM et al (2023) Uranium: an overview of physicochemical properties, exposure assessment methodologies, and health effects of environmental and occupational exposure. Environ Geochem Health 45(5):1183–1200
14. Piemonte V, De Falco M, Basile A (2013) Sustainable development in chemical engineering: innovative technologies. Wiley
15. Takeno N (2005) Atlas of eh-pH diagrams. Geological survey of Japan open file report no. 419(102): 285
16. Choppin GR (2006) Actinide speciation in aquatic systems. Mar Chem 99(1–4):83–92
17. Szabó A et al (2009) Geochemistry and determination possibilities of uranium in natural waters. Acta Technica Jaurinensis 2(1):19–34
18. Edwards SJ, Bowron DT, Baker RJ (2022) Insights into the solution structure of the hydrated uranyl ion from neutron scattering and EXAFS experiments. Dalton Trans 51(36):13631–13635
19. Kalintsev A et al (2023) Nature and coordination geometry of geologically relevant aqueous Uranium (VI) complexes up to 400 °C: a review and new data. J Hazard Mater 452:131309
20. Perry SN, Rodriguez VG, Burns PC (2023) Nanoscale calcium uranyl carbonate clusters in water. Chem Geol 641:121766
21. Rodríguez-Jeangros N, Seminario JM (2014) Density functional theory and molecular dynamics study of the uranyl ion (UO 2) 2+. J Mol Model 20:1–12
22. Olliia K (1988) Dissolution of UO 2 at various parametric conditions: a comparison between calculated and experimental results. MRS Online Proc Lib 127:337–342
23. Guilbert S et al (2002) Dissolution of uranium dioxide in simulated Boom clay water. Radiochim Acta 90(2):75–80
24. Lucchini J et al (2013) Uranium (VI) solubility in carbonate-free WIPP brine. Radiochim Acta 101(6):391–398
25. Stanley DM, Wilkin RT (2019) Solution equilibria of uranyl minerals: role of the common groundwater ions calcium and carbonate. J Hazard Mater 377:315–320
26. Dominé F, Velde BD (1985) Preliminary investigation of the processes governing the solubility of uranium in silicate melts. Bulletin de minéralogie 108(6):755–765
27. Banos A, Burrows R, Scott T (2021) A review of the mechanisms, reaction products and parameters affecting uranium corrosion in water. Coord Chem Rev 439:213899
28. Alam MS, Cheng T (2014) Uranium release from sediment to groundwater: influence of water chemistry and insights into release mechanisms. J Contam Hydrol 164:72–87
29. Lemire RJ, Garisto F (1992) The effect of ionic strength, groundwater composition and temperature on calculated radionuclide solubilities. Radiochim Acta 58(1):37–52
30. Kumar A et al (2011) Geochemical modelling of uranium speciation in the subsurface aquatic environment of Punjab State in India. J Geol Min Res 3(5):137–146
31. Deepika DN et al (2024) Studies on uranium concentration in groundwater samples and its associated health hazards to the residents of surrounding regions of Manchanabele reservoir, Bengaluru, Karnataka, India. Radiat Prot Dosimetry 200(11–12):1084–1089
32. Chahal A et al (2019) Study of uranium in drinking water around the Sohna fault line in Haryana. J Geol Soc India 94:428–436
33. Singh L et al (2013) Health risk assessments due to uranium contamination of drinking water in Bathinda region, Punjab state, India. Radioprotection 48(2):191–202
34. Winde F, Erasmus E, Geipel G (2017) Uranium contaminated drinking water linked to leukaemia—revisiting a case study from South Africa taking alternative exposure pathways into account. Sci Total Environ 574:400–421
35. Rudakov D et al (2023) A predictive assessment of the uranium ore tailings impact on surface water contamination: case study of the city of Kamianske, Ukraine. J Environ Radioact 268:107246
36. Ighalo JO et al (2024) A review of remediation technologies for uranium-contaminated water. Chemosphere 352:141322
37. Jha SK et al (2024) Uranium standards in drinking water: an examination from scientific and socio-economic standpoints of India. Environ Sci Pollut Res:1–14

38. Waseem A et al (2015) Distribution of natural uranium in surface and groundwater resources: a review. Crit Rev Environ Sci Technol 45(22):2391–2423
39. Gascoyne M (1992) Palaeoclimate determination from cave calcite deposits. Quat Sci Rev 11(6):609–632
40. Kumar P et al (2011) Dynamics and dispersion modelling of nanoparticles from road traffic in the urban atmospheric environment—a review. J Aerosol Sci 42(9):580–603
41. Hakonson-Hayes AC, Fresquez PR, Whicker F (2002) Assessing potential risks from exposure to natural uranium in well water. J Environ Radioact 59(1):29–40
42. Wrenn M, et al. (1983) Metabolism of ingested uranium and radium. In Conference on radioactivity in drinking water, Easton, MD
43. Diop L et al (2018) The influence of climatic inputs on stream-flow pattern forecasting: case study of Upper Senegal River. Environ Earth Sci 77:1–13
44. Kurttio P et al (2006) Kidney toxicity of ingested uranium from drinking water. Am J Kidney Dis 47(6):972–982
45. Bjørklund G et al (2020) Uranium in drinking water: a public health threat. Arch Toxicol 94:1551–1560
46. Cevirim-Papaioannou N (2018) Redox chemistry, solubility and hydrolysis of uranium in dilute to concentrated salt systems. Appl Geochem 98:286–300
47. Balaram V, Rani A, Rathore D (2022) Uranium in groundwater in parts of India and world: a comprehensive review of sources, impact to the environment and human health, analytical techniques, and mitigation technologies. Geosyst Geoenviron 1(2):100043
48. Mohapatra AK et al (2021) The role of manganese carbonate precipitation in controlling fluoride and uranium mobilization in groundwater. ACS Earth Space Chem 5(10):2700–2714
49. Redvers N et al (2021) Uranium exposure in American Indian communities: health, policy, and the way forward. Environ Health Perspect 129(3):035002
50. Haakonde T et al (2020) Preliminary assessment of uranium contamination in drinking water sources near a uranium mine in the Siavonga District, Zambia, and associated health risks. Mine Water Environ 39(4):735–745
51. Ough EA et al (2006) Determination of natural and depleted uranium in urine at the ppt level: an interlaboratory analytical exercise. Health Phys 90(5):494–499
52. Gieré R, Kaltenmeier R, Pourcelot L (2012) Uranium oxide and other airborne particles deposited on cypress leaves close to a nuclear facility. J Environ Monit 14(4):1263–1273
53. Wufuer R et al (2018) A survey of uranium levels in urine and hair of people living in a coal mining area in Yili, Xinjiang, *China*. J Environ Radioact 189:168–174
54. Reta G et al (2018) Environmental impact of phosphate mining and beneficiation: review. Int J Hydrol 2(4):424–431
55. Koppel DJ et al (2022) Current understanding and research needs for ecological risk assessments of naturally occurring radioactive materials (NORM) in subsea oil and gas pipelines. J Environ Radioact 241:106774
56. Yamazaki IM, Geraldo LP (2003) Uranium content in phosphate fertilizers commercially produced in Brazil. Appl Radiat Isot 59(2–3):133–136
57. Rahman MA, Hasegawa H (2011) High levels of inorganic arsenic in rice in areas where arsenic-contaminated water is used for irrigation and cooking. Sci Total Environ 409(22):4645–4655
58. Nolan J, Weber KA (2015) Natural uranium contamination in major US aquifers linked to nitrate. Environ Sci Technol Lett 2(8):215–220
59. Coyte RM et al (2018) Large-scale uranium contamination of groundwater resources in India. Environ Sci Technol Lett 5(6):341–347
60. Liu W et al (2017) Highly sensitive and selective uranium detection in natural water systems using a luminescent mesoporous metal–organic framework equipped with abundant Lewis basic sites: a combined batch, X-ray absorption spectroscopy, and first principles simulation investigation. Environ Sci Technol 51(7):3911–3921
61. Miyazaki I et al (2011) Astrocyte-derived metallothionein protects dopaminergic neurons from dopamine quinone toxicity. Glia 59(3):435–451

62. Ma M et al (2020) Emerging health risks and underlying toxicological mechanisms of uranium contamination: lessons from the past two decades. Environ Int 145:106107
63. Keith S et al (2013) Health effects, in toxicological profile for uranium. Agency for Toxic Substances and Disease Registry (US), Atlanta, GA
64. Aly MM, Hamza MF (2013) A review: studies on uranium removal using different techniques. Overview. J Dispers Sci Technol 34(2):182–213
65. Arzuaga X et al (2010) Renal effects of exposure to natural and depleted uranium: a review of the epidemiologic and experimental data. J Toxic Environ Health, Part B 13(7–8):527–545
66. Liu F et al (2015) Chemical and biological insights into uranium-induced apoptosis of rat hepatic cell line. Radiat Environ Biophys 54:207–216
67. Yue Y-C et al (2018) The toxicological mechanisms and detoxification of depleted uranium exposure. Environ Health Prev Med 23:1–9
68. WHO, Depleted uranium: sources, exposure and health effects. 2001, WHO
69. De Castro L et al (2023) Renal toxicity and biokinetics models after repeated uranium instillation. Sci Rep 13(1):4111
70. Guéguen Y, Frerejacques M (2022) Review of knowledge of uranium-induced kidney toxicity for the development of an adverse outcome pathway to renal impairment. Int J Mol Sci 23(8):4397
71. Sahoo S et al (2020) Scientific background and methodology adopted on derivation of regulatory limit for uranium in drinking water–a global perspective. Environ Adv 2:100020
72. Keith LS, Faroon OM (2022) Uranium. In: Handbook on the toxicology of metals. Elsevier, pp 885–936
73. Bergmann M, Graça MA (2020) Bioaccumulation and dispersion of uranium by freshwater organisms. Arch Environ Contam Toxicol 78(2):254–266
74. do Vale Lourenço JI (2012) Genotoxicity in humans and indicator species exposed to uranium wastes. Universidade de Aveiro (Portugal)
75. Gandhi TP, Sampath PV, Maliyekkal SM (2022) A critical review of uranium contamination in groundwater: treatment and sludge disposal. Sci Total Environ 825:153947
76. Frisbie SH, Mitchell EJ, Sarkar B (2013) World Health Organization increases its drinking-water guideline for uranium. Environ Sci: Processes Impacts 15(10):1817–1823
77. Yi L et al (2020) Characteristics and assessment of toxic metal contamination in surface water and sediments near a uranium mining area. Int J Environ Res Public Health 17(2):548
78. Mariki S et al (2023) Assessment of the effects of uranium exploration on wildlife, vegetation and tourism in Tanzania. Tanzania J Forest Nat Conserv 92(1):170–184
79. Gramss G, Voigt K-D (2014) Forage and rangeland plants from uranium mine soils: long-term hazard to herbivores and livestock? Environ Geochem Health 36:441–452
80. Li Z et al (2024) Translocation and transformation of uranium along the aquatic food chain: new insights into uranium risks to the environment. J Hazard Mater 478:135499
81. Kumar A et al (2011) Risk assessment for natural uranium in subsurface water of Punjab State, India. Human Ecol Risk Assess 17(2):381–393
82. Srivastava M et al (2022) A systematic study of uranium in groundwater and its correlation with other water quality parameters. Water Supply 22(3):2478–2492
83. Bairwa AK et al (2024) Seasonal variation and spatial distribution of uranium in sources of water in Tonk District of Rajasthan, India. Pollution 10(1):511–527
84. Lopez AM, Wells A, Fendorf S (2020) Soil and aquifer properties combine as predictors of groundwater uranium concentrations within the Central Valley, California. Environ Sci Technol 55(1):352–361
85. Devaraj N et al (2021) Spatio-temporal variations of Uranium in groundwater: implication to the environment and human health. Sci Total Environ 775:145787
86. Li C, Adeniyi EO, Zarzycki P (2024) Machine learning surrogates for surface complexation model of uranium sorption to oxides. Sci Rep 14(1):6603
87. Kim B et al (2024) Experimental analysis and prediction of radionuclide solubility using machine learning models: effects of organic complexing agents. J Hazard Mater 469:134012

88. Ganguly S, Bhan U (2023) Occurrences and mobility of uranium in soil profile due to groundwater–soil interaction. Hydrogeochem Aqua Ecosyst:181–198
89. Shi S et al (2021) Biological effects of uranium in water, soil and rice in uranium deposits in Southern China. J Radioanal Nucl Chem 328:507–517
90. Carvalho F et al (2023) The environmental behaviour of uranium. IAEA
91. Prasad M et al (2022) Uranium contamination in drinking water as a health concern in Uttarakhand, India. J Radioanal Nucl Chem 331(4):1933–1940

Water Quality Parameters Influencing Uranium Solubility

Impact of pH on Uranium Solubility

Manoj Kumar Jindal and Kiran Patni

Abstract Uranium is a naturally occurring radioactive element that has emerged as a significant environmental contaminant, particularly due to its increasing presence in water sources used for human consumption. It has been widely detected and reported in groundwater, surface water, and even public water supplies across various regions globally. The solubility of uranium in water and its mobility and bioavailability are governed primarily by the pH of the aqueous medium. Changes in water pH, influenced by both natural processes and anthropogenic activities, can significantly affect the chemical behavior of uranium, facilitating its release from rocks, soils, and other environmental matrices. This chapter focuses on the critical role of pH in understanding uranium solubility, with particular emphasis on uranium in its hexavalent oxidation state (U(VI)), which is highly soluble and environmentally mobile. This highlights various uranium-bearing minerals, their geochemical interactions, and the pH-dependent formation of uranium complexes, especially carbonate and hydroxide species. On the basis of global data on uranium concentrations and a review of the mechanisms that control its solubility and speciation, this chapter highlights how pH acts as a primary regulatory factor in uranium transport, distribution, and health risk. Furthermore, understanding these pH-dependent mechanisms is essential for developing effective environmental monitoring and remediation strategies to mitigate uranium contamination.

Keywords Uranium solubility · Water pH · Uranium speciation · Uranium complexes · Uranium mobility · pH-dependent solubility · Uranium minerals

M. K. Jindal (✉)
Center for Sustainable Drinking Water, Indian Institute of Technology Jodhpur, Jodhpur, Rajasthan, India

K. Patni
School of Allied Sciences, Graphic Era Hill University, Bhimtal Campus, Bhimtal, Uttarakhand, India

Centre for Promotion of Research, Graphic Era (Deemed to be) University, Dehradun, Uttarakhand, India

M. K. Jindal, K. Patni (eds.), *Water Quality-Uranium Solubility Nexus: Unlocking the Secrets of Uranium Solubility*, Earth and Environmental Sciences Library, https://doi.org/10.1007/978-3-031-98909-4_4

1 Why Uranium Solubility Is Important

Uranium is a naturally occurring radioactive element present in rocks and soil in the Earth's crust. It is a silvery-white heavy metal denoted by the symbol U and atomic number 92. The solubility of uranium, a prominent radioactive isotope, has recently attracted significant attention in the scientific community. This phenomenon remains complex and is intertwined with the intricate interplay between water chemistry and the multifaceted interactions of uranium with water. A detailed exploration of this field will shed light on the diverse factors influencing uranium solubility in various water sources. The global distribution of uranium has been documented across various regions worldwide. Reports show that uranium levels in groundwater vary from less than 0.01 to 3410 μg/L in European countries, including Finland, Sweden, Norway, Poland, Spain, Switzerland, and Greece. In North America and South America, including Canada, the United States, and Brazil, uranium levels range from less than 0.01 to 7780 μg/L. In African countries, including Morocco, Ethiopia, and Egypt, uranium levels vary from less than 0.001 to 519.4 μg/L. In Asia, including China, Iran, Jordan, Bangladesh, Japan, and India, uranium levels range from 0.0005 to 113.5 μg/L [1]. However, in India, a recent study by Srinivasan et al. [2] reported uranium concentrations in groundwater exceeding 1000 μg/L. Jindal et al. [3] reported even higher levels, exceeding 3000 μg/L, with a maximum value of 8649 μg/L reported in granitic aquifers of eastern Karnataka village, India.

In addition to groundwater, uranium has been reported in surface water worldwide. While surface water generally contains lower concentrations of uranium than groundwater, it is not entirely free of this element. The primary reason for the origin of uranium is that it is predominantly a geogenic contaminant, meaning that it originates naturally from rocks and soil, not directly from human activities. However, anthropogenic activities such as mining and industrial processes can indirectly contribute to uranium mobilization and release into surface water. Importantly, the human body efficiently excretes the majority (approximately 90–95%) of ingested uranium (by drinking or food) through urine and feces. While some portions may eventually reach surface water through improperly managed wastewater treatment systems or natural processes, this is not a major contributor to uranium levels in most surface water bodies. Natural activities such as the weathering of rocks and minerals containing uranium can also contribute to its presence in surface water in landslide areas or those areas where rocks are exposed at the surface. Additionally, specific geological formations might naturally have relatively high uranium concentrations, leading to elevated uranium levels in local water sources. Therefore, while surface water generally has lower uranium concentrations it is still important to monitor and study uranium levels in various scientific communities. This helps ensure the safety of drinking water and the understanding of the environmental behavior of uranium solubility. According to the United Nations Scientific Committee on the Effects of Atomic Radiation (UNSCEAR) report, Europe, North America, South America, and Asia have concentrations of <0.01 to 34 μg/L, 0.03 to 1737 μg/L, <0.01 to 50 μg/L, <0.01 to 50 μg/L, and 1.08 to 19.68 μg/L, respectively

[1]. Uranium has also been detected in public water supplies in many countries, including Finland, Germany, Austria, Greece, Poland, Spain, Italy, Norway, Sweden, Iceland, the USA, Argentina, Morocco, Iran, India, and Australia. The reported levels vary widely, ranging from below detectable levels (<0.001 μg/L) to 1770 μg/L. Even uranium has been found in bottled mineral water from some countries, including Germany, Spain, Slovenia, Italy, Poland, Croatia, Portugal, France, Austria, Norway, Sweden, Iceland, Argentina, Tunisia, Morocco, and Kuwait. The reported levels in bottled water also vary considerably, ranging from below detectable levels <0.006 to 698 μg/L [1].

All these reports highlight the importance of uranium in water, even though it is available in surface water, groundwater, a somewhat public water supply, and even in bottled water, which is generally purchased by every person in every country. Therefore, uranium solubility is an urgent need, and what is the correct scientific reason behind its solubility? The key aspect of understanding the availability of uranium in water resources is that it is a radioactive element and has a wide range of effects. As a result, all reports show that uranium is widely distributed, from below the detection limit to more than 8000 μg/L; thus, this report shows that uranium is widely distributed in water sources, and one of the main reasons for this is that the recommended value of uranium for drinking water according to the World Health Organization is only 30 μg/L, even in other countries. It varies from 30 μg/L to even a maximum of 1000 μg/L, as reported in the literature [4], but considering this fact, the quantity of uranium found is higher than the recommended values from different countries.

This high uranium dose intake by drinking water will have an impact on the health of persons who live in such places, and very high uranium doses can cause a variety of health problems. Health hazards may occur in these places owing to uranium chemotoxicity and radiotoxicity caused solely by natural conditions. Nephrotoxicity (kidney), bone toxicity (bone), hepatotoxicity (liver), neurotoxicity (brain), pulmonary toxicity (lung), and reproductive toxicity can all be caused by high uranium doses obtained through water. Ma et al. [5] highlighted all of these health consequences in their review study on the basis of actual events recorded in the Finland region and on mice and rats on the basis of a specified dose of uranium for certain time periods. Additional effects include oxidative stress (imbalance of redox homeostasis occurred when doses of 10, 20, or 40 μg/g per day of uranium nitrate were available in drinking water for 3 months), genotoxicity (observed in sperm DNA when doses of 40 mg/L in drinking water were obtained for 9 months), protein binding and abnormal function (long-term uranium exposure), metabolic disorders (disrupt normal metabolic processes and cellular homeostasis), and cell death.

According to Agency for Toxic Substances and Disease Registry (ATSDR) reports, 0.1–6% of the uranium that is consumed will enter the bloodstream [6]. If uranium is present in the bloodstream, which is conceivable since uranium is accessible in water in dissolved form, which more easily reaches the bloodstream, then the scenario is concerning/alarming. Although most ingested uranium is eliminated by urine or feces, deposition in various regions of the body for an extended length of time is conceivable owing to the large doses of uranium. According to ATSDR research, 66% of the deposited uranium detected in bones, which may persist for a very long period, has a half-life of between 70 and 200 days [6]. Other deposited sites in the body include the lung, liver, and kidneys (main target). Few other health effects are being studied on the basis of evidence from human and animal studies, such as neurobehavioral changes, decreased fertility, changes in the levels of certain chemicals in the brain, skin irritation, and mild skin damage; however, several health concerns associated with this emerging contaminant remain to be investigated in the future. Recently, Levin et al. [7] reported uranium as one of seven legacy and emerging drinking water contaminants or contaminant groups due to its exposure risk profiles and health effects. Therefore, it is important to consider uranium solubility, which widely depends on the pH of the water. Therefore, pH studies are more important and crucial for understanding the solubility of uranium.

A recent publication serves as a valuable asset for scientists by offering a comprehensive understanding of uranium solubility across different water sources and environmental settings. This chapter elucidates the factors influencing the complexity of uranium solubility in aquatic systems. This study provides a detailed overview of the uranium landscape, highlighting the diverse patterns of uranium occurrence in different aquatic environments with varying pH levels. This extensive study of uranium solubility and distribution is essential for scientists, researchers, and environmental professionals. This study provides crucial groundwork for understanding and addressing the multifaceted challenges associated with uranium in the environment.

2 Nature of Water

Water is a powerful but selective solvent. Water is often referred to as a "universal solvent" because of its exceptional ability to dissolve a wide range of polar and ionic compounds. This property arises from the unique structure of the molecule. The uneven sharing of electrons between oxygen and hydrogen creates a polar molecule with a slight negative charge on the oxygen end and a slight positive charge on the hydrogen end. These charged regions can interact with charged particles in other substances (ions) or the polar ends of other molecules, breaking them apart and allowing them to disperse throughout the water. While water can dissolve many elements and compounds found in the Earth's crust, it is not truly universal. Certain nonpolar substances, such as oils and fats, do not readily dissolve in water. The extent to which an element dissolves depends on its chemical properties and water characteristics, particularly its pH. Neutral water (pH 7) has a moderate dissolving capacity. However, changes in pH due to natural processes (such as weathering) or human activities (such as pollution) can significantly impact solubility. For example, iron is relatively insoluble in neutral water, but it becomes more soluble in acidic environments. This principle applies to many other elements found dissolved in water. The types of rocks and minerals that water interacts with (e.g., carbonate rocks) and the presence of anthropogenic contaminants (e.g., wastewater and industrial runoff) all play a role in influencing water pH and, consequently, its ability to dissolve various elements. The pH of the water influences the solubility. Highly acidic or basic water can alter the charge of some solutes, affecting their interaction with water molecules. Human activities such as waste dumping can introduce pollutants that alter water pH and composition, potentially impacting the solubility of certain elements. However, it is important to note that natural processes also play a significant role in changing water pH and influencing the elements it carries.

3 Oxidation State of Uranium

Uranium has a wide range of oxidation states, but the matter is whether this wide range of oxidation states is soluble and stable in aqueous solutions. To better understand this, you need to know about all the existing compound details under different conditions. Oxidation states of +3, +4, +5, and + 6 are present in uranium compounds [8]. The oxidation state +3 forms of uranium compounds U(III) are highly unstable and rarely soluble (generally insoluble) in water. This is because uranium readily oxidizes to a higher oxidation state, making it challenging to isolate and handle in the +3 oxidation state. Even in the absence of dissolved oxygen, U(IV) cannot form U(III). Moreover, when U(III) comes into contact with water, it reacts with water molecules, undergoes hydrolysis, and forms insoluble hydroxides, which precipitate. This depends on factors such as pH and other ions. Examples include UO(OH) and $U(OH)_3$, which are unlikely to find uranium in the +3 form as a complex in water, as their ability to further dissolve in water is limited. Generally, U(III) is insoluble, but some factors can slightly influence the solubility of U(III) in water. At low pH (acidic) conditions, U(III) solubility is slightly greater due to the formation of more soluble hydroxyl complexes. However, this increase is still quite limited compared with other oxidation states [9]. Strong complexing agents such as citrate or fluoride can form soluble complexes with U(III), but these complexes are specific and not commonly encountered in natural environments. Overall, the combination of instability and limited solubility makes U(III) an insignificant contributor to the overall mobility and environmental impact of uranium in aqueous environments.

With respect to the uranium(IV) oxidation state, Gayer and Leider [10] studied uranium(IV) hydroxide behavior in dilute basic and acidic media. Uranium(IV) is less stable and less common in aqueous solutions than U(VI) is due to its strong oxidizing tendency. However, U(IV) can exist and form specific compounds under certain conditions. Uranium(IV) hydroxide ($UO_2(OH)_2$) is a relatively insoluble black precipitate that readily forms in near-neutral pH solutions (6–8). This limited solubility explains the generally low concentration of U(IV) in water. Uranium(IV) halides (UX_4, where X is a halogen such as Cl, Br, or I) are more soluble than hydroxide and exist as green-colored salts. However, they are highly susceptible to oxidation by air or water and readily convert to the +6 state. While uranium can exist in the +4 oxidation state in water, forming specific compounds such as hydroxides and halides, its limited stability and strong tendency to oxidize in the presence of oxygen or other oxidizing agents convert it to the more stable U(VI) state [9]. The most common mobile form of uranium in water is typically U(VI). The uranium (V) state is unstable, disproportionate, and tends to be converted into +4 and +6 by reduction and oxidation in an aqueous medium, even as it reacts with itself.

4 Uranium Minerals

The vast list of minerals in which uranium occurs is as follows [8, 11]:

- Andersonite: $Na_2Ca(UO_2)(CO_3)_3{\cdot}6H_2O$
- Agrinierite: $K_2Ca[(UO_2)_3O_3(OH)_2]_2{\cdot}5H_2O$
- Althupite: $AlTh(UO_2)_7(PO_4)_4O_2(OH)_5{\cdot}15H_2O$
- Albrechtschraufite: $Ca_4Mg(UO_2)_2(CO_3)_6F_2{\cdot}17H_2O$
- Arsenuranospathite: $Al(UO_2)_2(AsO_4)_2F{\cdot}20H_2O$
- Betafite: $(Ca,U)_2(Ti,Nb,Ta)_2O_6(OH)$.
- Bayleyite: $Mg_2(UO_2)(CO_3)_3 18(H_2O)$.
- Becquerelite: $Ca(UO_2)_6O_4(OH)_6{\cdot}8(H_2O)$
- Belakovskiite: $Na_7(UO_2)(SO_4)_4(SO_3OH){\bullet}3H_2O$
- Bergenite: $Ca_2Ba_4[(UO_2)_3O_2(PO_4)_2]_3 16(H_2O)$ or $(Ba,Ca)_2(UO_2)_3(PO_4)_2(OH)_4 5.5(H_2O)$
- Bijvoetite-(Y): $(Y,REE)8(UO_2)16(CO_3)16O_8(OH)_8{\cdot}39H_2O$
- Bilietite: $Ba(UO_2)_6O_4(OH)_6{\cdot}8H_2$
- Boltwoodite: $(K_{0.56}Na_{0.42})[(UO_2)(SiO_3OH)]{\cdot}1.5(H_2O)$
- Braunerite: $(U, Ca, Y, Ce, La)(Ti, Fe)_2O_6$
- Clarkeite: $(Na,Ca,Pb)_2(UO_2)_2(O,OH)_3$ or $(Na,Ca,Pb)(UO_2)O(OH){\bullet}0\text{-}1H_2O$
- Carnotite: $K_2(UO_2)_2(VO_4)_2{\cdot}3H_2O$
- Coconinoite: $Fe_2Al_2(UO_2)_2(PO_4)4(SO_4)(OH)_2{\cdot}20H_2O$
- Cuprosklodowskite: $Cu(UO_2)_2(HSiO_4)_2{\cdot}6(H_2O)$
- $Pb_3[(UO_2)_4|O_4|(OH)_3]_2{\cdot}2\ H_2O$
- Davidite: $La_{0.7}Ce_{0.2}Ca_{0.1}Y_{0.75}U_{0.25}Ti_{15}Fe^{3+}5O_{38}$
- Demesmaekerite: $Pb_2Cu_5(UO_2)_2(SeO_3)_6(OH)_6{\cdot}2H_2O$
- Derriksite: $Cu_4(UO_2)(SeO_3)2(OH)6{\bullet}H2O$
- Euxenite: $(Y, Ca, Ce, U, Th)(Nb, Ta, Ti)_2O_6$
- Gummite: $(Pb{\gg}Ca,Ba)U_3Si012{*}6H_20$
- Mathesiusite: $K_5(UO_2)_4(SO_4)_4(VO_5){\cdot}4(H_2O)$
- Metatorbernite: $(Cu(UO_2)_2(PO_4)_2\ 8H_2O)$
- Mckelveyite-(Y): $Ba_3Na(Ca,U)Y(CO_3)_6{\cdot}3H_2O$
- Meyrowitzite: $Ca(UO_2)(CO_3)_2{\cdot}5H_2O$
- Margaritasite: $(Cs, K, H_3O)_2(UO_2)_2V_2O_8{\cdot}H_2O$
- Marthozite: $Cu(UO_2)_3(SeO_3)_3(OH)_2{\cdot}7H_2O$
- Masuyite: $Pb[(UO_2)_3O_3(OH)_2]{\cdot}3H_2O$

- Meisserite: $Na_5(UO_2)(SO_4)_3(SO_3OH)(H_2O)$
- Metazeunerite: $Cu(UO_2)_2(AsO_4)_2 \cdot 8H_2O$
- Mundite: $Al(UO_2)_3(PO_4)_2(OH)_3 \cdot 5(H_2O)$
- Polycrase: $(Y,Ca,Ce,U,Th)(Ti,Nb,Ta)_2O_6$
- Rameauite: $K_2CaU^{6+}6O_{20} \cdot 9H_2O$
- Rutherfordine: UO_2CO_3
- Steacyite: $K_{variable}(Ca,Na)_2(Th,U)Si_8O_{20}$
- Sklodowskite: $Mg(UO_2)_2(HSiO_4)_2 \cdot 5H_2O$
- Soddyite: $(UO_2)_2SiO_4 \cdot 2H_2O$
- Studtite: $[(UO_2)O_2(H_2O)_2] \cdot 2(H_2O)$ or $UO_4 \cdot 4(H_2O)$
- Uraninite: UO_2/coffinite ($USiO_4$)
- Umohoite: $(UO_2)MoO_4 \cdot 2H_2O$
- Uranocircite: $Ba(UO_2)_2(PO_4)_2 \cdot 10H_2O$
- Uranophane: $(Ca(UO_2)_2(SiO_3OH)_2 \cdot 5H_2O)$
- Uranopilite: $(UO_2)_6(SO_4)O_2(OH)_6 \cdot 14H_2O$
- Wyartite: $Ca_3U(UO_2)_6(CO_3)_2(OH)_{18} \bullet 3(H_2O)$
- Yttrogummite: Samarskite-(Y): $(Y,Ce,U,Fe)_3(Nb,Ta,Ti)_5O_{16}$, Euxenite-(Y): $(Y,Ce,U,Ca,Th)(Nb,Ta,Ti)_2O_6$, Polycrase-(Y): $(Y,Ca,Ce,U,Th)(Ti,Nb,Ta)_2O_6$.

Although several uranium minerals exist, as mentioned above, their diverse compositions, including various uranium compounds, contribute to this. Some of them are listed below [8, 11]:

Yellowcake (U_3O_8): U_3O_8
Uranium hexafluoride (UF_6): UF_6
Cerium uranium blue ($Ce(UO_2)_6$):$Ce(UO_2)_6$
Diuranium pentoxide (U_2O_5): U_2O_5
Ferro-Uranium (FeU_3): FeU_3
MOX fuel (mixed oxide fuel): $(U,Pu)O_2$
Pentavalent uranyl complexes: (UO_2)n(X)m (where X is a ligand such as Cl-, Br-, or I-)
Tetrauranium octadecafluoride (U_4F_{16}): U_4F_{16}
Uranium diboride (UB_2): UB_2
UC: Uranium carbide
Uranium oxide (UO_2): UO_2
Uranium pentachloride (UCl_5): UCl_5
Uranium pentafluoride (UF_5): UF_5
Uranium pentabromide (UBr_5): UBr_5
Uranium pentaiodide (UI_5): UI_5
Uranium rhodium germanium (UGeRh): UGeRh
Uranium ruthenium silicide (URuSi): URuSi
Uranous (UO_2): UO_2
Uranyl acetate, $UO_2(C_2H_3O_2)_2$
Uranium trioxide, UO_3
Uranyl ammonium carbonate, $UO_2CO_3 \cdot 2(NH_4)2CO_3$
Uranyl chloride, UO_2Cl_2

Uranyl carbonate, UO_2CO_3
Uranyl hydroxide, $UO_2(OH)_2$ or $(UO_2)_2(OH)_2$ in aqueous media
Uranyl sulfate, UO_2SO_4
Uranyl nitrate, $UO_2(NO_3)_2$
Uranyl zinc acetate, $ZnUO_2(CH_3COO)_4$

5 Uranium Compounds and Their Variation with pH

Rodden [12] described various methods for extracting uranium from minerals at different pH values. This highlights the importance of pH in both the solubility and extraction of uranium from its minerals or compounds. Additionally, extraction occurs at specific pH ranges, and reagents also play a crucial role in altering uranium chemistry in aqueous media. However, many reagents are now being reported in water, making water a host for various inorganic and organic materials. These materials, along with other interfering substances found in certain types of water, can affect uranium solubility. Therefore, understanding the chemistry of uranium solubility and extraction at specific pH values is crucial.

Uranyl salts such as uranyl nitrate, sulfate, chloride, acetate, and perchlorate are water-soluble yellow-colored salts. Uranyl fluoride and uranium hexafluoride are also soluble in water. Most uranium chlorides, bromides, and iodides either dissolve in water or react with water, resulting in simple solutions. Uranium carbides dissolve upon prolonged boiling with nitric acid but usually oxidize to U_3O_8 before complete dissolution. Uranium sulfates, acetates, and many other U(VI) salts are water soluble. However, uranium(IV) salicylate is soluble in water but readily soluble in organic solvents, as is uranium(IV) acetylacetonate. Conversely, uranium(IV) cupferrate, while insoluble in water, is soluble in ether and other organic solvents, making it important for uranium(IV) ion chemistry.

The sulfate ion forms complexes with uranium, where either two or three sulfate ions attach to a single uranyl ion. Examples include potassium disulfates and trisulfate compounds such as sodium uranyl trisulfate ($Na_4UO_2(SO_4)_3$). These sulfate complexes appear to inhibit the precipitation of UO_2 at a pH of 1.5, with the degree of interference increasing proportionally to the sulfate concentration (when the sulfate to uranyl ratio is greater than 1). Additionally, sulfate ions reduce the solubility of uranyl nitrate in ether, potentially due to a related effect.

Uranium(VI) forms complexes with organic compounds. Like uranium(VI), U(IV) also forms complexes with many organic compounds. For example, oxalate forms a stable, soluble complex of the type $(NH_4)_4U(C_2O_4)_4$ in neutral solutions. However, acidification causes U(IV) oxalate to precipitate, whereas the addition of ammonium hydroxide leads to the precipitation of U(IV) hydroxide or hydrated oxides. An acetate complex of U(IV) might exist, as U(IV) oxalate is soluble in ammonium acetate solutions. However, this solubility might simply be due to the solution being buffered to approximately pH 7, where U(IV) oxalate is soluble. Typically, peroxide precipitation for uranium occurs at a pH range of 0.5–3.5.

Solutions containing uranyl ions readily precipitate uranyl monohydrogen phosphate (UO_2HPO_4) when treated with phosphoric acid and ammonium uranyl phosphate ($NH_4UO_2PO_4$) when treated with diammonium phosphate. However, controlling pH is crucial, as incomplete precipitation occurs at lower pH values, and both salts become soluble in excess phosphoric or other strong acids. The optimal pH for precipitation is 1.7, with an acceptable range of 1.2 to 3.3. IRA-400 resin can be used directly with seawater at a pH range of 3–9. After washing with buffer at pH 5 and water, uranium was eluted with 0.8 N hydrochloric acid. To separate uranium from rare earth elements, a solution of 0.01 N in sulfuric acid is absorbed on Dowex-1 resin, and the uranium is then eluted with normal nitric acid. The emphasis here is that the pH at which the extraction is carried out is critical. Uranium is selectively absorbed from a sulfate solution at pH 2 by Dowex 2 and eluted with dilute hydrochloric acid. The pH range for forming and extracting the complex into chloroform is limited. The maximum extraction occurs from an aqueous solution at a pH of 7–8. Lower and higher pH values significantly reduce the extraction efficiency. Understanding how pH affects uranium solubility and complexation is crucial for various applications, including extraction from ores, environmental remediation, and nuclear fuel processing. Changing the pH can alter both the solubility of uranium minerals/compounds and the behavior of other materials that influence uranium solubility.

The uranium species (U-VI) present at different pH values are shown in Fig. 1. The acidity (pH) of a solution significantly impacts the form of uranium (VI) present. At low acidity (pH values below 5), uranium primarily exists as UO_2^{2+} ions. As the solution becomes less acidic (pH 5.0–7.0), UO_2^{2+} ions react with water molecules to form positively charged hydroxide complexes (UO_2OH^+) and larger multinuclear hydroxide complexes ($(UO_2)_3(OH)_5^+$) [13]. The x-axis of the graph

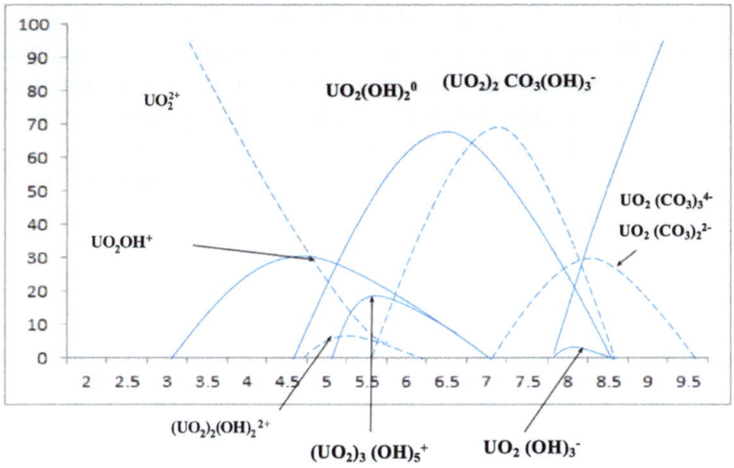

Fig. 1 Distribution of uranium species under different pH conditions ([13], reproduced with permission from Springer Nature)

represents the pH range, and the y-axis represents the mole fraction of the uranium species. UO_2^{2+} is the most soluble uranium species at all pH values. UO_2^{2+}, UO_2OH^+, $(UO_2)_2(OH)_2^{2+}$, $(UO_2)_3(OH)_5^+$, and $(UO_2)(OH)_2^0$, $UO_2(CO_3)$ favor acidic pH values in solution; however, $UO_2(CO_3)_3(OH)_3^-$, $UO_2(OH)_3^-$, $(UO_2)(CO_3)_3^{4-}$ and $UO_2(CO_3)_2^{2-}$ support the basic nature of water [13, 14].

Importantly, this graph illustrates the variation in uranium (VI) hydroxide carbonate complexes with pH. However, the actual solubility of uranium species in real water can also be affected by other factors, such as the presence of other ions in the solution and the reactive nature of aqueous solutions. The pH of the aqueous solution also plays a crucial role as an influencing factor that affects the remediation process of uranium. For example, in the metal ion adsorption process, an appropriate pH value is required for uranium(VI) biosorption onto a desired or suitable adsorbent. Various removal processes have been studied across acidic to alkaline pH ranges, mostly between 2.00 and 10.00, in water remediation [15].

6 Conclusion

The solubility of uranium in water is intricately controlled by pH, making it one of the most significant factors influencing uranium speciation, mobility, and bioavailability in the environment. As demonstrated in this chapter, uranium exists in multiple oxidation states and forms various complexes in a pH-dependent manner. Under acidic conditions, uranyl ions (UO_2^{2+}) and their hydrolyzed forms dominate, whereas alkaline conditions favor complexation with carbonate species, resulting in increased mobility and stability of uranium in solution. Uranium minerals and salts illustrate how both natural and anthropogenic influences alter water pH and impact uranium dissolution. Considering the widespread detection of uranium in global water systems, which often exceeds the WHO [16] and national safety thresholds, understanding pH-dependent solubility is vital for environmental monitoring, risk assessment, and water treatment strategies. Moreover, the documented health risks from chronic uranium exposure call for a precautionary approach to managing uranium-contaminated water, especially in regions where groundwater serves as the primary drinking source. Future work should focus on developing pH-based predictive models and implementing pH-controlled remediation techniques to mitigate uranium risk effectively.

References

1. UNSCEAR (2016) (United Nation Scientific Committee on the effect of Atomic Radiations). Sources, effects and risks of ionizing radiation. New York, United Nations, p 512. https://www.unscear.org/docs/publications/2016/UNSCEAR_2016_Report-CORR.pdf
2. Srinivasan R, Pandit SA, Karunakara N, Deepak Salim K, Sudeep Kumara M, Kumar R, Khatei G, Ramkumar KD (2021) High uranium concentration in groundwater used for drinking in parts of eastern Karnataka. India Curr Sci 121(11):1459

3. Jindal MK, Pandit SA, Karunakara N, Chandrashekara MS, Kumara S, Kumar V, Salim D, Srinivasan R (2023) High uranium dose from the groundwater in a granitic terrain in the eastern part of Karnataka, India. J Radioanal Nucl Chem. https://doi.org/10.1007/s10967-023-09053-6
4. Sahoo SK, Jha SK, Jha VN, Patra AC, Kulkarni MS (2021) Survey of uranium in drinking water sources in India: interim observations. Curr Sci 120(9):1482–1490
5. Ma M, Wang R, Xu L, Xu M, Liu S (2020) Emerging health risks and underlying toxicological mechanisms of uranium contamination: lessons from the past two decades. Environ Int 145:106107. https://doi.org/10.1016/j.envint.2020.106107
6. Keith S, Faroon O, Roney N, Scinicariello F, Wilbur S, Ingerman L, Llados F, Plewak D, Wohlers D, Diamond G (2013) Toxicological profile for uranium. United States Department of Health and Human Services, Agency for Toxic Substances and Disease Registry
7. Levin R, Villanueva CM, Beene D, Cradock AL, Donat-Vargas C, Lewis J, Martinez-Morata I, Minovi D, Nigra AE, Olson ED, Schaider LA (2024) US drinking water quality: exposure risk profiles for seven legacy and emerging contaminants. J Expo Sci Environ Epidemiol 34(1):3–22
8. Pownceby MI, Johnson C (2014) Geometallurgy of Australian uranium deposits. Ore Geol Rev 56:25–44
9. Hobart D (2013) Periodic table of elements: LANL, published on July 23, 2023. Retrieved from https://periodic.lanl.gov/92.shtml
10. Gayer KH, Leider H (1957) The solubility of uranium (IV) hydroxide in solutions of sodium hydroxide and perchloric acid at 25 C. Can J Chem 35(1):5–7
11. Moorhouse F (2023) Uranium (U) updated on August 15, 2023. Retrieved from https://thechemicalelements.com/uranium/
12. Rodden CJ (1964) Uranium, Chapter-1, Analysis of essential nuclear reactor materials (No. TID-21384). New Brunswick Lab., AEC, NJ (United States)
13. Deshmukh P, Sar SK, Jindal MK, Ray T (2023) Magnetite based green bio composite for uranium exclusion from aqueous solution. J Radioanal Nucl Chem 332:297–310. https://doi.org/10.1007/s10967-022-08723-1
14. Kimura T, Serrano G J, Nakayama S, Takahashi K, Takeishi H (1992) Speciation of uranium in aqueous solutions and in precipitates by photoacoustic spectroscopy. Radiochim Acta 58(1):173–178
15. Deshmukh P, Sar SK, Jindal MK (2023) Plant mediated magnetic nano composite as promising scavenger's radionuclides for the efficient remediation in aqueous medium. Chemosphere 312:137246. https://doi.org/10.1016/j.chemosphere.2022.137246
16. World Health Organization (2012) Uranium in drinking water. Background document for development of WHO guidelines for drinking-water quality. WHO, Geneva, pp 1–21

Temperature Influence on Uranium Solubility

Benjamin Uchechukwu Nwaka, Chukwuemeka Ehirim, I. Fabian Ezema, Emmanuel Obinna Ogueri, and K. Sudeep Kumara

Abstract Uranium species are various forms of uranium that exist in different environments, as observed in biological systems, rocks, soils, aquatic and marine environments. In aqueous environments such as groundwater, surface water, and marine systems, uranium can occur in solution as different chemical species. Temperature is an important factor that affects the speciation of uranium in various water bodies and environments. The study of the solubility, mobility and distribution of uranium among different chemical species, such as ions, complexes, ligands and colloids, is necessary. Temperature is an essential factor in controlling or influencing the solubility of uranium, and its effects are considered in conjunction with other geochemical parameters or water chemistry parameters. The relationship between the temperature and solubility of uranium is complex and depends on various factors, such as the minerology and sources of uranium, pH concentration, redox conditions, the presence of complexing ligands, the oxidation state, specia-

B. U. Nwaka (✉)
Environmental Radiation and Health Research Group, Department of Physics, Alvan Ikoku Federal University of Education, Owerri, Nigeria
e-mail: benjamin.nwaka@alvanikoku.edu.ng

C. Ehirim
Department of Physics, University of Port Harcourt, Port Harcourt, Rivers State, Nigeria
e-mail: chukwuemeka.ehirim@uniport.edu.ng

I. Fabian Ezema
Nano Research Group, Department of Physics and Astronomy, University of Nigeria, Nsukka, Enugu State, Nigeria
e-mail: fabian.ezema@unn.edu.ng

E. O. Ogueri
Environmental Radiation and Health Research Group, Department of Health Education, Alvan Ikoku Federal University of Education, Owerri, Nigeria
e-mail: emmmauael.ogueri@alvanikoku.edu.ng

K. Sudeep Kumara
Department of Physics, School of Engineering, Dayananda Sagar University, Bengaluru, Karnataka, India

M. K. Jindal, K. Patni (eds.), *Water Quality-Uranium Solubility Nexus: Unlocking the Secrets of Uranium Solubility*, Earth and Environmental Sciences Library, https://doi.org/10.1007/978-3-031-98909-4_5

tion, salinity and the strength of ions. The solubility of uranium increases, in general, with increasing temperature. High temperatures provide greater energy for uranium ions to travel and interact with surrounding water molecules, making it very easy for uranium ions to dissolve. High temperature also affects the distribution of uranium species in solution, as uranyl ions are more soluble than other uranium species. Temperature influences the oxidation of U(IV) to U(VI) and the reduction of U(VI) to U(IV). The stability of uranium complexes with ligands such as oxides, hydroxides, carbonates, phosphates, organic complexes and colloids are influenced by temperature. Furthermore, the rates of uranium precipitation reactions, such as the formation of uranium hydroxides and uranyl phosphate, are influenced by temperature.

Keywords Complexation · Ligands · Radionuclides · Uranium speciation · Temperature · Uranium solubility

1 Introduction

Water quality parameters are physical, chemical, biological and radiological features or characteristics employed in measuring, assessing and evaluating the quality of water in the hydrosphere. Hydrophysical studies are scientific investigations of the physical properties and behavior of water, including its flow pattern, mobility, transport, and interaction with the surrounding environment. It focuses on the physical parameters or physical aspects of water systems rather than the biological and chemical constituents. It involves the analysis of parameters such as solubility, temperature, pH, viscosity, surface tension, boiling and freezing point pressure, turbidity, odor, color, mobility, total dissolved solids (TDS) and conductivity for environmental media such as rock, soil, sediment, surface water and groundwater systems. Whereas hydrochemical study is the analysis of chemical composition of water in diverse or particular environment, examining several dissolved minerals, elements and ions present so as to understand the water quality. Identifying potential sources of pollution of surface and groundwater systems are very essential in geochemical and hydrochemical studies [1–5]. The chemical composition of water can result from natural, physical, chemical and biological processes. Some of the cations and anions found in water are uranyl ions (UO_2^{+2}), Na^+, K^+, Ca^{+2}, Mg^{+2}, Ca^{+2}, Cl^-, SO_4^{-2}, HCO_3^-, NO_2^-, NO_3^-, PO_3^-, PO_4^{-2}. Temperature, pH and redox potential play major roles in the solubility of uranium in water.

1.1 *Hydrosphere and Other Components of the Earth's System*

The hydrosphere is the total amount of water on planet Earth, which includes water on the surface of the planet, underground and in the air; this water can be liquid, vapor or ice. The hydrosphere is one of the four main spheres of the Earth among the atmosphere (which is a mixture of gases or air that surrounds the Earth), lithosphere (which is the Earth's solid outer layer or the terrestrial components of the biosphere), and biosphere, also called the ecosphere (the natural habitat of living organisms made up of the lithosphere's surface). The hydrosphere is a component of the Earth's system that comprises all of the water on planet Earth, which is interconnected with the atmosphere and lithosphere and includes atmospheric water vapor, ponds, lakes, streams, rivers, groundwater, ice caps, glaciers, estuaries, seas and oceans. It plays a vital role in regulating the climate of the Earth, as it helps to distribute heat around the globe and influences the patterns of weather. It is responsible for the continuous movement of water between the ocean, atmosphere and land. It provides habitats and support for a vast group or array of aquatic and marine ecosystems, such as coral reefs, estuaries and freshwater wetlands. The hydrosphere is a source of saltwater, freshwater, and groundwater for human consumption, agriculture, industry and medical purposes. Though hydrosphere plays a crucial role in climate regulation, water cycle, ecosystems and human activities, there are threats associated with hydrosphere, which include water pollution, climate change, over – extraction and rise in seas level. Human activities such as industrial water, agricultural runoff and sewage can contaminate freshwater sources. Rising temperatures and changing precipitation patterns can alter the distribution and availability of water resources. Over – extraction of ground can lead to decreased water tables and land resources. Rising sea levels can lead to coastal erosion, flooding and saltwater instruction into freshwater sources.

Temperature can strongly influence the hydrosphere, uranium solubility, mobility, rate of chemical reactions, oxygen solubility and other gases, some water quality parameters, toxicity and bioavailability, geochemical reactions and microbial activities. Temperature significantly impacts the hydrosphere by influencing the water cycle, including rates of evaporation, precipitation patterns, and the distribution of ice. An increase in temperature (warm temperature) results in increased evaporation, possibly causing more precipitation in some regions and areas while leading to drought in other regions and areas. As the Earth warms, seas and oceans warm as well, causing water to expand and resulting in a rise in sea level. A decrease in temperature (colder temperature) can result in ice formation and changes in ocean current [6, 7].

1.2 Hydrologic Cycle and Water Quality Parameters

In water cycle or hydrological processes, transpiration from trees and evaporation from oceans, lakes, streams and rivers enter the atmosphere, where it forms clouds and precipitation. Water vapor in the atmosphere condenses into clouds, fog and dew. Precipitation follows when water droplets in clouds become too heavy to remain suspended and fall to the Earth or ground as rain drops, snow, hails and sleets. The precipitation that falls as rain on land flows over the surface of the Earth as runoff, in due course returning to the oceans, seas, rivers, streams and lakes. The oceans and seas cover approximately 71% of the surface of the Earth and contain approximately 97% of the water in the Earth as one of the planets. Freshwater lake streams and rivers cover about 3% of the surface of the earth and contain about 2% of the planet's water. Groundwater is stored in aquifers beneath the surface of the Earth and accounts for approximately 1% of the water on the planet Earth. Water vapor in the atmosphere, as an important component of the hydrosphere, plays a key role in the Earth's climate system. Water quality parameters are used to evaluate the suitability of water for different purposes, such as for drinking, aquatic life support, irrigation and industrial uses. Some physical water quality parameters are temperature, pH, turbidity, color, TDS, and conductivity. Chemical water quality parameters include hardness, acidity, alkalinity, nutrients, dissolved oxygen (DO), biochemical oxygen demand (BOD), and stable elements such as zinc, chloride, sodium, magnesium, carbonate, nitrate, phosphate, and sulfate. In addition, heavy metals such as arsenic, cadmium, chromium, nickel, and lead are also important indicators. Biological water quality parameters include bacteria (such as *E. coli* and total coliforms, which can indicate fecal contamination and the risk of waterborne diseases), as well as viruses, protozoa, algae, and macroinvertebrates. Radiological parameters include the absorbed dose rate, annual effective dose rate and lifetime cancer risk in water samples.

1.3 Water Quality Guidelines and Standards

Various organization has established water quality guidelines and standards to protect the environments and human beings from health risks. Such organizations are United States Environmental Protection Agency (USEPA), World Health Organization (WHO), European Union (EU), Nigerian Standard for Drinking Water Quality (NSDWQ), Nigerian Industrial Standard (NIS), Standard Organization of Nigeria (SON) [8–11]. The guidelines and standards provide safety limits for various aforementioned water quality parameters, such as pH, TDS, carbonate, phosphate, heavy metals bacterial and protozoa, absorbed dose rate and the annual effective dose rate in water samples. Some of the water quality parameters that influence the solubility of uranium are temperature, pH, DO, redox potential or redox conditions, salinity effects, ionic strength, other physicochemical parameters, heavy metals, microorganisms and radioactive elements.

1.4 Role of Temperature in Influencing Uranium Solubility

Radioactive uranium, once dispersed, can have significant and long-lasting effects on ecosystems, soils, water sources, and human health. Variation in uranium concentrations in environmental media can lead to environmental impacts [12]. Some uranium radionuclides are embedded in soil, where they can remain for a long time, slowly decaying and emitting radiation. Others may be carried by air (wind) and water, spreading contamination over larger areas. The long-term effects include alterations in soil composition, contamination of water bodies and potential bioaccumulation in plants and animals, causing damage to the structure of deoxyribonucleic acid (DNA) during interactions, and these effects can last for decades or even centuries, depending on the isotopes involved, their concentrations and how they interact with environmental factors [12, 13].

The role of temperature in influencing uranium solubility is significant, as uranium solubility generally increases with increasing temperature. This may be attributed to the fact that, at elevated temperatures, uranium dissolves more readily, particularly in the presence of complexing ligands such as chloride and fluoride [14]. Higher temperatures provide uranium ions with increased kinetic energy, allowing them to move more freely and interact more effectively with surrounding water molecules, thereby enhancing dissolution. For instance, uranium salts such as uranyl nitrate are significantly more soluble at elevated temperatures than at lower temperatures. Temperature also affects the distribution of uranium species in solution and the stability of uranium–ligand complexes. Complexes with ligands such as hydroxide, carbonate, phosphate, and various forms of organic matter become more stable at higher temperatures, further contributing to increased uranium solubility.

2 Solute, Solvent (Universal Solvent) and Solution

A solute is a dissolved substance that be a solid, liquid or gas. A solvent is a substance (liquid or gas), usually a liquid that dissolves a solute. A solution is a mixture of two or more substances formed when a solid dissolves in a liquid such as water to produce a homogeneous mixture. A solid is called a solute, and a liquid is called a water solvent. Water is the most common solvent in nature and is usually referred to as a universal solvent. Solvents could be organic (e.g. acetone, benzene, toluene, methanol, ethanol, glycerin) or inorganic (e.g. water, ammonia, sulfuric acid, hydrochloric acid, and nitric acid). Organic solvents are substances typically composed of organic compounds capable of dissolving other substances, while inorganic solvents are substances composed of inorganic compounds that can also dissolve other materials. Specific solvents like petrol, kerosene, ammonia solution, and turpentine are effective for removing stains and paint, while a borax solution is useful for removing tea and coffee stains. The structure of a water molecule is polar in nature, as one side of the molecule is negatively charged, whereas the other side is positively charged. Water is a very good solvent for electrovalent solutes because of its polar nature or structure. Electrovalent solutes are inorganic solutes, including mineral salts, mineral acids and bases. Covalent compounds containing hydroxyl groups, such as organic solids such as sugars, and organic liquids such as alkanes and glycerol dissolve easily in water. Water, as a universal solvent, is a very suitable medium for most chemical reactions, especially in living systems. To a certain degree, gases can dissolve in water. Gases such as nitrogen (IV) oxides, ammonia and hydrogen chloride gas, which react with water or ionize in solution, are very soluble. Some other gases, such as chlorine, carbon(IV) oxides and sulphur(IV) oxides, that ionize slightly in water are fairly soluble in water. Gases such as hydrogen, nitrogen, oxygen, and carbon(II) oxides that do not ionize in water are sparingly soluble in water. The solubility of gases generally decreases with increasing temperature. Water can react with radioactive elements (^{238}U, ^{235}U, ^{226}Ra, ^{228}Ra), metals (sodium, potassium, calcium, magnesium), nonmetals (silicon, oxygen, chlorine), oxides of alkaline metals (calcium oxides. Magnesium oxides), oxides of nonmetals (carbon, nitrogen and sulphur). The reactions of water with organic and inorganic compounds are generally called hydrolysis. Familiar solutions are different from aqueous solutions, while familiar solutions are solutions where the solutes are dissolved in a liquid solvent, aqueous solutions are solutions whose liquid is water. Most substances are more soluble in water than others are.

3 Temperature and Temperature Variation

Temperature is a basic quantity that describes the extent of coldness or hotness of an object or a substance. Chemical reactions generally proceed faster at higher temperatures and slower at lower temperatures. An increase in the temperature of a system can lead to an increase in the rate of reaction, as energy in the form of heat is supplied to particles of the reactant so that the number of particles with energies equal to or greater than the activation energy increases. Additionally, the average velocity of all the reactant particles increases due to greater kinetic energy, resulting in a higher frequency of collision. As the number of effective collisions increases, the reaction proceeds at a faster rate. Variation in temperature is responsible for the formation of land and sea breezes. Temperature also influences physical properties such as viscosity, surface tension, density humidity, and relative humidity. As relative humidity is directly related to the air temperature, temperature affects the nature and type of cloud formation and precipitation as well as the moisture carrying capacity of the air. Temperature also controls the rates of evaporation and condensation, which are related to the degree of stability of the atmosphere. As the temperature increases, the pressure decreases, and the rate of evaporation increases. Since the atmosphere is primarily heated by conduction from the Earth's surface, areas closer to the ground are expected to be warmer than those at higher altitudes. Temperature, as a physical parameter, also influences the solubility of uranium [15].

4 Solubility of Substances and Solubility Curve

Solubility is a means of comparing the degree or extent to which different solutes can dissolve in a particular solvent at a specific temperature. In the laboratory, the solubility of a solute in a solvent at a particular or definite temperature is the maximum amount of solute in moles (mols.) or grams (g) that will saturate 1 dm^3 of the solvent at that temperature. If the solubility of a solute in a given solvent is plotted against their respective temperatures, a solubility curve or graph showing the effect of temperature on the solubility of the substance will be obtained. The solubilities of most solid solutes increase with increasing temperature. The solubilities of most common gases (except hydrogen chloride gas), however, decrease with increasing temperature. Gases that do not react with water are slightly soluble in water. Gases such as hydrogen chloride gas and ammonia, which are very soluble in water, form acidic and alkaline solutions, respectively. Air is sparingly soluble in water. The solubility of a gas is directly proportional to its pressure. When the aerated drink container or bottle is opened, the pressure is released, and bubbles of carbon (IV) oxide gas are released. Different uranium species have variable solubilities and mobilities, which affects their transport properties and fate in the environment. Among the uranium species, uranyl ions (UO_2^{+2}) and their complexes are normally more soluble and mobile uranium tetravalent ions (U^{+4}), and their complexes are less soluble and mobile in water. At high temperatures, more uranium is present as uranyl ions (UO_2^{+2}), which are more soluble than other uranium species (U^{+2}, U^{+3}, U^{+4}, U^{+5}).

5 Uranium Isotopes

Uranium is a radioactive element with an atomic or proton number of 92 (^{92}U), and its isotopes are found in natural and nuclear reactors. Natural uranium has two decay chains, the actinium series (^{235}U) and the uranium series (^{238}U), with relatively long half-lives. Uranium-234 (^{234}U) has a shorter half-life than do uranium-238 (^{238}U) and uranium-235 (^{235}U) [7, 16, 17]. Uranium–238 (^{238}U), uranium–235 (^{235}U) and uranium–234 (^{234}U) have natural uranium (abundance) contents of 99.275%, 0.720% and 0.0057%, respectively, for which uranium–238 (^{238}U) is regarded as the most stable of the three isotopes. The half-lives of isotopes of natural uranium–234(^{234}U), uranium–235(^{235}U) and uranium–238(^{238}U) are 2.47×10^5 years, 7.10×10^8 years and 4.51×10^9 years, respectively. These three isotopes of uranium are alpha emitters, and all the uranium parent isotopes decay to shorter-lived decay products referred to as daughters [7, 17]. Uranium–238 (^{238}U) is a very important isotope, as it is fertile, absorbing neutrons to produce radioisotopes that decay into fissile plutonium–239 (^{239}Pu). Long-lived alpha-emitting uranium–238 (^{238}U) decays through 18 members of the uranium series to stable lead–206 (^{206}Pb), whereas uranium–235 (^{235}U) decays through 15 members of the uranium series to lead–207 (^{207}Pb) [16]. Uranium-235 (^{235}U) is concentrated in a process called "enrichment" (enriched uranium). The enriched uranium would have more uranium—235 (^{235}U) and uranium—234 (^{234}U) and less uranium—238 (^{238}U). Depleted uranium would have more uranium—238 (^{238}U) and less uranium—235 (^{235}U) and uranium—234 (^{234}U). Other uranium isotopes exist, including uranium—214 (^{214}U), uranium—220 (^{220}U), uranium—232 (^{232}U), uranium—233 (^{233}U), uranium—236 (^{236}U) and uranium—242 (^{242}U) [18]. The uranium–233 (^{233}U) isotope is an artificial isotope produced in breeder reactors or nuclear reactors with high radioactivity. It is produced from thorium–232 (^{232}Th) through bombardment of neutrons. Except for uranium—220 (^{220}U), uranium—242 (^{242}U) and uranium—214 (^{214}U) isotopes have far shorter half–lives than primordial uranium isotopes (^{238}U and ^{235}U).

6 Uranium Oxidation States

The behavior of uranium at low and high temperatures in aqueous solutions is very important in the fields of environmental and radiation physics, geophysics, geochemistry, and water chemistry. Uranium exists in various oxidation states, including uranium(VI) or uranyl ions (U^{+6}), uranium(V) (U^{+5}), uranium(IV) or uranous ions (U^{+4}), uranium(III) (U^{+3}), and uranium(II) (U^{+2}) [3, 18–21]. Under oxidizing conditions in the marine environment, uranium is typically present in the hexavalent form of uranium (VI) as a large, divalent uranyl ion (UO_2^{+2}), which is highly mobile, forming soluble complexes primarily with phosphate and carbonate. Under reducing conditions in environments, uranium occurs in the tetravalent form of uranium(IV), forming relatively insoluble complexes with phosphates, hydrated fluorides, and hydroxides [18–21]. The oxidation state of uranium(V) (U^{+5}) seems relatively difficult to study, and data on its solubility almost do not exist because uranium(V) (U^{+5}) is always detected at equilibrium with uranium(IV) and uranium(VI) [19]. In nature, uranium forms complexes in one of these valence states: uranium (VI) or uranyl ions (U^{+6}) and uranium (IV) or uranous ions (U^{+4}). Solid uranyl ion (U^{+6}) compounds are soluble at room temperature and form stable species in aqueous solutions. Solid uranous ion (U^{+4}) compounds generally have very low solubilities in aqueous solutions at low temperatures (standard temperature of 25 °C) and pressures (1 paschal or bar) [19, 20].

7 Ligands

Ligands are molecules or ions that donate a pair of electrons to a central metal atom, thereby forming a coordinate covalent bond. They are atoms or molecules that bind to a central metal atom from a coordination compound. In the context of uranium geochemistry and chemistry, ligands play crucial roles in the formation of complexes with uranium ions, which can significantly influence uranium behavior, solubility, mobility and bioavailability. They are classified as simple ligands (carbon monoxide, water, ammonia, nitrogen, oxygen and halogens), complex ligands (ethylene dibromide and nitrile), unsaturated ligands (ammonia), anions (hydroxide, Cl^-, cyanide, CN^- and chloride Cl^-), and cations ($N_2H_5^+$, NO^+). Ligands used in uranium geochemistry and hydrochemical studies are grouped into organic ligands, inorganic ligands, chelating ligands and macromolecular ligands.

7.1 Types and Examples of Ligands

Four types of ligands are mentioned in this study, namely, monodentate ligands, bidentate ligands, polydentate ligands and chelating ligands. Ligands that are bound to the central metal atom or ion (uranium ions) through a single atom or group of atoms are called monodentate ligands. Examples of monodentate ligands include bromide (Br^-), chloride (Cl^-), and ammonia (NH_3^-). Ligands that are bound to the central metal atom or ion (uranium ions) through two atoms or groups are called bidentate ligands. Examples of bidentate ligands include ethylenediamine and oxalate ($C_2O_4^{-2}$). Polydentate ligands are ligands that are bound to the central metal atom or ion (uranium ions) through multiple atoms or groups. For example, ethylenediaminetetraacetic (EDTA) acid and ethylenediaminepentaacetitic (DTPA) acid have been used. Chelating ligands are ligands that form multiple bonds with the central atom or ion (uranium ions), resulting in a ring-like structure. Examples of chelating ligands include porphyrins, corrinoids, and crown ethers.

Some examples of ligands include the following:

- **Carbonate $(CO_3)_2^-$:** Carbonate is a common ligand in natural water bodies that forms complexes with uranium, as observed in $UO_2(CO_3)_2^{-2}$ and $UO_2(CO_3)_3^{-4}$. Carbonate has long been used as a typical transporting ligand for uranium in both ore deposit and waste repository contexts. [22]. Experiments evaluating the ability of carbonate-bearing liquids or fluids to dissolve and transport uranium at high temperatures revealed that at temperatures exceeding 100 °C, carbonate becomes completely irrelevant as a transporting ligand [22].
- **Hydroxide (OH^-):** a ligand that forms complexes such as $UO_2\ (OH)_2$ and $UO_2(OH)_3^-$. Since uranium is most effectively transported by hydrothermal fluid when it is present in its oxidation state of +6, its oxidation state of +6, such as uranyl ions (UO_2^{+2}), is highly soluble, forming stable complexes with a good range of inorganic ligands, such as OH^- and Cl^-. The presence of ligands such as OH^- and Cl^- serves as a major way to increase the uranium carrying capacity of hydrothermal fluids [22]. Under acidic conditions, uranyl complexes most readily with Cl^-; however, sea water, many groundwater systems and a number of uranium ore deposits are characterized by fluids with nearly neutral or slightly alkaline pH ranges.
- **Phosphate $(PO_4)_3^-$:** a ligand that forms complexes with uranium to yield $UO_2(PO_4)_2^-$ and $UO_2(PO_4)_3^{-5}$.
- **Organic ligands** include humic acids, fulvic acids and other organic molecules that can form complexes with uranium, influencing its behavior in natural environments.
- **Oxide ligands:** Oxide ions (O_2^-) can form strong bonds with uranium, resulting in the formation of uranyl ions (UO_2^{+2}).
- **Chelating ligands** can form strong complexes with uranium ions, making them useful for uranium separation and purification.

These ligands influence uranium geochemistry in various ways, such as changing uranium speciation and solubility, influencing uranium mobility and transport, affecting uranium bioavailability and toxicity, and impacting uranium geochemical reactions and fate.

7.2 *Characteristics of Ligands in Uranium Geochemistry and Hydrochemistry*

Some of the characteristics of ligands are discussed in terms of basicity, acidity, polarity, size and shape. Ligands can act as bases, donating electrons to the central metal atom or ion or uranium ions. Some ligands can act as acids, accepting electrons from the central metal atom or ion or uranium ions. Ligands can also be polar or nonpolar, affecting their ability to form bonds with central metal atoms, ions, or uranium ions. The size and shape of ligands can influence their ability to bind to central metal atoms, uranium ions, or ions.

7.3 *Importance of Ligands in Uranium Geochemistry and Diverse Environments*

Ligands can significantly impact the solubility and mobility of uranium in natural water bodies and other natural environments. The formation of complexes with ligands can affect the bioavailability and toxicity of uranium. Ligands can influence the reactivity of uranium ions with other elements and compounds, influencing geochemical processes such as adsorption, precipitation and complexation. Ligands can be used to separate and purify uranium from other elements and compounds.

8 Speciation in Uranium Geochemistry and Reactions

Speciation is the distribution of uranium among different chemical species, such as complexes, ions and colloids. It is the process of identifying and quantifying the different chemical species or forms of an element, such as uranium, in a given environment. In the context of uranium, various forms or species of uranium can exist in solution, including ions, complexes and colloids. While uranium species refer to various forms of uranium that can exist in different environments, including soils, rocks, and biological systems, uranium species in solution refer to the various forms of uranium that exist in groundwater, surface water, aquatic systems, sea water and ocean water (marine systems) [7].

8.1 Temperature-Dependent Reactions

Temperature is a vital factor that affects uranium speciation in different environments. The solubility of uranium increases with increasing temperature. This occurs because as the temperature increases, more energy is provided for uranium ions to move and interact with surrounding water molecules, thereby making it easier for uranium ions to dissolve. Additionally, temperature affects the distribution of uranium species in solution. There is a change in uranium speciation with respect to temperature. At high temperatures, more uranium is present as uranyl ions (UO_2^{+2}), which are highly mobile and more soluble than other uranium species are (U^{+3}, U^{+4} and U^{+5}). Temperature also influences complexation reactions, which involve the formation and stability of uranium complexes with ligands such as hydroxides, carbonate and organic matter. These complexes can enhance uranium solubility.

8.2 Temperature Effects on Uranium Speciation

Temperature influences uranium speciation through thermodynamic, kinetic, and solubility effects. These temperature-related factors play important roles in managing uranium behavior across diverse water bodies and environmental settings. In particular, the transport of uranium by hydrothermal fluids is key to understanding the formation of uranium ore deposits, assessing risks associated with proposed geological nuclear waste repositories, and, in some cases, informing nuclear reactor design [22]. Temperature encourages speciation of uranium in diverse environments by altering the thermodynamic effects, kinetic effects and solubility effects. For thermodynamic effects, temperature affects the equilibrium constant of uranium species, influencing their stability and distribution. In terms of kinetic effects, temperature influences the rates of uranium species reactions, such as complexation, precipitation and redox reactions. Temperature also affects the solubility of uranium species, influencing their concentration and distribution in solution. Complexation reactions, precipitation reactions and redox reactions are called temperature-dependent reactions.

In complexation reactions, the formation and stability of uranium complexes with ligands are influenced by temperature. Temperature also affects the rates of uranium redox reactions, such as the oxidation of uranium(IV) to uranium(VI) or the reduction of uranium(VI) to uranium(IV) ions. Temperature influences the rates of uranium precipitation reactions, such as the formation of uranyl hydroxide or uranyl phosphate. The relationship between the temperature and solubility of uranium depends on several factors, including but not limited to the pH concentration, redox potential or conditions, presence of complexing ligands, salinity and ionic strength, and uranium mineralogy and sources. The kinetics of mineral dissolution and desorption strongly influence the release of uranium in water. Owing to the high ionic strength and complex chemical composition of seawater and oceans, the uranium solubility of seawater and oceans is generally lower than that of freshwater systems, which are significantly influenced by organic matter and colloidal particles.

9 Complexation Reactions

Complexation reactions are the process of forming complexes that are molecules or ions that contain a central metal atom or ion surrounded by one or more ligands. In the context of uranium, complexation reactions involve the formation of complexes between uranium ions and various ligands. Examples of complexation reactions include uranium ions forming complexes with carbonates (uranium carbonate complexes), uranium ions forming complexes with hydroxide ions (uranium hydroxide complexes), uranium ions forming complexes with phosphate (uranium phosphate complexes), and organic complexes. These complexes are also regarded as some common uranium species, including uranyl ions (UO_2^{+2}), uranium (IV) ions or (U^{+4}) and uranium (VI) ions or (U^{+6}) [3, 18–21].

Analytical methods for studying uranium complexation include potentiometry, spectroscopy, chromatography and computational modeling. Potentiometry is a measure of the potential of a solution as a function of pH or ligand concentration. Spectroscopy uses techniques such as ultraviolet (UV), visible, infrared (IR) and nuclear magnetic resonance (NMR) spectroscopy to provide information about the structure and stability of uranium complexes. Chromatography separates the quantified uranium complexes on the basis of their interactions with a stationary phase. The computational modeling technique simulates the behavior of uranium complexes in various environments, assisting in predicting their formation and stability.

9.1 Types of Uranium Complexes

Two types of complexation reactions are inner–sphere complexation reactions and outer–sphere complexation reactions. In inner-sphere complexes, the complexes are formed when a ligand directly bonds to the uranium ion, replacing one or more water molecules in the uranium coordination sphere. Outer–sphere complexes are formed when a ligand interacts with a uranium ion through hydrogen bonding or electrostatic forces without directly bonding to the uranium.

9.2 Factors Influencing Uranium Complexation Reactions

- **pH:** pH can influence uranium complexation reactions. Changes in pH can affect the distribution of uranium complexes, with some complexes forming only within specific pH ranges. Different species dominate at different pH levels. The mobility of uranium species is most stable at intermediate pH values, and mobility increases as the pH approaches neutral conditions [21].
- **Temperature:** Temperature influences the rates of formation and solubility of uranium complexes. As the temperature increases, the rate of reactions typically accelerates. Additionally, at higher temperatures, the solubility of uranium in aqueous media increases due to the increased dissolution of uranium minerals and the formation of more soluble uranium species.
- **Ligand concentrations:** Layers such as hydroxide, carbonate, and organic matter can form complexes with uranium. The presence of complexing ligands can influence the rates of formation and solubility of uranium complexes and their speciation. In a previous study, uranium carbonate complexes demonstrated a severe decrease in stability at increased temperatures [22].
- **Redox conditions:** the oxidation state of uranium can be affected by changes in redox conditions, thus influencing the types of complexes that are formed. Reducing environments favor a lower oxidation state (U^{+4}), whereas oxidizing environments favor higher oxidation states (U^{+6}).
- **Competing ions:** The presence of other ions can compete with uranium for ligands, therefore affecting the formation and stability of uranium complexes.
- **Salinity and ionic strength: The** stability and distribution of uranium species and complexes are affected by changes in salinity and ionic strength.
- **Organic matter:** Organic matter contains functional groups that may bind to uranium ions. The interaction between organic matter and uranium may alter its transport and bioavailability.
- **Water chemistry:** Dissolved salts, metals, and other trace elements in water can affect the speciation and complexation of uranium.

9.3 Uranium Solubility in Water

The solubility of a substance is the amount of the substance that is required to form a saturated solution in a given amount of solvent at a specified temperature, and it is specific for a particular solvent (for example, water). Solubility is measured as the grams of solute per 100 grams of solvent. The solubility of sodium chloride in water is 36.0 grams (g) per 100 grams (g) of water at 20 °C. With respect to the solubility of substances, whether liquid or gas, the temperature must be specified because the solubility varies with temperature [21]. For gases, the pressure must also be specified. Uranium solubility is the ability of uranium to dissolve in water and other liquids. The solubility of uranium in water is a vital factor in determining its potential impact on water quality, and it is typically measured in units of concentration, such as micrograms per liter (*μg/L*), milligrams per liter (*mg/L*), and molarity (M). Uranium is relatively insoluble in water, with a solubility of approximately 20–30 *mg/L* at room temperature. However, its solubility can increase in the presence of ligands such as carbonate, phosphate and hydroxide ions. Generally, uranium–238 (^{238}U) is less soluble than uranium–235 (^{235}U) and uranium–234 (^{234}U) are because of its lower reactivity and tendency to form less soluble compounds.

10 Solubility and Mobility of Uranium Species

The solubility of uranyl minerals is very important because it controls the transport and distribution of uranium in many oxidizing environments [23]. Temperature influences the solubility of uranium species, encouraging and impacting the distribution of uranium species in solution. Uranium speciation helps in predicting the behavior, transport and fate of uranium in the environment as well as its potential impacts on human health and the ecosystem. However, uranium can exist in several species or oxidation states, including +2, +3, +4, +5, and +6, as can uranium (II), uranium (III), uranium (IV), uranium (V), and uranium (VI). However, in natural environments, uranium is typically found in the +4 and +6 oxidation states. The specific role of uranium(IV) in the +5 oxidation state is difficult to elucidate in solubility studies since, as an intermediate valence state, uranium(V) is systematically accompanied by uranium(IV) and uranium(VI) [24, 25]. Some of the common uranium species include uranyl ions (UO_2^{+2}), uranium (IV) ions (U^{+4}) with an oxidation state of +4, uranium (VI) ions (UO_2^{+2}) with an oxidation state of +6 (U^{+6}), carbonate complexes ($UO_2(CO_3)_2^{-2}$; $UO_2(CO_3)_3^{-4}$), hydroxide complexes ($UO_2\ (OH)_2$; $UO_2(OH)_3^{-}$), chloride complexes ($UO_2\ Cl_2$; $UO_2CL_3^{-}$), oxide complexes (UO_2^{+2}), phosphate complexes ($UO_2(PO_4)_2^{-}$; $UO_2(PO_4)_3^{-5}$), organic complexes and colloids. Uranium can bind to organic matter, such as humic acids and fulvic acids, which can influence the solubility and bioavailability of uranium. Uranium can also exist as colloids, which are small particles that can remain suspended in solution. While some uranium species are more soluble and mobile, others are less soluble and mobile. Examples of more soluble and mobile as well as less soluble and mobile materials are highlighted as follows:

10.1 More Soluble and Mobile Uranium Species

- Uranyl ion (UO_2^{+2}): **A** uranyl ion (UO_2^{+2}) is a linear molecule with oxygen atoms bonded to the uranium atom. It is the most common and stable uranium species in oxidizing environments and is highly soluble and mobile in water, especially in the presence of complexing ligands such as hydroxide, carbonates, and phosphate.
- The uranium (VI) ion (U^{+6}): uranium (VI) ion (U^{+6}) is similar to or related to the uranyl ion (UO_2^{+2}); however, it has a different oxidation state, which is hexavalent (U^{+6}). It is more soluble and mobile in water than is the uranium(IV) oxidation state.
- Uranium hydroxide complexes: Uranium hydroxide complexes such as $UO_2\,(OH)_2$ and $UO_2\left(OH\right)_3^-$ are soluble and mobile in water, especially in alkaline media or environments.
- Uranyl carbonate complexes: Uranyl carbonate complexes such as $UO_2\left(CO_3\right)_2^{-2}$ and $UO_2\left(CO_3\right)_3^{-4}$ are highly soluble and mobile in water, and these complexes are stable over a wide range of pH and redox conditions.

10.2 Less Soluble Uranium Species

- Uranium (IV) ions (U^{+4}): uranium (IV) ions are in a reduced state or form of uranium that can exist in reducing environments. In particular, in the absence of complexing ligands, uranium(IV) ions are less soluble and mobile in water than are uranium(IV) oxidation states and uranyl ions (UO_2^{+2}).
- Uranium(II) oxide and uranium hydroxide complexes: Compared with uranyl ions and their complexes, uranium(II) oxide and uranium hydroxide complexes are less soluble and mobile in water.
- Uranium phosphate complexes: Uranium can form complexes with phosphate ions. Compared with uranium ions and their complexes, uranium phosphate complexes, such as $UO_2\left(PO_4\right)_3^{-5}$ and $UO_2\left(PO_4\right)_2^-$, are less soluble and mobile in water.

Generally, the solubility and mobility of uranium species decreases with increasing pH and decreasing redox potential. The presence of complexes, namely, carbonate, phosphate and hydroxide, can also influence the solubility and mobility of uranium species. The groundwater is relatively high due to the formation of carbonate complexes [26].

11 Uranium Oxides (UO_2, UO_3, U_3O_8)

Uranium (II) oxide (UO_2) is known as brown oxide or uraninite, and it has low solubility in water. The ore body is primarily in a tetravalent state (U^{+4}). Uranium(II) oxide (UO_2) reacts with carbonate in the presence of oxygen to produce a mixture of mono uranate and di urinate, and high temperatures favor mono urination. One hundred percent oxidation of uranium(II) oxide (UO_2) is possible at 723 *K* in 1 h [27]. Uranium(III) oxide (UO_3), known as orange oxide, is moderately soluble in water. When UO_3 is heated in air, it decomposes to form $U_3O_{8.}$ In addition, when UO_3 is heated to a temperature greater than 710 °C in the presence of $H_{2,}$ it is converted to UO_2. Triuranium octoxide (U_3O_8) is the most stable form of uranium and is used for disposal and storage. It is greenish black in color with low solubility in water. Other compounds include ammonium diuranate [$(NH_4)_2U_2O_7$] and uranium peroxide ($UO_4.nH_2O$), which are pale yellowish compounds that are moderately soluble in water. High-temperature reactions both in melts and in the solid-statestate of uranium oxides ($UO_{2,}$ UO_3 and U_3O_8) that result in the formation of insoluble alkali metals of lithium (Li) and cesium (Cs) uranates have been reported [27].

12 Influence of Temperature on Solubility of Solids, Liquids and Gas

The solubility of the majority of solid substances increases as the temperature increases; however, the effects are difficult to predict and vary widely from one solute to another. The solubility of solids in water increases with increasing temperature, whereas the gas solubility decreases as the temperature increases. When a solvent (liquid) with a gas (solute) dissolved in it is heated, the kinetic energy of both the solvent and the solute increases. As the kinetic energy of the gaseous solute increases, its molecules tend to escape the attraction of the solvent molecules and return to the gas phase. Therefore, the solubility of a gas decreases as the temperature increases. The temperature dependence of solubility can be visualized with the help of a graph of solubility against temperature called the solubility curve. Solubility curves can be used to determine if a given solution is saturated or unsaturated. If the graph of the solubility against the temperature of a substance is fairly flat, an increase in temperature has relatively little effect on the solubility of the substance or compound (NaCl, for example). If the solubility of a substance or a compound is very steep, an increase in temperature dramatically increases the solubility of that substance or compound ($KNO_{3,}$ for example). Substances such as NH_3, SO_2 and HCL have solubilities that decrease as the temperature increases. These compounds are gases at standard pressure. A report by Peiffert et al. [28] on the solubility of uranium in both carbonated fluid and granitic melt in equilibrium in a haplogranitic-uranium oxide H_2O-Na_2CO_3 system at 720–770 °C 2Kbar revealed that temperature variation in the range of 720–770 °C had no significant effects on uranium solubility in either aqueous fluid or silicate melt.

13 Uses of Uranium Isotopes and Damages Caused by Uranium Isotopes

Primordial uranium–238 (^{238}U) and uranium–235 (^{235}U) isotopes as well as the decay products of uranium–234 (^{234}U) are found in significant amounts in the Earth's crust (water, seas, oceans, soil, and rocks), with different solubilities [20, 29]. Uranium is soluble in acidic solutions [30]. Some quantities of uranium are present in magma, and because of the chemical properties of uranium, it can be used as a means of understanding magmatic processes [31]. Uranium isotopes are extensively used in the study of sources of dissolved solutes in streams, rivers, seas and oceans and as radioactive tracers in the dating of several oceanographic processes [7, 17, 20]. Depleted uranium (uranium containing mostly ^{238}U) can be used as projectiles in armor—piercing weapons—and in the shielding of radiation. Uranium-235 (^{235}U) is very importantly used as fuel for nuclear power plants and nuclear reactors that run submarines and naval ships. It is also used as a nuclear weapon because it is fissile.

The ingestion of high concentrations of uranium and its progenies can cause severe health effects, such as liver or bone cancer. Because uranium decays by alpha particles during external exposure, inhaling elevated concentrations can cause lung cancer. Owing to its chemical properties, uranium is a toxic element; therefore, when elevated quantities of uranium are ingested, kidney damage occurs sooner than the radioactive effects that can cause cancers of the liver and bones [7, 32, 33].

14 Summary

The solubility of uranium generally increases with increasing temperature, as more uranium can dissolve in solution due to the formation of more stable uranium complexes at elevated temperatures, particularly with carbonate ions, which enhances the mobility of uranium in aqueous environments. Therefore, the relationship between the temperature and solubility of uranium is essential for evaluating potential hazards or risks associated with uranium contamination in surface water and groundwater systems, particularly in areas or locations with fluctuating temperatures. The increased acidity of water with increasing temperature enhances the hydrolysis and solubility of uranium. The temperature can increase the solubility of uranium in water and nitric acid solutions because increasing the temperature increases the acidity of water and the hydrolysis of uranium. As the temperature increases, the water becomes more acidic, resulting in an increase in uranium solubility. The solubility of uranium also increases with increasing temperature because higher temperatures enhance the dissolution of uranium minerals and increase the formation of soluble complexes.

15 Conclusion

Temperature can influence the solubility of uranium in a number of ways, including increased complexation, hydrolysis reactions, redox conditions, chemical speciation, the influence of pH, and geological compositions, among other conditions. As the temperature increases, the formation of soluble uranium complexes with ligands such as carbonate ions becomes more favorable, leading to increased solubility in aqueous solutions or environments. Elevated temperatures can promote hydrolysis reactions, which can generate more soluble uranium species. Temperature also influences redox conditions by influencing the oxidation state of uranium. The oxidation state of uranium can be influenced by temperature, with reducing conditions favoring less soluble uranium (IV) species. The dominant chemical form of uranium in solution significantly impacts its solubility, with uranium(IV) species being more soluble than uranium(VI) oxidation states. Uranium solubility tends to be greater at alkaline pH values or under alkaline conditions. The constituents of the surrounding rock and groundwater system can influence uranium solubility, with certain minerals facilitating uranium dissolution.

References

1. Balaram V, Rani A, Rathore DPS (2022) Uranium in groundwater in parts of India and world: a comprehensive review of sources, impact to the environment and human health, analytical techniques, and mitigation technologies. Geosyst Geoenviron 1(2):100043
2. Deshmukh P, Sar SK, Jindal MK (2023) Plant mediated magnetic nano composite as promising scavenger's radionuclides for the efficient remediation in aqueous medium. Chemosphere 312:137246
3. Jindal MK, Pandit SA, Karunakara N, Changrashekara SN, Salim SKD, Srinivasan R (2023) High uranium dose from groundwater in a granitic terrain in the eastern part of Karnatakar, Indian. J Radioanal Nucl Chem. https://doi.org/10.1007/s10967-023-09053-6
4. Sajwan RS, Joshi V, Ahamad T, Kumar N, Parmar P, Jindal MK (2024) Assessment of radon transportation and uranium content in the tectonically active zone of Himalaya, India. Sci Total Environ 926:171823
5. Sar SK, Diwan V, Biswas S, Singh S, Sahu M, Jindal MK, Arora A (2018) Study of uranium level in groundwater of Balod district of Chhattisgarh state, India and assessment of health risk. Hum Ecol Risk Assess Int J 24(3):691–698
6. Kundzewicz ZW (2008) Earth and planetary sciences: hydrosphere. Encyclopedia of Ecology. https://www.sciencedirect.com/topics/earth-and-planetary-sciences/hydrosphere

7. Nwaka BU, Ogueri EO, Jindal MK, Amala D, Avwiri GO, Jibiri NN, Amakom CM, Ezema FI (2025) Chapter 6: Impact of radioactive elements and oceanographic factors on marine ecosystems. In: Radiation status in the marine world. Springer, Cham. https://doi.org/10.1007/978-3-031-82403-6
8. Nigerian Standard for Drinking Water Quality (NSDWQ) (2007) Nigerian Industrial Standard (NIS) approved by Standard Organization of Nigeria Governing Council, ICS 13.060.20: 15–19
9. UNSCEAR (2012) Sources, effects and risks of ionizing radiation. United Nations Scientific Committee on the Effects of Atomic Radiation; Report to the General Assembly with Scientific Annexes A and B, United Nations, New York
10. World Health Organization (WHO) (2017) Guideline for drinking—water quality fourth (4th) edition incorporating the first addendum. World Health Organization Cataloguing in publication data, Geneva
11. Nwaka BU, Avwiri GO (2021) Physicochemical water quality within Okpoi Okwu and Uburu Salt lakes, Ebonyi State, Nigeria. IOSR J Environ Sci Toxicol Food Technol 15(6):7–11
12. Nwaka BU, Avwiri GO, Jindal MK, Kiran P (2025) Assessment of radiation hazard from radioactive elements 226Ra, 232Th, 40K in local and industrially refined salt from consumption in Nigeria. J Radioanal Nucl Chem. https://doi.org/10.1007/s10967-025-10037-x
13. Najar M, Maglas NNM, Wang H, Qiang Z, Ali MMM (2025) Enhancing radiological risk evaluation through AI and hotspot code integration: a comparative study of LOCA and SGTR. Radiat Phys Chem 230(112580). https://doi.org/10.1016/j.radphyschem.2025.112580
14. Mishra P, Sati M, Krishnamurthi R (2024) A review on uranium mineralization related to Na-metasomatism: Indian and international examples. Geosciences 14(11):304. https://doi.org/10.3390/geosciences14110304
15. Ella E, Granfors M, Schild D, Kastriot S (2020) The effects of temperature and full dissolution kinetics under H_2 atmosphere. J Nucl Mater 531:151981. https://doi.org/10.1016/j.jnucmat.2019.151981
16. Nwaka BU (2019) Introduction to ionizing radiation measurement and quantities, LAMBERT Academic Publishing, 17 Meldrum Street Beau Bassin 71504, Mauritius. ISBN-13:978-613-9-87463-7, www.amazon.com
17. Zhao C, Zhang P, Li X, Ning Y, Tan L, Edwards RL, Yao X, Cheng H (2020) Distribution characteristics and influencing factors of uranium isotopes in saline lakes water in the Northeast of Qaidam Basin. Minerals 10(1):74. https://doi.org/10.3390/min10010074
18. Fujii Y, Higuchi N, Haruno Y, Nomura M, Suzuki T (2006) Temperature dependence of isotope effects in uranium chemical exchange reaction. J Nucl Sci Technol 43(4):400–406. https://doi.org/10.1080/18811248.2006.9711111. http://www.tandfonline.com/loi/tnst20
19. Olivier P, Laurent T, Annabelle L, Etienne D (2022) Solubility of uranium oxide in ternary aluminosilicate glass melts. J Non-Cryst Solids 595:121845. https://doi.org/10.1016/j.jnoncrysol.2022.121845
20. Stirling HS, Andersen MB, Potter E, Halliday AN (2007) Low—temperature isotopic fractionation uranium. Earth Planet Sci Lett 264(1–2):208–225. https://doi.org/10.1016/jepsi.2007.09.019
21. van Hartesveldt NF (2020) Uranium Solubility in high temperature reduced systems. MSc thesis and dissertation, 4856, submitted to the Faculty of Mississippi State University, Mississippi. https://scholarsjunction.msstate.edu/td/4846
22. Kalintsev A, Migdisov A, Alcorn C, Baker J, Brugger J, Mayanovic RA (2021) Uranium carbonates complexes demonstrate drastic decrease in stability at elevated temperature. Commun Chem 4:120. https://doi.org/10.1038/S42004-021-00558-3
23. Gorman-Lewis D, Burns PC, Fein JB (2008) Review of uranyl mineral solubility measurements. J Chem Thermodynam 40:335–352

24. Bailey EH, Ragnarsdottir KV (1994) Uranium and thorium solubilities in seduction zone fluids. Earth Planet Sci Lett 124(1–4):119–129. https://doi.org/10.1016/0012-821X(94)0071-9
25. Chevreux P, Lourent T, Annabelle L, Tanya V, Sebastian B, Le GF et al (2021) Uranium solubility and speciation in reductive soda—lime aluminosilicate glass melts. J Nucl Mater 554:152666. https://doi.org/10.1016/j.jnucmat.2020.152666
26. Kumar A, Rout S, Marayanana U, Mishra MK, Tripathi RM, Singh J (2011) Geochemical modeling of uranium specialtion in the subsurface aquatic environment of Punjab State in India. J Geol Mining Res 3(5):137–146. ISSN: 2006 – 9766 https://www.academia.otg/jgmr
27. Griffiths TR, Volkovich VA (1999) A review of high temperature oxidation of uranium oxides in molten salts and in solid states to form alkali metal urinates, and their composition and properties. J Nucl Mater 274(3):229–251. https://doi.org/10.1016/S0022-3115(99)00091-4
28. Peiffert C, Cuney M, Nguyen-Trung C (1994) Uranium in granitic magama: Part 1 Experimental determination of uranium solubility and fluid-melt partition coefficient in the uranium oxides-haplogranite-H_2O-Na_2CO_3 system at 720–770 °C 2Kbar. Geochim Cosmochim Acta 58(11):2495–2507. https://doi.org/10.1016/0016.7037(94)90026-4
29. Michael AK, Hooyu L, Michael JE, Seth GJ, Rob M, Jess FA, Françoise LHT (2022) ^{238}U, ^{235}U and ^{234}U in sea water and deep-sea corals: a high precision reappraisal. Geochim Cosmochim Acta 336:231–248. https://doi.org/10.1016/j.gca.2022.09.018
30. Alexander K, Artos M, Joel B, Hongwu X (2024) Uranium (VI) hydrolyses up to 250 °C and its geological implications. Geochim Cosmochim Acta 277:68–83. https://doi.org/10.1016/j.gca.2024.05.014
31. Florent D, Bruce V (1985) Preliminary investigation of the processes governing the solubility of uranium in silicate melts. Bull Mineral 108:755–765
32. John SOG, Olukokun SF, Mathuthu M (2025) Assessment of radioactivity concentrations and associated radiological health risk in natural spring mineral bottled drinking water from South Africa. Water 17(2):156. https://doi.org/10.3390/w17020156
33. Singh KP, Chandra S, Prasad M, Panwar P, Joshi A, Sharma S et al (2025) Measurement of natural radionuclides and health risk assessment of soil samples of the main central thrust region in Garhwal Himalaya, India. Environ Monit Assess 197(2):128. https://doi.org/10.1007/s10661-024-13569-z

Dissolved Oxygen and Its Role in Uranium Solubility

Brian Youden, Andrew Carrier, Ken Oakes, Mark Servos, Runqing Jiang, and Xu Zhang

Abstract Redox reactions involving uranium (U) are key processes governing its environmental mobility, accumulation in ore bodies, isotopic fractionation, remediation, and its leaching, extraction, and recovery from mines. Notably, the influx of oxidants, such as dissolved molecular oxygen (DO) or nitrate (NO_3^-), into aquifers and storage systems threatens long-term sequestration of tetravalent uranium (U^{4+}) minerals via particulate or solid surface dissolution as mobile hexavalent uranium (U^{6+}) species, which can present a significant health risk to those reliant on groundwater. Although nitrate influx is often attributed to anthropogenic inputs, DO naturally infiltrates reduced sediments through advective oxic groundwater flow, rainwater infiltration, and capillary fringe transport. Aquifers also exhibit substantial variations in DO laterally, vertically, and temporally due to variations in soil and rock systems, remediation treatments, and seasonal watershed dynamics. Thus, transitions between oxic and anoxic conditions can result in substantial geochemical changes and stimulation or repression of microbial activity. These reactions with DO are significantly altered by the presence of biomass, metal cations and minerals, (bi)carbonate, and other inorganic ligands. In this chapter, the interactions of DO with uranium species are reviewed and discussed. The influence of DO in practical applications, i.e., uranium extraction and bioremediation, is also examined.

B. Youden · M. Servos
Department of Biology, University of Waterloo, Waterloo, ON, Canada

A. Carrier · X. Zhang (✉)
Department of Chemistry, Cape Breton University, Sydney, NS, Canada
e-mail: xu_zhang@cbu.ca

K. Oakes
Department of Biology, Cape Breton University, Sydney, NS, Canada

R. Jiang
Department of Biology, University of Waterloo, Waterloo, ON, Canada

Department of Medical Physics, Grand River Regional Cancer Centre, Kitchener, ON, Canada

M. K. Jindal, K. Patni (eds.), *Water Quality-Uranium Solubility Nexus: Unlocking the Secrets of Uranium Solubility*, Earth and Environmental Sciences Library, https://doi.org/10.1007/978-3-031-98909-4_6

Keywords Dissolved oxygen · Uranium solubility · Uranium extraction

1 Interaction of Dissolved Oxygen with Uranium Species

Uranium is a toxic and radioactive element with two main stable valence states: oxidized hexavalent U^{6+} and reduced tetravalent U^{4+}, although other oxidation states, e.g., pentavalent U^{5+}, also occur in solid minerals. Aqueous U^{6+} is stable as dissolved uranyl ions (UO_2^{2+}) over a wide pH range but is most often complexed with carbonate ions or other ligands in natural waters. The high mobility of U^{6+} is correlated with its high toxicity. In contrast, U^{4+} is significantly soluble only at very low pH values (<2) and thus occurs primarily as uraninite (UO_2) precipitates or other uranium-bearing solids, e.g., U_3O_8, β–U_3O_7, U_4O_9, or amorphous U^{4+} phases [1]. The reduction of soluble U^{6+} to immobile U^{4+} is considered an effective strategy for the immobilization, remediation, and long-term storage of uranium in soil and groundwater. However, reduced U^{4+} species are readily re-oxidized and re-released into the environment due to changes in redox conditions or during extraction.

Reduced U^{4+} precipitates are subject to oxidative dissolution or remobilization by various compounds, such as dissolved molecular oxygen (DO), Fe ions, nitrate (NO_3^-), denitrification intermediates (NO_2^-, NO, and N_2O), and Fe- or Mn-oxides, or through complexation with organic ligands, such as siderophores, citrate, and humic substances [2]. Typically, the principal oxidizing agent in natural groundwaters is DO. During the dissolution of UO_2 particles and spent fuel, O_2 increases the surface oxidation state from UO_2 to UO_{2+x} ($0 < x < 0.25$), producing soluble U^{6+} on the surface [3]. Peroxides and other reactive oxygen species (ROS) generated from radioactive decay are also oxidants but generally only form in appreciable concentrations in intermediate- to high-level radioactive wastes.

Many studies have demonstrated a clear dependence between the oxidation and dissolution rates of UO_2 on DO concentrations, although the trends have not been consistent, with large variations observed within some datasets [3, 4]. This likely depends on several factors, such as variations in uranium ore and fuel composition, oxygen concentration and pressure, pH, temperature, diffusion limitations [5, 6], and the existence of biomass, inorganic ions, or other minerals in the system that competitively consume oxygen [7, 8]. Casella et al. [4], for instance examined the effect of DO on the dissolution of pure and Gd_2O_3-doped UO_2 (to simulate fission products in used fuel). Dissolution by DO increased as the temperature rose; however, a significantly weaker dependence was found in the doped fuels for all the examined temperatures and DO ranges. This effect was most significant at higher temperatures (50 and 75 °C). These results indicate that fuel transformation during fission and the formation of secondary uranium phases at low temperatures and carbonate concentrations likely inhibit dissolution by DO [9]. Furthermore, in a study of Mn^{2+}-doped biogenic UO_2, the equilibrium solubility was half that of conventional biogenic UO_2, and oxidation by DO was 38-fold slower [10]. This observation suggests that environmentally relevant cations incorporated into the UO_2

structure may also provide a stabilizing effect against DO. When considering the effect of oxygen pressure, Xu et al. [11] reported that increased oxygen pressure significantly enhanced the uranium leaching rate of sandstone ore under acidic conditions. When the oxygen pressure increased from 0.5 to 1.0 to 1.5 MPa (pH 2), the total uranium leaching increased from 77% to 84% to 97%, respectively; however, when the leaching solution pH increased to 3.0, the uranium dissolution rate decreased significantly, i.e., to ~45% and 75% at 1.0 and 1.5 MPa, respectively. In contrast, Kacham and Suri reported that the leaching of uranium from ore with a high percentage of pyrite in alkaline solutions occurred faster at oxygen partial pressures below 1 atm and decreased significantly with increasing pressure because of competitive oxidation of the pyrite [12]. Further work is therefore needed to understand the complex interplay of factors modifying uranium dissolution by DO.

The oxidation of dissolved U^{4+} by O_2 at low pH can be described by the following reaction:

$$U^{4+} + \frac{1}{2}O_2 + H_2O \rightleftharpoons UO_2{}^{2+} + 2H^+ \quad (1)$$

In comparison, the oxidation of solid U^{4+} (UO_2) in the presence of O_2 and bicarbonate (HCO_3^-) can be described as follows [13]:

$$2UO_2 + O_2 \rightleftharpoons 2UO_3 \quad (2)$$

$$UO_3 + 2HCO_3^- \rightleftharpoons UO_2(CO_3)_2{}^{2-} + H_2O \quad (3)$$

$$UO_3 + CO_3^{2-} + 2HCO_3^- \rightleftharpoons UO_2(CO_3)_3{}^{4-} + H_2O \quad (4)$$

Or simply:

$$UO_2 + \frac{1}{2}O_2 + 2HCO_3^- \rightleftharpoons UO_2(CO_3)_2{}^{2-} + H_2O \quad (5)$$

As carbonate is a strong complexing agent for U^{6+}, it quickly removes oxidized uranium from particle surfaces that would otherwise accumulate and passivate the surface. Additionally, because carbonate is present in most natural waters, the rate-limiting step for uranium dissolution is often the oxidation of U^{4+} to U^{6+} by O_2, not the accumulation of oxidation products on the solid surface (Fig. 1a).

The presence of dissolved co-ions at contamination sites can also significantly impact the effects of O_2 on uranium dissolution. Wang et al. [14] examined the effects of Mn^{2+}, a redox-active cation, on UO_2 dissolution under both oxic and anoxic conditions using batch and continuous-flow reactors (Fig. 1b). Soluble Mn^{2+} is found in groundwater at appreciable concentrations at uranium-contaminated sites, e.g., up to 20 μM at the Rifle field research site (Colorado, USA; a former uranium processing site). A significant inverse correlation between dissolved Mn and DO concentrations was also observed at the Rifle site, indicating dynamic

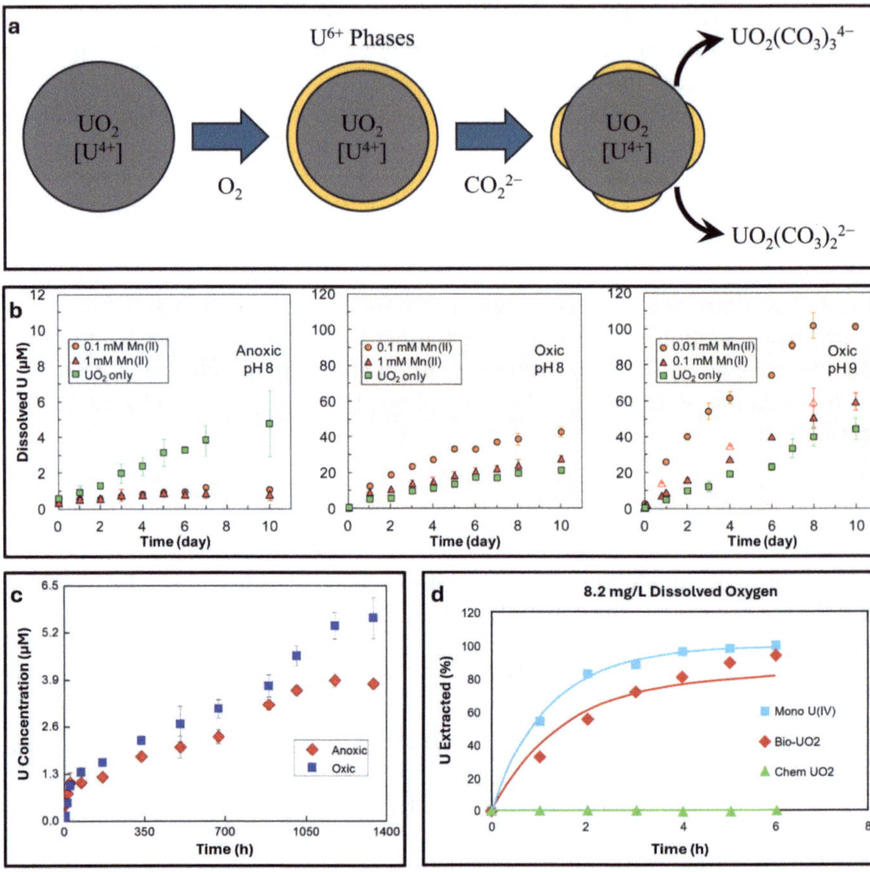

Fig. 1 (**a**) Schematic of uraninite (UO_2) dissolution by dissolved oxygen (DO) and carbonate. (**b**) Dissolution of UO_2 in anoxic and oxic conditions in the presence of Mn^{2+} cations [14]. (**c**) Comparative dissolution of studtite under anoxic and oxic conditions [15]. (**d**) Comparative dissolution of monomeric U^{4+} species (Mono U(IV)), biogenic UO_2 (Bio-UO_2), and chemogenic UO_2 (Chem UO_2) in the presence of DO [16]

seasonal manganese redox cycling. Under anoxic conditions, Mn^{2+} inhibited UO_2 dissolution, which was attributed to the blocking of access to U^{4+} surface sites by oxidants through the adsorption of Mn^{2+} and $MnCO_3$. Another investigation revealed that Ca^{2+} and Zn^{2+}, which are also common at contamination sites, inhibit UO_2 dissolution under both oxic and anoxic conditions via the precipitation of a calcium-uranyl and zinc carbonate (hydrozincite, $Zn_5(CO_3)_2(OH)_6$) phase that passivates UO_2 surfaces [17]. This inhibition was most substantial under oxic conditions, where the dissolution rates were 7 and 1450 times lower in the presence of Ca^{2+} and Zn^{2+}, respectively, than in cation-free water. In contrast, Mn^{2+} promotes UO_2 dissolution under oxic conditions through redox cycling, producing reactive manganese species that dissolve UO_2 faster than O_2 alone. With rapid Mn turnover during redox cycling, Mn concentrations as low as 10 μM increase UO_2 dissolution rates in the presence of O_2.

Notably, most studies limit their focus to uraninite-like minerals, but the importance of the crystal structure and the interaction of oxygen with other uranium deposits are less well understood. For instance, Kim et al. [15] reported only a relatively small increase in uranium dissolution from studtite (UO_4) in oxic groundwater relative to anoxic groundwater (Fig. 1c). Cerrato et al. [16] reported that biomass-associated monomeric U^{4+} species were more labile than biogenic uraninite in the presence of O_2, which in turn was more labile than chemogenic (crystalline) uraninite (Fig. 1d). These results suggest that amorphous U^{4+} is more labile than uraninite in subsurface environments. A concentration of 8.2 mg/L DO and 100 mM $NaHCO_3$ was sufficient to oxidize all biogenic uraninite (620 nmol/g U/s) and monomeric U^{4+} (990 nmol/g U/s) after 6 h. In contrast, the dissolution rate of chemogenic uraninite was only 1 nmol/g U/s, approximately an order of magnitude lower than that previously reported by Ulrich et al. [18] using continuous flow tank reactors, whereas the rate constant for the dissolution of biogenic uraninite was an order of magnitude greater. This may be due to differences in the reactor types (batch vs continuous flow) that affect the accumulation of U^{6+}. Another important difference is that the biogenic uraninite in Cerrato et al. [16] was associated with biomass. More work is needed to understand the interaction of O_2 with different forms of UO_2 and non-uraninite minerals.

2 Dissolved Oxygen Levels in Aqueous Environments

Anoxic sediments are major sinks for uranium in the environment alongside biogenic carbonates and oceanic basalts. Under suboxic or anoxic conditions, uranium can be removed from sedimentary porewater via the reduction of U^{6+} to U^{4+} and replenished by diffusion at the sediment–water interface. Importantly, the DO concentrations of different water sources and aquifers exhibit significant spatial and temporal heterogeneity, ranging from <1 to >20 mg/L, depending on several factors. Two similar 100% air-saturated bodies of water will not necessarily contain the same DO concentration because of differences in temperature, elevation, salinity, the presence of (micro)organisms, and pollution. For example, in fresh water at 1 atm of pressure (sea level), the DO concentration reaches ~14.6 mg/L at 0 °C but only 9.1, 8.3, and 7.0 mg/L at 20, 25, and 35 °C, respectively. In addition, temporal DO variations arise following cessation of remediation treatments (O_2 rebound), mining, and seasonal watershed dynamics. The DO levels in surface waters also follow a daily cycle due to the influence of photosynthetic plants and algae that photosynthesize during the day and respire at night.

Seasonal variations in DO affect uranium mobility. For example, at the Rifle field research site, DO varies from <0.1 mg/L during fall and winter to ~1.5 mg/L during spring and summer meltwater runoff. Lezama-Pacheco et al. [19] thus examined seasonal UO_2 dissolution by O_2 using a diffusible gel puck at the Rifle research site.

Under low O_2 conditions (<0.3–0.6 mg/L) in the P-103 well, no significant uranium dissolution was observed over a 674-day period. In comparison, in the B-02 well, the DO concentration was 1.2 mg/L at the start of the experiment (July 16, 2009), which occurred during the spring/summer meltwater event. After 102 d, the gels had leached 55% of their original uranium. The groundwater composition in the B-02 well was similar to that in the P-103 well, except for the higher DO concentrations, indicating that DO caused oxidative uraninite dissolution to occur during the spring/summer. During the subsequent fall–winter period (October 26, 2009–April 13, 2010), the DO level decreased to <0.6 mg/L, and no uraninite leaching was recorded over this interval. During the remaining year of the experiment (to May 21, 2011), involving the 2010 and first portion of the 2011 spring-summer meltwater events, the uranium loss ultimately climbed to 90%. This comparison highlights the seasonal dependence of DO on uranium dissolution during the spring/summer period. Under intermediate O_2 conditions in the Rifle site groundwater, i.e., 0.6–1.5 mg/L during spring–summer, biogenic uraninite is expected to persist for only a few years.

Several studies have highlighted the role of chemical oxygen demand in uranium dissolution. The addition of organic compounds, especially domestic and animal sewage, industrial waste, and crop wastewater, dramatically reduces DO in water through redox processes and the presence of aerobic bacteria. Bioreduced inorganic compounds may also be preferentially oxidized by DO. For example, Moon et al. [20] examined the oxidation of bioreduced uranium by oxygen and nitrate. While nitrate does not significantly re-oxidize reduced species, such as U^{4+}, Fe^{2+}, and sulfide, denitrification intermediates, such as nitrite, can. Both injected oxygen (20% gas mixture, 270 μM) and nitrate (1.6 mM) resolubilized the majority (88 and 97%, respectively) of the uranium precipitate within 54 d. Although oxygen is a more thermodynamically favorable oxidant than nitrate, nitrate-dependent uranium oxidation was significantly faster than oxygen-dependent uranium oxidation because oxygen preferentially reacts with other reduced compounds, i.e., iron sulfides. Excess nitrogen is converted to nitrate in oxygen-rich soils, which then percolate into groundwater, especially in shallow unconfined aquifers [2]. Bone et al. [21] investigated uranium dissolution from reduced sediments upon exposure to artificial groundwater containing O_2 or NO_3^- under diffusion-limited conditions. Little uranium was leached during the 85-day reaction, despite rapid diffusion of groundwater within the sediments and the presence of non-uraninite U^{4+} species. The production of ferrous iron and sulfide in conjunction with rapid oxidant consumption suggested that the sediments harbored high concentrations of bioavailable organic carbon that fuel anaerobic microbial respiration and stabilized U^{4+}. These results suggest that seasonal influxes of O_2 and NO_3^- may cause only localized uranium leaching.

In addition to temporal variations in DO, limitations in the penetration depth and transport of DO significantly impact uranium mobility. Compared with those at the surface, bottom water and deeply buried sediments typically have low DO concentrations. An investigation at the Rifle mining site identified potential effects of both groundwater composition and transport limitations on biogenic UO_2 stability [5]. The Rifle site has a shallow, unconfined alluvial aquifer with uranium concentrations between 0.1 and 0.4 mg/L (0.4–1.8 μM), a circumneutral pH, and low but measurable DO concentrations. The uranium dissolution rate was 50–100 times greater in laboratory experiments than in field experiments in which permeable sample cells were used to maintain groundwater contact with UO_2 solids. The difference between the laboratory and field dissolution rates was attributed primarily to the diffusive limitations exerted by the permeable sample cells on DO and U^{6+} transport; however, slight differences in the compositions of the actual vs. artificial groundwater used for laboratory experiments may also have contributed to the different dissolution rates [5, 6]. A column study by Moon et al. [22] indicated that the addition of nitrate re-oxidized more bioreduced U^{4+} than DO because DO reacts faster with organic compounds and biogenic iron sulfides, slowing the advancement of the reaction front through the column and protecting U^{4+} from DO contact. The authors observed that 50–80% of the influent nitrate concentration broke through the columns within 5 h of the initiation of the oxidation phase, whereas no DO was detected in the effluent until after 40 d. Iron sulfides were more effective at protecting biogenic uranium from DO than nitrate, even when the two oxidants had the same electron acceptor equivalence (0.27 mM oxygen and 0.25 mM nitrate). Unlike well-mixed flow-through reactor systems, where the presence of mackinawite ($(Fe, Ni)_{1+x}S$ where x = 0–0.11) or organic matter keeps DO levels low throughout the reaction period, DO gradients may exist in column reactors and natural sediments. As a result, the uranium accumulation rate in sediments is nonlinearly related to the depth of oxygen penetration.

3 Isotope Fractionation during Uranium Oxidation

Uranium isotopes in ancient sedimentary rocks, i.e., shales and carbonate rocks, are widely used as tools to reconstruct paleo-redox conditions [23], but the behavior of uranium isotopes under modern non-sulfidic anoxic vs. oxic conditions remains poorly understood [24]. Different uranium isotopes also have different uses as nuclear fuels. The most abundant isotopes of uranium are primordial ^{238}U (99.2745%) and ^{235}U (0.72%), with half-lives of 4.5 and 0.7 billion years, respectively. Because of their long half-lives, variations in $^{238}U/^{235}U$ are largely the result of chemical reactions rather than radioactive decay [25]. Several field and laboratory studies have shown that the reduction in U^{6+} causes a notable shift in the isotopic distribution of uranium [26, 27], favoring heavier isotopes, which appears to be caused by a nuclear volume effect [28, 29]. This allows the use of $^{238}U/^{235}U$ ratios ($\delta^{238}U$) to monitor and potentially quantify redox conditions. Andersen et al. [30] presented a model where uranium isotopic compositions of modern sediments up to 1.2‰ higher than those of seawater indicate anoxic or euxinic (both anoxic and sulfidic) seawater conditions, whereas values comparable to those of seawater indicate oxic conditions. The reduction of uranium diffusing into porewater is thought to be fast under fully sulfidic conditions, leading to large uranium isotope fractionation toward higher $\delta^{238}U$

values. However, since U^{4+} is quickly re-oxidized by natural oxidants, such as oxygen and nitrate, the application of $\delta^{238}U$ measurements to indicate reduction may be misleading [31]. If oxygen from oxic bottom water penetrates the upper porewater, the distance between the sediment–water interface and the uranium reduction zone increases, decreasing isotopic fractionation. Bioturbation at oxic sites also contributes to maintaining the supply of uranium with low $\delta^{238}U$ values to several cm deep.

The effect of the oxidation reaction by DO on isotope fractionation has recently been examined for both dissolved and solid U^{4+} [31]. Experiments focusing on dissolved U^{4+} oxidation at low pH indicated that the initial U^{6+} product was isotopically lighter by about 1.1‰, with the remaining U^{4+} progressively enriched in ^{238}U with a constant equilibrium fractionation factor. In contrast, solid U^{4+} in an aerated $NaHCO_3$ solution (20 mM, pH 9.4) experienced only weak fractionation (~0.1–0.3‰). This small shift was explained by the adsorption of isotopically light U^{6+} onto the U^{4+} solid. More specifically, the authors attributed the lack of isotopic fractionation to a "rind effect," where each surface layer of the solid particles must be completely oxidized before the next layer degrades. As such, the oxidation of each layer but not the first results in minimal isotope fractionation. If the oxidative dissolution process involves kinetic fractionation, this would cause an initial release of isotopically distinct uranium from the surface. However, this would be short-lived as the exposed surface becomes increasingly fractionated in the opposite direction via adsorption, negating the initial fractionation. Furthermore, because only surface uranium atoms exchange isotopes with dissolved U^{6+}, the impact of equilibrium fractionation is limited by the small percentage of surface-active U^{4+}. This depends on the crystal structure, morphology, concentration, and particle size of the U^{4+} minerals. Therefore, while isotopic fractionation occurs during U^{4+} oxidation, isotope fractionation may be limited if solid U^{4+} dissolution occurs in a congruent manner and/or if the contact time between the U^{6+} and U^{4+} phases is brief. For instance, in natural weathering environments, the weathering fluid is flushed away at rates dependent on local hydrogeological conditions, such as permeability and precipitation, which may prevent significant isotope exchange. Measurements of $\delta^{238}U$ in various natural weathering environments should therefore be undertaken to better understand potential isotope fractionation during the oxidation of U^{4+}-bearing minerals.

4 Effect of Dissolved Oxygen on Uranium Remediation

Uranium remediation is classified as the use of either abiotic (chemical) or biotic (biological) processes. Microbial (biotic) reduction of water-soluble U^{6+} species to U^{4+} species, such as uraninite UO_2 precipitates, has been proposed as a promising technology to bioremediate uranium-contaminated groundwater and soil. Several metal- and sulfate-reducing bacteria, including *Geobacter*, *Shewanella, and Desulfovibrio* species, are a focus of uranium reduction research because of their prevalence in the environment, ease of culture, and ability to reduce uranium enzymatically; however, because uranium is not an essential nutrient, there are no known dedicated U^{6+}-reducing enzymes. Following the application of bioavailable organic compounds, e.g., glucose, acetate, or ethanol, to contaminated water, metal-reducing microorganisms utilize these substrates by consuming oxidants such as oxygen and nitrate [32]. Then, dissimilatory metal-reducing microorganisms oxidize the remaining organics to CO_2 and reduce metals simultaneously [33, 34]. Importantly, the aqueous speciation of U^{6+} affects the bioavailability of uranium for microbial reduction. Most metal-reducing bacteria can reduce carbonate-complexed uranium, but uranyl-carbonate and organic-uranium complexes may limit several sulfate-reducing bacterial strains, although the mechanism of this inhibition is not well understood [35].

Because of its ability to re-oxidize biogenic uranium, DO presents a key challenge for biotic remediation. The Oak Ridge Field Research Center, for example, has investigated the bioreduction of U^{6+} in the Y-12 aquifer following the injection of organic electron donors to stimulate metal- and sulfate-reducing bacteria to reduce U^{6+} to U^{4+} to submicromolar levels [36]. This process produces iron sulfides and organic matter that can act as DO sinks. However, the cessation of donor amendments and the reintroduction of DO were accompanied by a rapid rebound in dissolved U^{6+} concentrations, highlighting that sustained amendments are necessary for long-term remediation. At pH <4, U^{4+} is more soluble and can be kinetically stabilized against abiotic oxidation by DO. However, under these conditions, microbially mediated aerobic oxidation is possible and has been observed in several species. For example, *Acidithiobacillus ferrooxidans*, a well-studied autotrophic organism commonly found in acidic rock drainage, catabolically couples U^{4+} oxidation to DO reduction, comparable to its ability to grow via Fe^{2+} oxidation. This is because the U^{6+}/U^{4+} redox couple has a similar potential to the $Fe(OH)_3/Fe^{2+}$ couple [35]. Importantly, although *A. ferrooxidans* is often used in laboratory studies because it is easily cultured, it is not necessarily the dominant organism in acid rock wastes or engineered bioleaching systems.

In addition to biotic remediation, several recent studies have examined the abiotic immobilization of soluble U^{6+} using iron, i.e., iron ions [1, 37], zero-valent iron [38–40], iron sulfides [41, 42], ferrihydrite [43–46], goethite [47], magnetite [48], and hematite [49], as an efficient and eco-friendly method to mitigate uranium migration at many nuclear sites and surface environments. There are three primary mechanisms of uranium immobilization by iron-based materials: (1) adsorption [47, 49, 50], (2) reductive precipitation to uranium minerals [37–40, 42, 48], and (3) coprecipitation into iron minerals [43–46]. As with biotic reduction, the presence of DO significantly impacts these processes. Xie et al. [1], for example, compared U^{6+} sedimentation rates with or without Fe^{2+} and DO under neutral conditions and reported that the precipitation efficiency was much greater under oxygen-rich conditions. Specifically, the sedimentation rates at Fe/U molar ratios of 0, 0.85 and 8.52 were ~60%, 94%, and 96%, respectively, under oxygen-rich conditions and only ~16%, 79%, and 82%, respectively, under oxygen-free conditions. Thus, the U^{6+} sedimentation rates with and without Fe^{2+} under oxygen-rich conditions were nearly 16% and 44% higher, respectively. Notably, this method resulted in uranium partially precipitating as U^{5+}, which was somewhat resistant to reoxidation by DO and acid dissolution.

In many cases, DO inhibits uranium immobilization via the oxidation of reduced UO_2, although some iron minerals partially counter these effects [51]. Mackinawite, for instance, rapidly immobilizes U^{6+} to reduced UO_2, U_4O_9, or U_3O_8 solid phases when it is added or naturally formed through the biostimulation of sulfate-reducing bacteria [52]. Given that bacteria capable of reducing uranium, iron, and SO_4^{2-} are ubiquitous in natural systems, reduced U^{4+} solids and FeS can be simultaneously produced at uranium-contaminated sites where Fe^{3+} minerals and sulfate are frequently encountered. Mackinawite also acts as a redox buffer against the re-oxidation of reduced crystalline uranium phases by preferentially consuming DO. In addition, iron sulfide oxidation produces iron (oxyhydr)oxides, e.g., nanogoethite and lepidocrocite, which are sorptive surfaces for additional U^{6+} [53]. However, when a large amount of DO is available, mackinawite oxidation products conversely catalyze the oxidation of non-crystalline U^{4+} species [54, 55]. Bi et al.[56], for example, performed abiotic batch experiments with synthetic UO_2 in the presence or absence of synthetic mackinawite (FeS) under simulated groundwater conditions (pH = 7, P_{O2} = 0.02 atm, and P_{CO2} = 0.05 atm). The kinetic profiles of dissolved uranium indicated that FeS inhibited UO_2 dissolution for about 51 h by effectively scavenging DO and forming iron oxyhydroxides and elemental sulfur. In the batch experiments, after FeS was depleted via oxidation, the DO level increased, and UO_2 dissolution occurred at an initial rate of 12 ± 4 nmol/g/s, which was higher than the value of 5.4 ± 0.3 nmol/g/s in the control experiment where FeS was absent.

5 Applications of Dissolved Oxygen for Uranium Extraction

Currently, several common techniques are used to extract uranium from ore, including open-pit, underground, borehole, and in situ leaching (ISL, also known as in situ recovery (ISR)) mining techniques.[57] ISL/ISR mining is the most common extraction technique, with DO often playing an important role. ISL/ISR involves injecting a solvent/leaching solution, i.e., lixiviant (typically O_2-saturated water containing sulfuric acid, sodium bicarbonate or CO_2), which is subsequently pumped out of the ground for processing. The ideal lixiviant will oxidize the uranium in the ore and contains a complexing agent that aids uranium dissolution and forms strong aqueous complexes that remain dissolved with few interactions with the host rocks [58].

Acid leaching, which was predominantly used in former Soviet republics and Australia, involves the injection of a dilute strong acid, e.g., sulfuric or hydrochloric acid. For solid ore containing U^{6+}, such as UO_3, the acid-based ISL/ISR reaction is expressed as:

$$UO_3 + 2H^+ \rightleftharpoons UO_2^{2+} + H_2O \quad (6)$$

Depending on the composition of the acid solution, the uranyl ion then combines with sulfate, nitrate, or chloride ions, forming soluble complexes, such as UO_2SO_4, $UO_2(NO_3)_2$, or UO_2Cl_2. For the more common U^{4+} ore, the reaction proceeds via [58]:

$$UO_2 + \frac{1}{2}O_2 + 2H^+ \rightleftharpoons UO_2^{2+} + H_2O \tag{7}$$

Although lixiviant can be an alkaline, neutral, or acidic solution, alkaline and circumneutral lixiviants are most commonly used in the United States because of the high concentrations of carbonate minerals, such as calcite or limestone, which increase acid consumption [35]. Additionally, acid leaching has the added environmental risk of dissolving associated toxic heavy metal(loid)s. Early (1970s–1980s) alkaline ISL/ISR mining of uranium involved the use of oxygen, sodium chlorate or hypochlorite, or potassium permanganate as oxidants, which are often mixed with sodium or ammonium carbonate/bicarbonate to increase uranium solubility by forming aqueous uranyl carbonate complexes. However, because of the difficulty of restoring ammonia-based sites, the current practice in the US involves the use of O_2 and CO_2 gas in the initial injection solution (Fig. 2). The neutral ISL/ISR method minimizes groundwater pollution during leaching and is considered to be more environmentally friendly than either acidic or alkaline leaching [58, 59].

The injection of dissolved CO_2 and O_2 substantially increases the redox potential and concentration of dissolved HCO_3^-, which oxidizes U^{4+} to U^{6+} and stimulates

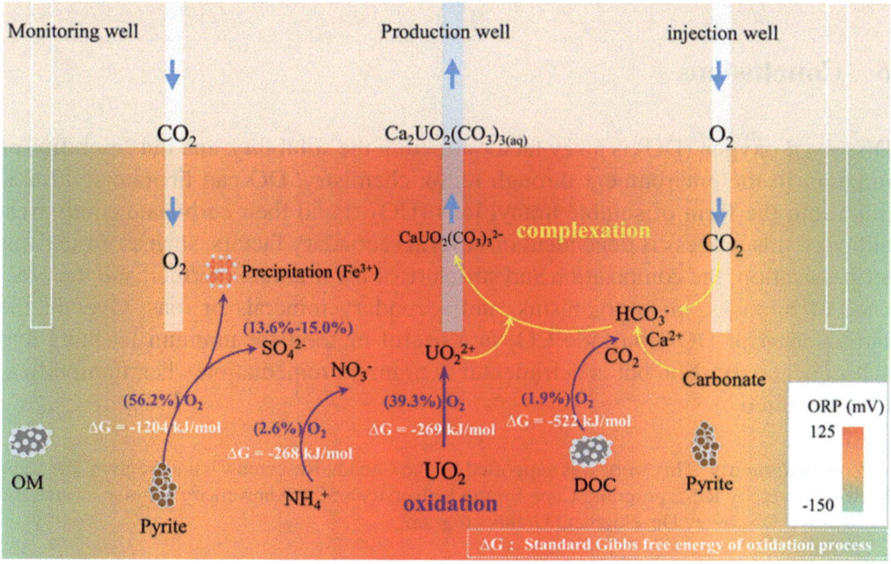

Fig. 2 In situ leaching/recovery (ISL or ISR) using CO_2 and O_2 injections in a sandstone-type UO_2-containing deposit. OM: Organic matter. DOC: Dissolved organic carbon. ORP: oxygen reduction potential [8]

complexation with bicarbonate, forming uranyl bi and tricarbonate ($[UO_2(CO_3)_2^{2-}]$ and $[UO_2(CO_3)_3^{4-}]$), or ternary Ca/Mg-uranyl carbonate complexes [13]. However, many factors affecting uranium leaching under CO_2 and O_2 conditions are not yet fully understood, especially the relative contributions of multiple O_2 consumption processes. Generally, the leaching of uranium can be described by Eqs. 2, 3, 4, 5, 8, and 9 [58, 60]:

$$UO_2 + \frac{1}{2}O_2 + 2CO_3^{2-} + H_2O \rightleftharpoons UO_2\left(CO_3\right)_2^{2-} + 2OH^- \tag{8}$$

$$UO_2 + \frac{1}{2}O_2 + 3HCO_3^- \rightleftharpoons UO_2\left(CO_3\right)_3^{4-} + H^+ + H_2O \tag{9}$$

The resulting carbonate complexes are subsequently removed from the production wells and processed through an ion exchange resin [35]:

$$2R\left(Cl\right) + 2Na^+ + UO_2\left(CO_3\right)_2^{2-} \rightleftharpoons R_2UO_2\left(CO_3\right)_2 + 2NaCl \tag{10}$$

or

$$4\left[R\text{-}N\text{-}\left(CH_3\right)_3HCO_3\right] + UO_2\left(CO_3\right)_3^{4-} \rightleftharpoons \left[\left(R\text{-}N\text{-}\left(CH_3\right)_3\right)_4\right] UO_2\left(CO_3\right)_3 + 4HCO_3^- \tag{11}$$

6 Conclusions

Dissolved oxygen (DO) is a key factor affecting the solubility, and thus mobility, of uranium in the environment through redox chemistry. DO can liberate solid uranium into the form of soluble uranyl ions (UO_2^{2+}) and their carbonate complexes, although this process is dependent on several secondary factors, such as groundwater passivation; the composition and structure of the uranium ore/fuel; and the presence of biomass, microorganisms, and secondary minerals or ions. Overall, the introduction of DO alongside CO_2 is a useful method for uranium leaching and subsequent extraction but is detrimental to uranium remediation when immobilization is desired.

Acknowledgments This work was supported by the Canada Research Chairs program, the Cape Breton University RISE program, Cape Breton University's 50th Anniversary Scholar Award, and the NSERC Discovery Development Grants program.

References

1. Xie Y, Fang Q, Li M, Wang S, Luo Y, Wu X, Lv J, Tan W, Wang H, Tan K (2020) Low concentration of Fe(II) to enhance the precipitation of U(VI) under neutral oxygen-rich conditions. Sci Total Environ 711:134827. https://doi.org/10.1016/j.scitotenv.2019.134827
2. Luo W, Gu B (2011) Dissolution of uranium-bearing minerals and mobilization of uranium by organic ligands in a biologically reduced sediment. Environ Sci Technol 45:2994–2999. https://doi.org/10.1021/es103073u
3. Oversby VM (1999) Uranium dioxide, SIMFUEL, and spent fuel dissolution rates - a review of published data. Sweden
4. Casella A, Hanson B, Miller W (2016) The effect of fuel chemistry on UO_2 dissolution. J Nucl Mater 476:45–55. https://doi.org/10.1016/j.jnucmat.2016.04.025
5. Campbell KM, Veeramani H, Ulrich K-U, Blue LY, Giammar DE, Bernier-Latmani R, Stubbs JE, Suvorova E, Yabusaki S, Lezama-Pacheco JS, Mehta A, Long PE, Bargar JR (2011) Oxidative dissolution of biogenic uraninite in groundwater at old rifle, CO. Environ Sci Technol 45:8748–8754. https://doi.org/10.1021/es200482f
6. Giammar DE, Cerrato JM, Mehta V, Wang Z, Wang Y, Pepping TJ, Ulrich K-U, Lezama-Pacheco JS, Bargar JR (2012) Effect of diffusive transport limitations on UO_2 dissolution. Water Res 46:6023–6032. https://doi.org/10.1016/j.watres.2012.08.034
7. Abdelouas A, Lutze W, Nuttall HE (1999) Oxidative dissolution of uraninite precipitated on Navajo sandstone. J Contam Hydrol 36:353–375. https://doi.org/10.1016/S0169-7722(98)00151-X
8. Lu C, Xiu W, Guo H, Lian G, Yang B, Zhang T, Bi E, Shi Z (2023) Multi-isotope based identification and quantification of oxygen consuming processes in uranium hosting aquifers with CO_2 + O_2 in situ leaching. Water Resour Res 59:e2022WR033980. https://doi.org/10.1029/2022WR033980
9. Shoesmith DW, Tait JC, Sunder S, Gray WJ, Steward S, Russo RE, Rudnicki JD (1996) Factors affecting the differences in reactivity and dissolution rates between UO_2 and spent nuclear fuel. Canada
10. Veeramani H, Schofield EJ, Sharp JO, Suvorova EI, Ulrich K-U, Mehta A, Giammar DE, Bargar JR, Bernier-Latmani R (2009) Effect of Mn(II) on the structure and reactivity of biogenic uraninite. Environ Sci Technol 43:6541–6547. https://doi.org/10.1021/es900556k
11. Xu Y, Wang L, Su X, Chen M, Liao W, Ren Y, Du Z, Ding Y (2023) Study on mechanism of oxygen oxidation leaching with low acid for high acid consumption sandstone uranium deposit. Processes 11
12. Kacham AR, Suri AK (2014) Application of a topochemical reaction model to predict leaching behavior of high carbonate uranium ores in alkaline solutions: an experimental case study. Hydrometallurgy 141:67–75. https://doi.org/10.1016/j.hydromet.2013.10.005
13. Yang Y, Qiu W, Liu Z, Song J, Wu J, Dou Z, Wang J, Wu J (2022) Quantifying the impact of mineralogical heterogeneity on reactive transport modeling of CO_2+O_2 in-situ leaching of uranium. Acta Geochim 41:50–63. https://doi.org/10.1007/s11631-021-00502-1
14. Wang Z, Tebo BM, Giammar DE (2014) Effects of Mn(II) on UO_2 dissolution under anoxic and oxic conditions. Environ Sci Technol 48:5546–5554. https://doi.org/10.1021/es5002067
15. Kim J, Kim H, Kim W-S, Um W (2018) Dissolution of studtite $[UO_2(O_2)(H_2O)_4]$ in various geochemical conditions. J Environ Radioact 189:57–66. https://doi.org/10.1016/j.jenvrad.2018.01.010
16. Cerrato JM, Ashner MN, Alessi DS, Lezama-Pacheco JS, Bernier-Latmani R, Bargar JR, Giammar DE (2013) Relative reactivity of biogenic and chemogenic uraninite and biogenic noncrystalline U(IV). Environ Sci Technol 47:9756–9763. https://doi.org/10.1021/es401663t

17. Cerrato JM, Barrows CJ, Blue LY, Lezama-Pacheco JS, Bargar JR, Giammar DE (2012) Effect of Ca^{2+} and Zn^{2+} on UO_2 dissolution rates. Environ Sci Technol 46:2731–2737. https://doi.org/10.1021/es203751t
18. Ulrich K-U, Ilton ES, Veeramani H, Sharp JO, Bernier-Latmani R, Schofield EJ, Bargar JR, Giammar DE (2009) Comparative dissolution kinetics of biogenic and chemogenic uraninite under oxidizing conditions in the presence of carbonate. Geochim Cosmochim Acta 73:6065–6083. https://doi.org/10.1016/j.gca.2009.07.012
19. Lezama-Pacheco JS, Cerrato JM, Veeramani H, Alessi DS, Suvorova E, Bernier-Latmani R, Giammar DE, Long PE, Williams KH, Bargar JR (2015) Long-term in situ oxidation of biogenic uraninite in an alluvial aquifer: impact of dissolved oxygen and calcium. Environ Sci Technol 49:7340–7347. https://doi.org/10.1021/acs.est.5b00949
20. Moon HS, Komlos J, Jaffé PR (2007) Uranium reoxidation in previously bioreduced sediment by dissolved oxygen and nitrate. Environ Sci Technol 41:4587–4592. https://doi.org/10.1021/es063063b
21. Bone SE, Cahill MR, Jones ME, Fendorf S, Davis J, Williams KH, Bargar JR (2017) Oxidative uranium release from anoxic sediments under diffusion-limited conditions. Environ Sci Technol 51:11039–11047. https://doi.org/10.1021/acs.est.7b02241
22. Moon HS, Komlos J, Jaffé PR (2009) Biogenic U(IV) oxidation by dissolved oxygen and nitrate in sediment after prolonged U(VI)/Fe(III)/SO42− reduction. J Contam Hydrol 105:18–27. https://doi.org/10.1016/j.jconhyd.2008.10.014
23. Kendall B, Brennecka GA, Weyer S, Anbar AD (2013) Uranium isotope fractionation suggests oxidative uranium mobilization at 2.50Ga. Chem Geol 362:105–114. https://doi.org/10.1016/j.chemgeo.2013.08.010
24. Bruggmann S, Gilleaudeau GJ, Romaniello SJ, Severmann S, Canfield DE, Anbar AD, Scholz F, Frei R (2022) Uranium isotope cycling on the highly productive Peruvian margin. Chem Geol 590:120705. https://doi.org/10.1016/j.chemgeo.2021.120705
25. Weyer S, Anbar AD, Gerdes A, Gordon GW, Algeo TJ, Boyle EA (2008) Natural fractionation of $^{238}U/^{235}U$. Geochim Cosmochim Acta 72:345–359. https://doi.org/10.1016/j.gca.2007.11.012
26. Basu A, Sanford RA, Johnson TM, Lundstrom CC, Löffler FE (2014) Uranium isotopic fractionation factors during U(VI) reduction by bacterial isolates. Geochim Cosmochim Acta 136:100–113. https://doi.org/10.1016/j.gca.2014.02.041
27. Bopp CJIV, Lundstrom CC, Johnson TM, Sanford RA, Long PE, Williams KH (2010) Uranium 238U/235U isotope ratios as indicators of reduction: results from an in situ biostimulation experiment at rifle, Colorado, USA. Environ Sci Technol 44:5927–5933. https://doi.org/10.1021/es100643v
28. Lau KV, Lyons TW, Maher K (2020) Uranium reduction and isotopic fractionation in reducing sediments: insights from reactive transport modeling. Geochim Cosmochim Acta 287:65–92. https://doi.org/10.1016/j.gca.2020.01.021
29. Schauble EA (2007) Role of nuclear volume in driving equilibrium stable isotope fractionation of mercury, thallium, and other very heavy elements. Geochim Cosmochim Acta 71:2170–2189. https://doi.org/10.1016/j.gca.2007.02.004
30. Andersen MB, Romaniello S, Vance D, Little SH, Herdman R, Lyons TW (2014) A modern framework for the interpretation of 238U/235U in studies of ancient ocean redox. Earth Planet Sci Lett 400:184–194. https://doi.org/10.1016/j.epsl.2014.05.051
31. Wang X, Johnson TM, Lundstrom CC (2015) Isotope fractionation during oxidation of tetravalent uranium by dissolved oxygen. Geochim Cosmochim Acta 150:160–170. https://doi.org/10.1016/j.gca.2014.12.007
32. Green SJ (2011) Microorganisms and processes linked to uranium reduction and immobilization. In: Microbial metal and metalloid metabolism. Wiley, pp 117–138
33. Lovley DR, Phillips EJP, Gorby YA, Landa ER (1991) Microbial reduction of uranium. Nature 350:413–416. https://doi.org/10.1038/350413a0

34. Vodyanitskii YN (2011) Chemical aspects of uranium behavior in soils: a review. Eurasian Soil Sci 44:862–873. https://doi.org/10.1134/S1064229311080163
35. Campbell KM, Gallegos TJ, Landa ER (2015) Biogeochemical aspects of uranium mineralization, mining, milling, and remediation. Appl Geochem 57:206–235. https://doi.org/10.1016/j.apgeochem.2014.07.022
36. Wu W-M, Carley J, Luo J, Ginder-Vogel MA, Cardenas E, Leigh MB, Hwang C, Kelly SD, Ruan C, Wu L, Van Nostrand J, Gentry T, Lowe K, Carroll S, Luo W, Fields MW, Gu B, Watson D, Kemner KM, Marsh T, Tiedje J, Zhou J, Fendorf S, Kitanidis PK, Jardine PM, Criddle CS (2007) In situ bioreduction of uranium (VI) to submicromolar levels and reoxidation by dissolved oxygen. Environ Sci Technol 41:5716–5723. https://doi.org/10.1021/es062657b
37. Taylor SD, Marcano MC, Rosso KM, Becker U (2015) An experimental and ab initio study on the abiotic reduction of uranyl by ferrous iron. Geochim Cosmochim Acta 156:154–172. https://doi.org/10.1016/j.gca.2015.01.021
38. Li Z-J, Wang L, Yuan L-Y, Xiao C-L, Mei L, Zheng L-R, Zhang J, Yang J-H, Zhao Y-L, Zhu Z-T, Chai Z-F, Shi W-Q (2015) Efficient removal of uranium from aqueous solution by zero-valent iron nanoparticle and its graphene composite. J Hazard Mater 290:26–33. https://doi.org/10.1016/j.jhazmat.2015.02.028
39. Tsarev S, Collins RN, Fahy A, Waite TD (2016) Reduced uranium phases produced from anaerobic reaction with nanoscale zerovalent iron. Environ Sci Technol 50:2595–2601. https://doi.org/10.1021/acs.est.5b06160
40. Tsarev S, Collins RN, Ilton ES, Fahy A, Waite TD (2017) The short-term reduction of uranium by nanoscale zero-valent iron (nZVI): role of oxide shell, reduction mechanism and the formation of U(v)-carbonate phases. Environ Sci Nano 4:1304–1313. https://doi.org/10.1039/C7EN00024C
41. Liu H, Zhu Y, Xu B, Li P, Sun Y, Chen T (2017) Mechanical investigation of U(VI) on pyrrhotite by batch, EXAFS and modeling techniques. J Hazard Mater 322:488–498. https://doi.org/10.1016/j.jhazmat.2016.10.015
42. Yang Z, Kang M, Ma B, Xie J, Chen F, Charlet L, Liu C (2014) Inhibition of U(VI) reduction by synthetic and natural pyrite. Environ Sci Technol 48:10716–10724. https://doi.org/10.1021/es502181x
43. Marshall TA, Morris K, Law GTW, W. Mosselmans JF, Bots P, Roberts H, Shaw S (2015) Uranium fate during crystallization of magnetite from ferrihydrite in conditions relevant to the disposal of radioactive waste. Mineral Mag 79:1265–1274. https://doi.org/10.1180/minmag.2015.079.6.02
44. Massey MS, Lezama-Pacheco JS, Jones ME, Ilton ES, Cerrato JM, Bargar JR, Fendorf S (2014) Competing retention pathways of uranium upon reaction with Fe(II). Geochim Cosmochim Acta 142:166–185. https://doi.org/10.1016/j.gca.2014.07.016
45. McBriarty ME, Kerisit S, Bylaska EJ, Shaw S, Morris K, Ilton ES (2018) Iron vacancies accommodate uranyl incorporation into hematite. Environ Sci Technol 52:6282–6290. https://doi.org/10.1021/acs.est.8b00297
46. Roberts HE, Morris K, Law GTW, Mosselmans JFW, Bots P, Kvashnina K, Shaw S (2017) Uranium(V) incorporation mechanisms and stability in Fe(II)/Fe(III) (oxyhydr)oxides. Environ Sci Technol Lett 4:421–426. https://doi.org/10.1021/acs.estlett.7b00348
47. Guo Z, Li Y, Wu W (2009) Sorption of U(VI) on goethite: effects of pH, ionic strength, phosphate, carbonate and fulvic acid. Appl Radiat Isot 67:996–1000. https://doi.org/10.1016/j.apradiso.2009.02.001
48. Skomurski FN, Ilton ES, Engelhard MH, Arey BW, Rosso KM (2011) Heterogeneous reduction of U^{6+} by structural Fe^{2+} from theory and experiment. Geochim Cosmochim Acta 75:7277–7290. https://doi.org/10.1016/j.gca.2011.08.006
49. Zhao D, Wang X, Yang S, Guo Z, Sheng G (2012) Impact of water quality parameters on the sorption of U(VI) onto hematite. J Environ Radioact 103:20–29. https://doi.org/10.1016/j.jenvrad.2011.08.010

50. Wei J, Zhang X, Liu Q, Li Z, Liu L, Wang J (2014) Magnetic separation of uranium by $CoFe_2O_4$ hollow spheres. Chem Eng J 241:228–234. https://doi.org/10.1016/j.cej.2013.12.035
51. Zhang G, Burgos WD, Senko JM, Bishop ME, Dong H, Boyanov MI, Kemner KM (2011) Microbial reduction of chlorite and uranium followed by air oxidation. Chem Geol 283:242–250. https://doi.org/10.1016/j.chemgeo.2011.01.021
52. Hyun SP, Davis JA, Sun K, Hayes KF (2012) Uranium(VI) reduction by iron(II) monosulfide mackinawite. Environ Sci Technol 46:3369–3376. https://doi.org/10.1021/es203786p
53. Gallegos TJ, Fuller CC, Webb SM, Betterton W (2013) Uranium(VI) interactions with mackinawite in the presence and absence of bicarbonate and oxygen. Environ Sci Technol 47:7357–7364. https://doi.org/10.1021/es400450z
54. Loreggian L, Sorwat J, Byrne JM, Kappler A, Bernier-Latmani R (2020) Role of iron sulfide phases in the stability of noncrystalline tetravalent uranium in sediments. Environ Sci Technol 54:4840–4846. https://doi.org/10.1021/acs.est.9b07186
55. Smedley PL, Kinniburgh DG (2023) Uranium in natural waters and the environment: distribution, speciation and impact. Appl Geochem 148:105534. https://doi.org/10.1016/j.apgeochem.2022.105534
56. Bi Y, Hyun SP, Kukkadapu RK, Hayes KF (2013) Oxidative dissolution of UO_2 in a simulated groundwater containing synthetic nanocrystalline mackinawite. Geochim Cosmochim Acta 102:175–190. https://doi.org/10.1016/j.gca.2012.10.032
57. Mkandawire M (2013) Biogeochemical behaviour and bioremediation of uranium in waters of abandoned mines. Environ Sci Pollut Res 20:7740–7767. https://doi.org/10.1007/s11356-013-1486-3
58. Saunders JA, Pivetz BE, Voorhies N, Wilkin RT (2016) Potential aquifer vulnerability in regions down-gradient from uranium in situ recovery (ISR) sites. J Environ Manag 183:67–83. https://doi.org/10.1016/j.jenvman.2016.08.049
59. Su X, Liu Z, Yao Y, Du Z (2020) Petrology, mineralogy, and ore leaching of sandstone-hosted uranium deposits in the Ordos Basin, North China. Ore Geol Rev 127:103768. https://doi.org/10.1016/j.oregeorev.2020.103768
60. Qiu W, Yang Y, Song J, Que W, Liu Z, Weng H, Wu J, Wu J (2023) What chemical reaction dominates the CO_2 and O_2 in-situ uranium leaching? Insights from a three-dimensional multicomponent reactive transport model at the field scale. Appl Geochem 148:105522. https://doi.org/10.1016/j.apgeochem.2022.105522

Redox Potential: A Key Player in Uranium Fate

Yvon Ukwishaka, Kashyap Komal, Ayush Kumar Singh, Manish Kumar, Biswa Mohan Sahoo, Pardeep Singh, and Preetismita Borah

Abstract Redox potential (oxidation–reduction potential, ORP), a measure of a system's reducing or oxidizing capacity, plays a vital role in determining the fate of uranium in different environments and its impact on human health and ecosystem stability. The ORP influences uranium's oxidation state, solubility and mobility, bioavailability, and contamination potential in natural ecosystems. Uranium, a naturally occurring radionuclide heavy metal found in soil, rock, and water, exists primarily in both mobile and immobile phases depending on its two oxidation states: highly soluble and mobile uranium(VI) and less soluble and less mobile uranium(IV). Under oxidizing conditions (oxic, high redox potential), uranium is present mainly in the form of uranium(VI), increasing the risk of environmental and health issues. Conversely, under reducing conditions (suboxic to anoxic, low redox potential), uranium tends to favor the formation of uranium(IV), leading to lower solubility and mobility. Other factors, such as pH, organic matter, and microbial activity, can also influence uranium behavior in the environment, hence providing different implications concerning the contamination of water resources. A broadened understanding of uranium redox chemistry in water resources and the impact of redox potential on uranium mobility are crucial for the development of mitigation frame-

Y. Ukwishaka
Nebraska Public Health Environmental Laboratory, Lincoln, NE, USA

K. Komal · A. K. Singh
Chandigarh University, Mohali, Punjab, India

M. Kumar · P. Borah (✉)
CSIO-Central Scientific Instruments Organization, Chandigarh, India
e-mail: preetismita@csio.res.in

B. M. Sahoo
School of Pharmacy and Life Sciences, Centurion University of Technology & Management, Khurda, Odisha, India

P. Singh
Department of Environmental Studies, PGDAV College, University of Delhi, New Delhi, India

M. K. Jindal, K. Patni (eds.), *Water Quality-Uranium Solubility Nexus: Unlocking the Secrets of Uranium Solubility*, Earth and Environmental Sciences Library, https://doi.org/10.1007/978-3-031-98909-4_7

works and effective remediation strategies. This chapter explores the complex chemistry of uranium redox and its solubility in various water bodies, as well as the speciation of uranium at the soil–water interface, offering a broad understanding of ORP dynamics and their relationship with uranium chemistry to address uranium contamination and safeguard environmental and human health.

Keywords Uranium redox chemistry · Redox potential · Radionuclide · Remediation · Uranium contamination

1 Uranium Contamination and Redox Potential in Natural Environment

Uranium contamination has emerged as a significant environmental and public health issue, particularly in regions with high anthropogenic activities, including uranium mining, nuclear weapon production, nuclear power generation, and waste disposal [1, 2]. The tenacity of uranium in the natural environment, especially in water systems (groundwater and surface water), poses a global threat because of its toxicity and radioactivity, making it vital to understand the processes that influence uranium behavior in these systems [2]. One of the primary factors controlling the mobility of uranium in water is redox potential (Eh). The redox potential of an environment measures its ability to either accept or donate electrons, a critical aspect in driving oxidation–reduction (redox) reactions [3]. These reactions are central to determining uranium's oxidation state, which in turn controls its solubility, mobility, and potential to bioaccumulate in water systems.

Uranium exists in two primary oxidation states: a reduced form (U(IV)) and an oxidized form (U(VI)), which behave very differently. U(VI) is highly soluble and mobile, whereas U(IV) is insoluble and tends to precipitate; thus, uranium speciation occurs. Under reducing conditions, U(VI) can be reduced to U(IV), leading to its immobilization in sediments. Conversely, in oxidizing environments, U(IV) can be oxidized to U(VI), increasing its mobility and resulting in groundwater contamination [4, 5]. The redox-based transition between U(VI) and U(IV) represents one of the most critical factors affecting the environmental fate and transport of uranium. The objective of this chapter is to explore how redox potential plays a vital role in uranium's fate in different environmental matrices. This chapter also delves into how different factors, such as pH, dissolved oxygen, and microbial activity, influence redox potential and how manipulating redox potential can be useful in remediation strategies designed to reduce uranium contamination.

1.1 Why Is Uranium Contamination a Looming Global Concern?

Uranium (atomic number = 92) is a naturally occurring radioactive element, although its concentration in some situations has increased dramatically because of human activity. When uranium is discharged into water systems (groundwater and surface water), it poses a variety of environmental and health problems because of the following factors:

1. **Chemical Toxicity**: Uranium is a heavy metal that, like other heavy metals, may cause serious chemical toxicity when consumed or breathed. The chemical toxicity of uranium mostly affects the kidneys, where it accumulates after it enters the circulation [6].
2. **Radioactivity**: In addition to its chemical toxicity, uranium is a radioactive element whose radioactive isotopes pose long-term health risks. The most abundant uranium isotope, U-238, decays slowly over billions of years, emitting different particles (alpha particles, beta particles, and gamma rays). While U-238 is relatively low in radioactivity, its disintegration chain creates a sequence of more radioactive isotopes, such as radon-222 and radium-226, which release harming radiation. This radiation can lead to an increased risk of cancer [7].
3. **Mobility in Water**: Uranium mobility in water is most concerning when uranium exists in its hexavalent state (U(VI)), which is very soluble. In oxygen-rich environments, uranium forms stable uranyl ion (UO_2^{2+}) complexes that dissolve readily in water. This makes uranium contamination more dangerous since it may travel long distances through groundwater and surface water systems, spreading far beyond its source of contamination [7].

1.2 The Role of Redox Potential (Eh)

The redox potential (Eh) is the chemical environment's capacity to gain or lose electrons, influencing the oxidation state of elements such as uranium. A more positive Eh corresponds to oxidizing conditions, where uranium tends to lose electrons and exists in higher oxidation states (U(VI)). A more negative Eh indicates reducing conditions, where uranium gains electrons and exists in lower oxidation states (U(IV)) [8]. The importance of redox potential in determining uranium fate and speciation arises from its direct role in oxidation–reduction reactions across numerous environmental matrices. When uranium is oxidized (U(VI)), it becomes soluble and mobile in water. However, when reduced (U(IV)), it is insoluble and precipitates out of solution, making it immobile. The standard redox reaction equation that represents the interchange between uranium species and governs uranium mobility is summarized as follows:

$$UO_2^{2+} + 4H^+ + 2e^- \rightleftharpoons UO_{2(s)} + 2H_2O_{(l)} \quad (1)$$

Equation reaction 1, shows how U(VI) is reduced to U(IV) in the form of insoluble UO_2 (as the uranyl ion, UO_2^{2+}). On the other hand, when the oxidizing conditions are relatively high, a reverse reaction takes place, converting UO_2 back into UO_2^{2+} and increasing uranium solubility.

2 Uranium Redox Chemistry

The behavior of uranium in aqueous environments mostly depends on its oxidation state, which is impacted by the environment's redox potential (Eh). In natural water, uranium mostly exists in two oxidation states: uranium(IV) and uranium(VI). Redox reactions govern the equilibrium between these oxidation states and play a role in determining whether uranium will remain immobile in sediments or spread into groundwater, where it may cause contamination. This section examines the chemistry of these two oxidation states, the redox reactions that govern their transformation, and the environmental factors that drive these processes [5].

2.1 Reduced Uranium: U(IV)

In reducing environments, a state with low or reduced redox potential (more negative), uranium exists in the form of uranium(IV). In general, U(IV) occurs as uraninite (UO_2), which is sparingly soluble or immobile. Under these conditions, U(IV) tends to precipitate out of solution via complexes with other compounds. These formed complexes are stable and immobile, thus posing a reduced risk of moving to the groundwater system [5]. The following redox equation (2) represents the reduction of U(VI) (the soluble form) to U(IV) (the insoluble form).

$$UO_2^{2+} + 4H^+ + 2e^- \rightarrow UO_{2(s)} + 2H_2O_{(l)} \quad (2)$$

Equation 2 involves the reduction of uranyl ions (UO_2^{2+}), the predominant form of U(VI) in aqueous solutions, to uraninite (UO_2), a solid and insoluble form of uranium that precipitates under reducing conditions. This reduction process is favored in environments with reducing agents (Fig. 1).

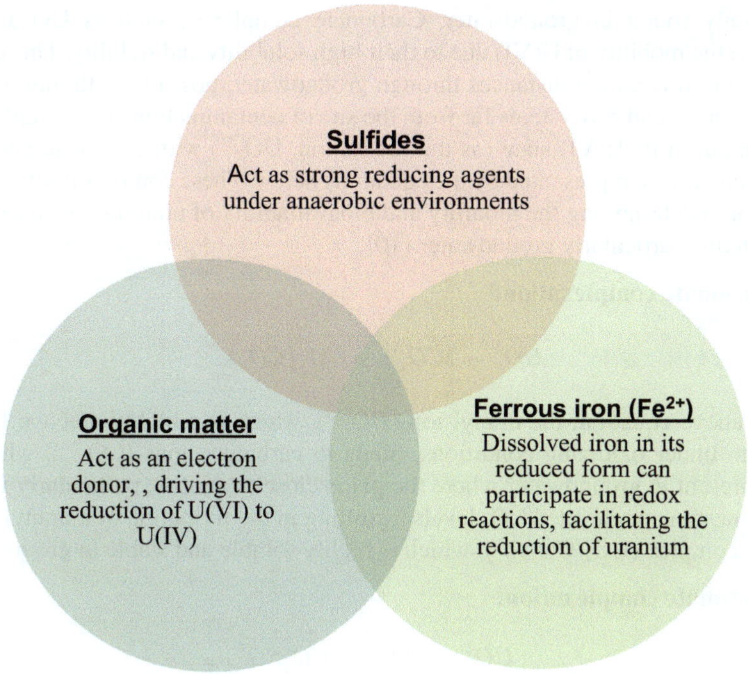

Fig. 1 Reduction process

2.2 *Oxidized Uranium: (VI)*

In oxidizing environments (higher oxidation state—greater redox potential), uranium exists as uranium(VI), which is more soluble and mobile than is the reduced form (U(IV)). Naturally, the most common form of U(VI) in water is the uranyl ion (UO_2^{2+}), which readily forms complexes with other ions, such as carbonate (CO_3^{2-}), phosphate (PO_4^{3-}), and sulfate (SO_4^{2-}). These complexes are highly soluble, meaning that U(VI) can remain in solution and migrate through soil–groundwater systems [9]. In general, U(IV) to U(VI) oxidation occurs in the presence of oxygen, as shown in the equation below:

$$UO_{2(s)} + \frac{1}{2}O_{2(g)} + 2H^+ \rightarrow UO_2^{2+} + H_2O_{(l)} \quad (3)$$

In Eq. 3, uraninite (UO_2), the solid form of U(IV), oxidizes in the presence of dissolved oxygen, resulting in soluble uranyl ions (UO_2^{2+}). This reaction is especially essential in oxygen-rich settings, such as surface water or aquifers, which experience periodic oxygen infiltration and fluctuations. In the oxidized form of U(VI), uranium can form stable aqueous complexes, especially with carbonate, which is commonly found in groundwater. Carbonate complexes, such as $UO_2(CO_3)_3^{4-}$, increase the mobility of U(VI) due to their high solubility and stability. This enables uranium to travel long distances through groundwater, possibly polluting drinking water sources and ecosystems far from the site of contamination. The complexation of uranium in its U(VI) state (as the uranyl ion, UO_2^{2+}) with carbonate and phosphate ions can be represented by the equations below. These complexation reactions are vital in determining the mobility and bioavailability of uranium in environmental systems, particularly groundwater [10].

- **Carbonate complexation:**

$$UO_2^{2+} + 3CO_3^{2-} \rightarrow UO_2\left(CO_3\right)_3^{4-} \tag{4}$$

In the above equation, the uranyl ion (UO_2^{2+}), which is a highly reactive form of uranium under oxidizing conditions, binds to carbonate ions (CO_3^{2-}), which are often present in groundwater where the pH is close to neutral, particularly in areas with limestone or elevated CO_2 levels, resulting in the formation of a uranyl–tricarbonate complex ($UO_2(CO_3)_3^{4-}$), which is highly soluble and stable in groundwater.

- **Phosphate complexation:**

$$UO_2^{2+} + PO_3^{2-} \rightarrow UO_2PO^{4-} \tag{5}$$

The reaction between the uranyl ion (UO_2^{2+}) and the phosphate ion (PO_4^{3-}) produces a uranyl-phosphate complex ($UO_2PO_4^-$). The uranyl ion U(VI) is an oxidized form of uranium that is highly reactive in nature. Phosphate ions (PO_4^{3-}) can bind to uranium and form stable complexes. The uranyl-phosphate complex ($UO_2PO_4^-$) produced during this reaction is less soluble than uranyl-carbonate complexes. As a result, phosphate can reduce uranium mobility in groundwater, lowering its ability to spread, hence reducing the risk of contamination, which is why phosphate is often used in remediation strategies to immobilize uranium in contaminated environments.

2.3 Dynamics Between U(IV) and U(VI)

The dynamic redox chemistry of uranium can be explained by the reversible reaction between U(IV) and U(VI), which is influenced by a variety of environmental factors. Under reducing conditions, the Eh is low, and uranium exists mostly as U(IV), which is insoluble and persists in sediments and water. In oxidizing environments, the Eh is relatively high, and U(VI) predominates, remaining in soluble forms that can be transported by water [11]. This dynamic can be summarized in a simplified redox cycle:

1. **Reduction**: In the presence of organic matter, acidic/slightly acidic conditions, ferrous iron, or sulfides, U(VI) is reduced to immobile U(IV), limiting uranium mobility.

$$UO_2^{2+} + 4H^+ + 2e^- \rightarrow UO_{2(s)} + 2H_2O_{(l)} \quad (6)$$

2. **Oxidation**: In the presence of oxygen and alkaline conditions, U(IV) is oxidized to U(VI), increasing the solubility and mobility of uranium.

$$UO_{2(s)} + \frac{1}{2}O_{2(g)} + 2H^+ \rightarrow UO_2^{2+} + H_2O_{(l)} \quad (7)$$

Importantly, uranium redox processes are controlled not only by the presence of dissolved oxygen or reducing species but also by pH, temperature, and microbial activity. Certain bacteria can catalyze the reduction of U(VI) to U(IV), which enhances the immobilization process in sediment-water matrices. Similarly, pH can affect uranium speciation by shifting the balance and stability of U(VI) complexes, where higher pH conditions promote carbonate complex formation, increasing mobility.

3 Interplay Between Factors Influencing Redox Potential and Hence Uranium Behavior

The redox potential (Eh) of an environment influences the oxidation state, solubility, and mobility of uranium in nature. Several environmental conditions may affect Eh and, as a result, uranium redox behavior, including pH, carbonate chemistry, organic matter, microbial activity, iron and manganese oxides, and dissolved oxygen. All of these variables interact in intricate ways to either increase uranium mobility or contribute to uranium immobilization in various environmental matrices.

3.1 pH and Carbonate Chemistry

The pH of a solution is one of the most essential factors controlling uranium redox behavior and speciation. Depending on the pH, uranium may form a variety of complexes in natural water systems, determining whether it remains mobile or becomes immobilized. pH affects uranium's ability to transition between its oxidation states, U(IV) and U(VI).

3.1.1 Impact of pH on Uranium Solubility and Redox Behavior

Under acidic conditions, uranium exists as U(VI) in its soluble uranyl ion form (UO_2^{2+}) and may form complexes with different ligands. As the pH increases, uranium interacts with carbonate ions (CO_3^{2-}), resulting in stable and soluble uranyl carbonate complexes. This can significantly increase the mobility of U(VI) in groundwater, particularly in oxidizing environments [12].

For example, at higher pH values (alkaline conditions), U(VI) produces the following stable carbonate complex:

$$UO_2^{2+} + 3CO_3^{2-} \rightarrow UO_2\left(CO_3\right)_3^{4-} \tag{8}$$

This uranyl carbonate complex, $UO_2(CO_3)_3^{4-}$, is highly soluble and remains in solution, allowing uranium to travel long distances in groundwater. The formation of these complexes depends on both the pH and the concentration of dissolved carbonate ions. In contrast, under acidic conditions (lower pH), uranium mobility and solubility decrease significantly, and U(IV) becomes the more stable form. U(IV) tends to form insoluble precipitates such as uraninite (UO_2), reducing the mobility of uranium in the environment.

$$UO_2^{2+} + 4H^+ + 2e^- \rightarrow UO_{2(s)} + 2H_2O_{(l)} \tag{9}$$

3.1.2 Alkaline Conditions and Carbonate Complexation

In high-pH (alkaline) environments, carbonate chemistry plays an important role in uranium mobility. Stable uranyl carbonate complexes such as $UO_2(CO_3)_3^{4-}$ can hinder the reduction of U(VI) to U(IV), even in environments that would otherwise favor reduction. The carbonate ions stabilize U(VI), preventing it from precipitating as UO_2 under reducing conditions (low Eh). This is particularly problematic in environments with abundant carbonate ions because it allows uranium to remain in its more soluble and mobile U(VI) form even in conditions that would ordinarily favor uranium immobilization [12].

3.2 Organic Matter and Microbial Activity

Organic matter and microbial activity are also among the driving forces for redox reactions in natural environments, predominantly in reducing uranium from U(VI) to U(IV). OM acts as an electron donor, facilitating redox processes, whereas certain microbes mediate the reduction of U(VI) to U(IV) during anaerobic respiration reactions.

3.2.1 Role of Organic Matter

Organic matter is often abundant in sedimentary environments such as wetlands, peat bogs, and other areas rich in decaying vegetation. When organic matter undergoes microbial decomposition, it generates reducing conditions by releasing electrons. These electrons are then available to reduce U(VI) to U(IV), promoting uranium immobilization (eq. 10). The presence of organic matter is a key factor in natural attenuation processes, where uranium contamination is gradually immobilized through microbial and geochemical interactions [13].

$$UO_2^{2+} + 4H^+ + 2e^- + Org \rightarrow UO_{2(s)} + 2H_2O_{(l)} + OrgH^+ \tag{10}$$

3.2.2 Microbial-Mediated Uranium Reduction

Certain microorganisms, such as *Geobacter* and *Shewanella* species, can utilize uranium as an electron acceptor during anaerobic respiration. In this bioreduction process, these microbes reduce soluble U(VI) to insoluble U(IV), precipitating uranium as uranium dioxide (UO_2) (eq. 11). This microbially driven redox process is a critical mechanism for the immobilization of uranium in reducing environments. The electrons required for this reduction reaction are sourced from organic matter or hydrogen. In ecosystems with high microbial activity, bioreduction can significantly decrease uranium mobility by converting uranium into insoluble U(IV), which then becomes trapped in sediments. This transformation not only enhances the stability of uranium but also mitigates its potential for contamination in groundwater and surface water, highlighting the essential role of microbial processes in biogeochemical cycling and environmental remediation efforts [14–16].

$$UO_2^{2+} + 4H^+ + 2e^- + microbes \rightarrow UO_{2(s)} + 2H_2O_{(l)} \tag{11}$$

3.3 *Iron and Manganese Oxides*

Iron (Fe) and manganese (Mn) oxides are abundant in soils and sediments, and they play dual roles in uranium redox chemistry by acting as both electron donors and electron acceptors, depending on their oxidation state, mediating uranium redox transformations and controlling uranium mobility.

3.3.1 Iron Oxides

Iron generally exists in two main oxidation states: ferrous iron (Fe^{2+}), which is reduced and soluble, and ferric iron (Fe^{3+}), which is oxidized and can form insoluble oxides. Fe^{2+} can reduce U(VI) to U(IV), whereas Fe^{3+} can oxidize U(IV) back to

U(VI), thereby controlling the cycle of uranium among its mobile, intermediate and immobile forms. In anoxic environments, where ferrous iron is present in significant amounts, Fe^{2+} donates electrons to U(VI), reducing it to insoluble UO_2 and becoming oxidized to Fe^{3+} in the process (Eq. (12)) [17–20].

$$UO_2^{2+} + Fe^{2+} \rightarrow UO_{2(s)} + Fe^{3+} \tag{12}$$

Conversely, ferric iron (Fe^{3+}) can oxidize U(IV) back to U(VI), mainly in environments where redox conditions fluctuate/alternate, leading to alternating dominance of Fe^{2+} and Fe^{3+} (Eq. (13)).

$$UO_{2(s)} + Fe^{3+} \rightarrow UO_2^{2+} + Fe^{2+} \tag{13}$$

3.3.2 Manganese Oxides

Manganese oxides, particularly manganese(IV) oxide (MnO_2), play a significant role in uranium redox chemistry, similar to the role of iron. MnO_2 serves as a potent oxidizing agent that is able to oxidize uranium from its tetravalent state (U(IV)) to its hexavalent state (U(VI)). This transformation increases the solubility and mobility of uranium in aqueous environments, which is particularly crucial in contaminated sites or regions with high uranium concentrations (Eq. 14) [20].

$$UO_{2(s)} + MnO_2 \rightarrow UO_2^{2+} + MnO \tag{14}$$

This environmental scenario most commonly occurs in environments comprising both iron and manganese oxides, which alter the fate of uranium in natural and anthropogenic systems by cycling uranium between its soluble and insoluble forms. Impact of DO on the fate of uranium in natural and anthropogenic environments is shown in Table 1.

4 Uranium Redox in Remediation Strategies

Understanding the redox behavior of uranium is critical for designing effective remediation strategies to manage uranium contamination in groundwater and soils. The manipulation of redox conditions to control uranium speciation is a key aspect of many remediation technologies since uranium solubility and mobility heavily depend on its oxidation state. By promoting the reduction of uranium from its soluble U(VI) form to its less soluble U(IV) form, remediation strategies can significantly reduce the spread of contamination.

Table 1 Impact of dissolved oxygen (DO) on the fate of uranium in natural and anthropogenic environments

Condition	Description	Equation
Oxidizing conditions	In oxic environments, where dissolved oxygen is abundant, U(IV) is rapidly oxidized to U(VI), which is more soluble and mobile oxygen acts as an electron acceptor, oxidizing insoluble U(IV) to the soluble uranyl ion (UO_2^{2+}). The presence of dissolved oxygen in groundwater or surface waters is a major factor controlling uranium mobility in these environments, as U(VI) can travel with the water, leading to the contamination	$UO_{2(s)} + \frac{1}{2}O_{2(g)} + 2H^+ \rightarrow UO_2^{2+} + H_2O_{(l)}$
Reducing conditions	In anoxic environments, where oxygen is absent, U(VI) is reduced to U(IV), which precipitates as uraninite (UO_2). These reducing conditions are often created by microbial activity, organic matter decomposition, or the presence of reducing agents such as ferrous iron or sulfide.	$UO_2^{2+} + 4H^+ + 2e^- \rightarrow UO_{2(s)} + 2H_2O_{(l)}$

4.1 *Reductive Immobilization*

Reductive immobilization is a remediation technique that focuses on reducing harmful mobile U(VI) to less harmful immobile U(IV), where it precipitates as an insoluble compound or complex, typically uraninite (UO_2). The technique manipulates the redox process by creating reducing conditions in contaminated environments, which leads to the immobilization of uranium in sediments and soils, effectively limiting its mobility and preventing further spread of contamination in groundwater (Table 2) [15].

4.2 *Bioremediation*

Bioremediation harnesses the natural ability of microorganisms to reduce U(VI) to U(IV), effectively immobilizing uranium by leveraging biological processes. This technique is increasingly favored and promising because it is cost effective and sustainable and takes advantage of naturally occurring microbial communities present in the subsurface environment (Table 3) [15].

Table 2 Summarized techniques and implementation methods

Section	Description
4.1.1 Mechanism of reductive immobilization	The goal of reductive immobilization is to introduce reducing agents into contaminated groundwater or soil to lower the redox potential (eh). Common reducing agents include hydrogen (H_2), sulfides (S^{2-}), or organic carbon sources (such as acetate or lactate), which serve as electron donors in the redox reaction. The reduction of U(VI) to U(IV) is a two-electron transfer process following the equation below: $UO_2^{2+} + 4H^+ + 2e^- \rightarrow UO_{2(s)} + 2H_2O_{(l)}$ The uranyl ion (UO_2^{2+}) is reduced by gaining electrons, leading to the formation of insoluble uraninite (UO_2), which precipitates out of solution. The reduction is facilitated by the lowering of the redox potential, which stabilizes the U(IV) form, which is then trapped in sediments, where it remains immobile.
4.1.1.1 Field implementation	To achieve reductive immobilization in contaminated sites, various techniques can be used to introduce reducing agents into the subsurface: - **Direct injection of reducing agents**: With. This method, reducing agents such as hydrogen gas, or sulfide are directly injected into the groundwater to create reducing conditions. - **In situ chemical reduction (ISCR)**: This technique involves the use of chemical reductants like zero-valent iron (ZVI), which releases electrons as it corrodes, reducing U(VI) to U(IV). - **Organic carbon addition**: Adding organic substrates, such as acetate, stimulates microbial activity, which indirectly creates reducing conditions by consuming oxygen and producing reduced metabolites that drive uranium reduction. Overall, reductive immobilization is useful for long-term uranium management in aquifers and contaminated sediments since it aims to trap uranium in a stable, insoluble form that resists remobilization under reducing conditions and it can take long time to show positive outcome.

4.3 Permeable Reactive Barriers (PRBs)

Permeable reactive barriers (PRBs) are in situ remediation technologies employed to treat contaminated groundwater as it flows through a reactive material. PRBs are typically installed perpendicular to the direction of groundwater flow and allow the water to pass through while reacting with contaminants, including uranium, immobilizing them (Table 4) [21].

The performance of PRBs depends on factors such as the rate of groundwater flow, the concentration of uranium, and the longevity of the reactive materials. Regular monitoring is required to ensure that the PRB remains effective and that the reactive materials have not been depleted or become clogged with precipitates as time passes. In few cases, remediation strategies that manipulate redox conditions, such as reductive immobilization, bioremediation, and PRBs, are crucial for controlling uranium contamination. By reducing U(VI) to U(IV), these techniques immobilize uranium, preventing its spread in groundwater and reducing its global health and environmental impacts. Understanding the redox behavior of uranium

Table 3 Summarized bioremediation process and long-term sustainability assessment

Section	Description
4.2.1 Microbial uranium reduction	Certain bacteria, such as *Geobacter* and *Shewanella*, are known to use uranium as an electron acceptor during anaerobic respiration. These bacteria's adaptability to thrive in environments where oxygen is limited or absent (anaerobic conditions) and can reduce U(VI) to U(IV), plays role in precipitating uranium as solid uraninite (UO_2) based on the following equation: $UO_2^{2+} + 4H^+ + 2e^- + microbes \rightarrow UO_{2(s)} + 2H_2O_{(l)}$ The bacteria derive energy by transferring electrons to U(VI), reducing it to the immobile U(IV) form, which precipitates as UO_2 in sediments. Microorganisms capable of uranium reduction often use organic matter as electron donors, which fuels their metabolic processes.
4.2.2 Stimulation of microbial activity	In contaminated environments, bioremediation is enhanced by adding organic substrates that stimulate the growth and spike activity of uranium-reducing bacteria. The addition of organic carbon sources like acetate increases microbial respiration, which promotes the reduction of uranium. This approach can be deployed by: 1. In situ bioremediation: Organic substrates are injected into the groundwater through wells, providing a food source for native microbial communities that can reduce U(VI). 2. Ex situ bioremediation: Contaminated groundwater is pumped to the surface, treated in bioreactors containing uranium-reducing bacteria, and then reinjected into the subsurface. However, the success of bioremediation processes depends on several factors, including the availability of electron donors, the presence of active microbial communities, and the redox conditions of the subsurface environment.
4.2.3 Long-term sustainability of bioremediation	One of the major advantages of bioremediation is its potential for long-term sustainability. Once successfully established, microbial communities can continue to function in the subsurface, maintaining reducing conditions and keeping uranium immobilized as long as the appropriate conditions are sustained. Nevertheless, the longevity of bioremediation systems depends on the manipulation of redox conditions, and periodic and frequent monitoring may be required to ensure the ongoing effectiveness of bioremediation process.

allows for the design of more effective and sustainable remediation strategies that can provide long-term protection for affected ecosystems and human health [22].

5 Conclusion

Redox potential (oxidation–reduction potential, ORP) plays a pivotal role in governing the fate and mobility of uranium in different environmental matrices. Through alteration of the oxidation state of uranium, it directly impacts its solubility,

Table 4 Summarized permeable reactive barrier structure, function, types and advantages

Section	Description	Key Reaction
Structure and function of PRBs	PRBs are passive treatment system consisting of a trench filled with reactive materials such as zero-valent iron (ZVI), organic carbon, or other reductants that promote reducing conditions. While contaminated groundwater flows through the barrier, redox reactions occur, reducing U(VI) to U(IV) and precipitating uranium as an immobile solid, typically uraninite (UO_2).	$UO_2^{2+} + Fe^0 \rightarrow UO_{2(s)} + Fe^{2+}$
Types of reactive materials used in PRBs	Several types of reactive materials can be used in PRBs, depending on the specific contaminants and environmental conditions. Common materials include: **Zero-valent iron (ZVI):** Creates strong reducing conditions by releasing electrons. **Organic carbon:** Promotes microbial activity and indirectly facilitates uranium reduction. **Activated carbon:** Adsorbs uranium and removes it from groundwater.	
Advantages of PRBs	PRBs offer several advantages: Passive operation: Operate without ongoing energy input or maintenance. Long-term effectiveness: Can reduce U(VI) for extended periods. Cost-effective: More economical than active remediation technologies.	

interaction with mineral rock surfaces, and potential bioavailability. Under reducing environments, uranium primarily exists in the less soluble U(IV) form and is often immobilized as uraninite or adsorbed onto organic matter. Conversely, under oxidizing environments, uranium predominantly occurs in highly soluble and mobile U(VI) forms, posing greater health and environmental risks of contamination in water surface and groundwater systems. There is an urgent need to understand and monitor redox conditions for effective remediation strategies, including bioremediation and permeable reactive barriers (PRBs). In addition, incorporating redox potential into large-scale experiments followed by predictive models can enhance our understanding and assessment of environmental risks, increasing our ability to design targeted interventions to mitigate uranium contamination in natural and engineered settings.

Acknowledgments Authors would like to extend their sincere acknowledgment and encouragement from the director of CSIR-Central Scientific Instruments Organization, Chandigarh, India. YU would like to also acknowledge all coauthors for their commitment and contribution towards the completion of this chapter.

References

1. Ma M, Wang R, Xu L, Xu M, Liu S (2020) Emerging health risks and underlying toxicological mechanisms of uranium contamination: lessons from the past two decades. Environ Int 145:106107. https://doi.org/10.1016/j.envint.2020.106107
2. Bjørklund G, Semenova Y, Pivina L et al (2020) Uranium in drinking water: a public health threat. Arch Toxicol 94:1551–1560. https://doi.org/10.1007/s00204-020-02676-8
3. Peiffer S, Kappler A, Haderlein SB, Schmidt C, Byrne JM, Kleindienst S, Vogt C, Richnow HH, Obst M, Angenent LT, Bryce C, McCammon C (2021) A biogeochemical–hydrological framework for the role of redox-active compounds in aquatic systems. Nat Geosci 14(5):264–272. https://doi.org/10.1038/s41561-021-00742-z
4. Pan Z, Bártová B, LaGrange T, Butorin SM, Hyatt NC, Stennett MC, Kvashnina KO (2020) Nanoscale mechanism of UO2 formation through uranium reduction by magnetite. Nat Commun 11(1):1–12. https://doi.org/10.1038/s41467-020-17795-0
5. Deng C, Liang J, Sun R, Wang Y, Fu P, Wang B, Gao S, Huang W (2023) Accessing five oxidation states of uranium in a retained ligand framework. Nat Commun 14(1):1–7. https://doi.org/10.1038/s41467-023-40403-w
6. Asic A, Kurtovic-Kozaric A, Besic L, Mehinovic L, Hasic A, Kozaric M, Hukic M, Marjanovic D (2017) Chemical toxicity and radioactivity of depleted uranium: the evidence from in vivo and in vitro studies. Environ Res 156:665–673. https://doi.org/10.1016/j.envres.2017.04.032
7. Zhang L, Chu J, Xia B, Xiong Z, Zhang S, Tang W (2022) Health effects of particulate uranium exposure. Toxics 10(10):575. https://doi.org/10.3390/toxics10100575
8. Patterson CG (1998) Oxidation–reduction. In: Geochemistry. Encyclopedia of earth science. Springer, Dordrecht. https://doi.org/10.1007/1-4020-4496-8_232
9. Keener M, Hunt C, Carroll TG, Kampel V, Dobrovetsky R, Hayton TW, Ménard G (2020) Redox-switchable carboranes for uranium capture and release. Nature 577(7792):652–655. https://doi.org/10.1038/s41586-019-1926-4
10. Lezama-Pacheco JS, Cerrato JM, Veeramani H, Alessi DS, Suvorova E, Bernier-Latmani R, Giammar DE, Long PE, Williams KH, Bargar JR (2015) Long-term in situ oxidation of biogenic uraninite in an alluvial aquifer: impact of dissolved oxygen and calcium. Environ Sci Technol 49(12):7340–7347. https://doi.org/10.1021/acs.est.5b00949
11. Stewart BD, Amos RT, Fendorf S (2011) Effect of uranium(VI) speciation on simultaneous microbial reduction of uranium(VI) and iron(III). J Environ Qual 40(1):90–97. https://doi.org/10.2134/jeq2010.0304
12. Kalintsev A, Migdisov A, Alcorn C, Baker J, Brugger J, Mayanovic RA, Akram N, Guo X, Xu H, Boukhalfa H, Caporuscio FA, Viswanathan H, Wang Y, Matteo E, Roback R (2021) Uranium carbonate complexes demonstrate drastic decrease in stability at elevated temperatures. Commun Chem 4(1):1–8. https://doi.org/10.1038/s42004-021-00558-3
13. Wu L, Song Z, Wu Y, Xia S, Kuzyakov Y, Hartley IP, Fang Y, Yu C, Wang Y, Chen J, Guo L, Li Z, Zhao X, Yang X, Zhang Z, Liu S, Wang W, Ran X, Liu C et al (2024) Organic matter composition and stability in estuarine wetlands depending on soil salinity. Sci Total Environ 945:173861. https://doi.org/10.1016/j.scitotenv.2024.173861
14. Hu Y, Wang Y, Han X, Shan Y, Li F, Shi L (2021) Biofilm biology and engineering of Geobacter and Shewanella spp. for energy applications. Front Bioeng Biotechnol 9:786416. https://doi.org/10.3389/fbioe.2021.786416
15. You W, Peng W, Tian Z, Zheng M (2021) Uranium bioremediation with U(VI)-reducing bacteria. Sci Total Environ 798:149107. https://doi.org/10.1016/j.scitotenv.2021.149107
16. Zou L, Huang Y, Long Z et al (2019) On-going applications of *Shewanella* species in microbial electrochemical system for bioenergy, bioremediation and biosensing. World J Microbiol Biotechnol 35:9. https://doi.org/10.1007/s11274-018-2576-7
17. Pérez-Guzmán L, Bogner KR, Lower BH (2010) Earth's ferrous wheel. Nat Educ Knowl 3(10):32

18. Kappler A, Bryce C, Mansor M, Lueder U, Byrne JM, Swanner ED (2021) An evolving view on biogeochemical cycling of iron. Nat Rev Microbiol 19(6):360–374. https://doi.org/10.1038/s41579-020-00502-7
19. Cain TJ, Smith AT (2021) Ferric iron reductases and their contribution to unicellular ferrous iron uptake. J Inorg Biochem 218:111407. https://doi.org/10.1016/j.jinorgbio.2021.111407
20. Li F, Yin H, Zhu T, Zhuang W (2024) Understanding the role of manganese oxides in retaining harmful metals: insights into oxidation and adsorption mechanisms at microstructure level. Eco-Environ Heal 3(1):89–106. https://doi.org/10.1016/j.eehl.2024.01.002
21. Obiri-Nyarko F, Grajales-Mesa SJ, Malina G (2014) An overview of permeable reactive barriers for in situ sustainable groundwater remediation. Chemosphere 111:243–259. https://doi.org/10.1016/j.chemosphere.2014.03.112
22. Budania R, Dangayach S (2023) A comprehensive review on permeable reactive barrier for the remediation of groundwater contamination. J Environ Manag 332:117343. https://doi.org/10.1016/j.jenvman.2023.117343

Effect of Salinity on Uranium Solubility

Divya Patel

Abstract One important aspect of aquatic chemistry is salinity, which has significant effects on the solubility of heavy metals and other physicochemical factors, such as pH, TDS, and EC. Salinity also has a significant effect on uranium (U) solubility, according to numerous studies. This chapter explores the relationship between salinity and uranium speciation and how it affects uranium solubility. While several oxidation states of uranium exist in nature, U(IV) and U(VI), which exhibit notably distinct behaviors in aqueous environments, are the most common oxidation states. The uranyl ion (UO_2^{2+}) is the main form of U(VI), and it is more soluble and may form a variety of complexes than U(IV), which is usually insoluble and precipitates in reducing circumstances. Salinity increases influence the ionic strength, complexation dynamics, and anion competitiveness, all of which affect uranium solubility. This chapter focuses on evaluating the mobility and solubility of uranium in saline environments.

Keywords Uranium · Salinity · Solubility · Mobility · Aquatic chemistry

1 Introduction

Both surface and underground aquatic systems contain uranium, which is radioactively and chemically toxic [1]. The three most common uranium isotopes found in nature are ^{234}U, ^{235}U, and ^{238}U. ^{235}U and ^{238}U are the most prevalent and have long half-lives; the content ratio of $^{238}U/^{235}U$ is 137.832 [2]. There are several oxidation states of uranium in nature, including +2, +3, +4, +5, and + 6. However, there are two primary oxidation states for uranium: +4 and + 6 under naturally occurring

D. Patel (✉)
National Sugar Institute, Kanpur, UP, India

Department of Environmental Studies, The Maharaja Sayajirao University, Vadodara, India

M. K. Jindal, K. Patni (eds.), *Water Quality-Uranium Solubility Nexus: Unlocking the Secrets of Uranium Solubility*, Earth and Environmental Sciences Library, https://doi.org/10.1007/978-3-031-98909-4_8

geochemical circumstances [3]. Under oxidizing conditions, U is mostly present in natural waters as dissolved uranyl carbonates ($UO_2(CO_3)_3^{4-}$), and as a result, it does not strongly associate with particles or colloids. Chemical weathering is principally responsible for introducing ^{238}U, ^{235}U, and ^{234}U into the ocean as dissolved forms. U is usually transported by rivers, and its concentration is influenced by the degree of chemical weathering and the lithology of the drainage basin. As ^{238}U breaks down in the parent material during the weathering cycle, α-recoil mechanisms produce excess ^{234}U (compared with its parent ^{238}U), resulting in an excess of ^{234}U at the mineral interface and in river runoff relative to secular equilibrium [4]. The most water-soluble U oxidation state is U(VI) (uranyl, UO_2^{2+}). When U(VI) is reduced to U(IV) (uranous), it is thought to precipitate as an insoluble oxide ($UO_2(s)$) under extremely reducing conditions [5]. A wide range of other chemical factors, including pH, salinity, hardness, and alkalinity, significantly influence uranium speciation and solubilization in water [6]. Even the mobility of uranium in the soil environment is dependent upon the pH, texture, redox potential (Eh), and composition (mineral and organic content) of the soil solution [7]. Soluble forms have the ability to move with soil water, be absorbed by aquatic or plant life, leach into groundwater, and ultimately enter the food chain. Moreover, the amount of salt present in the soil influences uranium behavior. Two potential mechanisms that could be involved in this process are (1) competition for sorption sites on the solid phase between positively charged U species and salt-derived cations and (2) the ability of salt-derived anions to combine with uranium species. This chapter covers a comprehensive examination of the relationships between uranium solubility and salinity, providing information that is crucial for both theoretical understanding and practical applications in uranium-based waste management and the behavior of uranium in a saline environment.

2 Uranium Behavior in the Environment

Uranium is a naturally occurring element that is present in trace amounts in all compartments of the environment [8]. All of the uranium isotopes are radioactive. Naturally occurring uranium isotopes such as ^{234}U, ^{235}U, and ^{238}U decay radioactively, emitting an alpha particle followed by moderate gamma radiation. The dominant isotope, ^{238}U, generates a wide range of decay products, including the significant radionuclides ^{226}Ra and ^{222}Rn. The decay process continues until a stable, nonradioactive decay product is produced [9]. Uranium in shallow or surface natural waters is usually oxidized (U^{6+}), and its aquatic chemistry includes soluble uranyl ions (UO_2^{2+}) and aqueous complexes. Uranium is present mostly in solution as UO_2^{2+} at low pH values. However, at higher pH values, uranyl produces a variety of hydrolysis species, including UO_2OH^+, $(UO_2)_3(OH)^{5+}$, and $UO_2(OH)_2^0$ [10]. The uranyl ion (UO_2^{2+}) is a large, divalent, positively charged ionic species comparable in size to potassium, calcium, and cesium. In natural waters, uranium normally forms complexes with phosphate, fluoride, chloride, sulfate and carbonate, hydroxide, and possibly silicate ions and organic compounds, with fluoride complexes dominating below pH 5 and phosphate complexes dominating between pH 4 and 7.5 [11].

3 Uranium and Salinity Relationship

The uranium–salinity relationship in brine and sea ice demonstrates that U behaves conservatively in regard to salinity variations induced by sea ice melting and formation, which appear to match the basic concentration and dilution of seawater. This cautious predisposition appears to be maintained over salt levels ranging from low to rather high, up to 135 g/kg (Fig. 1). Compared with the nonconservative behavior of U during saltwater or brackish water mixing, fluctuations in U content in brackish waters from estuaries most likely reflect intricate chemical reactions and water–sediment interactions in the mixing zone [4].

TDS, salinity, and uranium content were reported to have correlation values of 0.69 and 0.68, respectively, by Kumar et al. [12]. Under reducing conditions, uranium was found as insoluble complexes, whereas under oxidizing conditions, it was found as soluble complexes. When the salinity and TDS levels are high, soluble complexes are discharged into groundwater under high ionic strength circumstances [12].

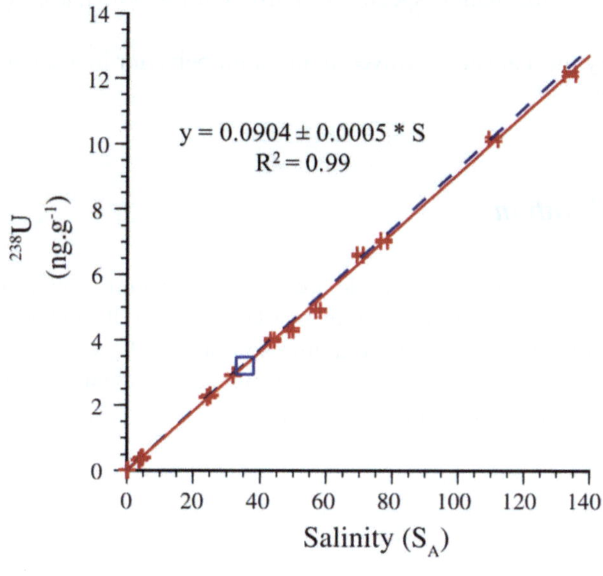

Fig. 1 Uranium salinity relationship (Reproduced from Not et al. [4] with permission from Elsevier Publishing)

4 Effect of Salinity on Uranium Solubility, Speciation, and Mobility

The solubility and speciation of uranium may be impacted by higher salinity, as it increases the ionic strength of water. Competing ions can change the chemical forms of uranium, promoting the formation of specific complexes [13–15]. The oxidation–reduction potential (ORP), pH, and complexing agents, such as carbonates, phosphates, fluorides, sulfates, and silicates, are the primary determinants of uranium concentration and mobility in aqueous systems [15]. A multivariate study revealed substantial correlations between U concentrations and salinity, TDS, NO_3^-, K, Na, HCO_3^-, Cl^-, and F^-. Among these factors, uranium was highly correlated with salinity and TDS. According to this correlation, U mobilization in the alluvial plains of northwest India and Punjab groundwater is caused mostly by the salt effect, which results from intense ion competition [16]. The groundwater in the Punjab district contains an elevated level of uranium as a result of mineral weathering. Uraninite is likely to be transformed into carnotite (U(VI)) minerals in an oxidizing environment, promoting uranium solubility and mobility in groundwater [17]. A prior study revealed that uranium concentrations in shallow water are more strongly correlated with salinity, TDS, NO_3^-, K, Na, HCO_3^-, Cl^-, and F^- than those in deep water. The TDS and salinity remained strongly related (U-TDS: $\rho = 0.5$ to 0.6, U-salinity: $\rho = 0.5$). This association demonstrates that uranium mobilization is mostly induced by the salt effect, which arises from enhanced ion competition. This effect is often observed when uranium levels rise in conjunction with an increase in mixed water species (e.g., Mg-Cl, Na-HCO_3 and Na-Cl). $Ca_2UO_2(CO_3)_2^-$ and $CaUO_2(CO_3)_3^-$ are the main U species responsible for mobility, according to speciation data [18].

The primary mechanisms involved in uranium behavior in a saline environment are listed below.

4.1 *Complexation*

Halides such as chlorine and bromide are common in saline environments. These anions can form stable complexes with U(VI), such as $(UO_2)(O_2)(Cl)(H_2O)_2^-$ and $(UO_2)(O_2)(Br)(H_2O)_2^-$, increasing uranium solubility [19]. Other species, such as carbonate, sulfate, phosphate, calcium, and magnesium, make complexes of uranium in saline aquatic environments [20, 21].

4.2 *Ionic Strength*

Depending on the prevailing chemical species, ions can increase or decrease uranium solubility in an aqueous solution. Li et al. examined the behavior of Studtite (a uranium mineral) in a saline environment and reported that ionic strength has a substantial influence on the rate of Studtite formation. Dissolution studies further demonstrated that the rate of studtite dissolution increased with increasing ionic strength, especially in the presence of Cl^- and Br^- ions [22]. Many studies have shown that uranium transport and fate in groundwater environments are influenced by geochemical factors such as pH, redox conditions, and the availability of ligand-forming species, including dissolved carbonate, Ca^{2+}, and Mg^{2+} ions [23].

5 Uranium Behavior in the Natural Saline Environment

For the first time, Not et al. [4] investigated the relationship between salinity and uranium in Arctic Sea ice using an extended salinity scale. They noticed the conservative behavior of uranium in ocean water [4]. However, it is not valid in brackish waters, such as those found in estuaries, where both conservative and nonconservative behaviors have been documented [24, 25]. Zhao et al. studied four saline lakes in the northeastern Qaidam Basin to determine how uranium isotopes were spread and influenced. The ^{238}U concentration and activity ratios of $^{234}U/^{238}U$ ($[^{234}U/^{238}U]AR$) did not fluctuate significantly within the same lake; however, there was some variation in the distribution across lakes. Moreover, aqueous ^{238}U concentrations within a specific range increased as total dissolved solids (TDS) and salinity levels increased, as did pH. The adsorption of aqueous ^{238}U is controlled by TDS and salinity, similar to the pH in natural waters [2].

6 Some Case Studies and Field Observations

Groundwater-dominated salts, including sodium, magnesium, sulfate, chloride, and others, also affect uranium concentrations, especially in arid and semiarid zones [26]. In a meta-analysis of uranium contamination in Punjab's alluvial plains, Sahoo et al. reported that mixed water species containing Na–Cl, Mg–Cl, and $Na–HCO_3$ increased the uranium concentration in groundwater. They also noted that uranium mobility in groundwater is caused by a few main uranium species, including $Ca_2UO_2(CO_3)^{2-}$ and $CaUO_2(CO_3)^{3-}$[16]. Another field observation revealed a substantial association between Na and Cl in the arid zone, reflecting the effect of saltwater on uranium. The same study also revealed that the majority of the samples with high uranium contents were from the Na–Cl hydrological facies [27]. Total salinity has little effect on uranium in groundwater in some humid wet and dry regions, such as Deoghar in Jharkhand, but certain parameters, such as nitrate, sulfate, and phosphate, are positively correlated with uranium [28, 29]. In addition to these findings, numerous other investigations have demonstrated that elevated salinity contributes to uranium mobility and solubility in water [30–33].

7 Conclusion

Salinity significantly affects uranium solubility, determining its mobility and potential environmental damage. Salinity can increase uranium solubility by modifying ionic strength and complexation dynamics, particularly when specific anions are present. Understanding these interactions is critical for managing uranium in aquatic systems, as well as establishing appropriate environmental and waste disposal techniques. However, few studies have investigated the effect of salinity on uranium solubility in the environment. As a result, more research is recommended, particularly case studies of uranium behavior in saline environments. Future research should also emphasize the combined impacts of salinity, pH, redox conditions, and temperature to construct comprehensive models of uranium behavior in various aquatic systems.

Acknowledgments The author thanks the NSI, Kanpur, and MSU Baroda, Vadodara.

References

1. Rout D, Ravi P, Kumar A, Tripathi R (2015) Study on speciation and salinity-induced mobility of uranium from soil. Environ Earth Sci 74
2. Zhao C, Zhang P, Li X, Ning Y, Tan L, Edwards RL, Yao X, Cheng H (2020) Distribution characteristics and influencing factors of uranium isotopes in Saline Lake waters in the Northeast of Qaidam Basin. Minerals 10:74
3. Romanchuk AY, Vlasova IE, Kalmykov SN (2020) Speciation of uranium and plutonium from nuclear legacy sites to the environment: a mini review. Front Chem 8:630
4. C. Not, K. Brown, B. Ghaleb, C. Hillaire-Marcel, Conservative behavior of uranium vs. salinity in Arctic Sea ice and brine, Mar Chem, 130–131 (2012) 33–39
5. Duff MC, Amrhein C (1996) Method for the separation of uranium (IV) and (VI) oxidation states in natural waters. J Chromatogr A 743:335–340
6. Goulet RR, Thompson PA, Serben KC, Eickhoff CV (2015) Impact of environmentally based chemical hardness on uranium speciation and toxicity in six aquatic species. Environ Toxicol Chem 34:562–574
7. Malikova IN, Strakhovenko VD, Ustinov MT (2020) Uranium and thorium contents in soils and bottom sediments of lake Bolshoye Yarovoye, western Siberia. J Environ Radioact 211:106048
8. Zavodska L, Kosorinova E, Scerbakova L, Lesny J (2008) Environmental chemistry of uranium. HV ISSN:1418–7108

9. UNSCEAR (2017) Radiation, Sources, effects and risks of ionizing radiation, united nations scientific committee on the effects of atomic radiation (UNSCEAR) 2016 report: report to the general assembly, with scientific annexes, United Nations
10. IAEA (2023) The environmental behaviour of uranium. International Atomic Energy Agency
11. Bird GA (2012) Uranium in the environment: behavior and toxicity. In: Meyers RA (ed) Encyclopaedia of sustainability science and technology. Springer New York, New York, pp 11220–11262
12. Kumar M, Kaushal A, Sahoo B, Sarin A, Mehra R, Jakhu R, Bhalla A, Sharma N (2017) Measurement of uranium and radon concentration in drinking water samples and assessment of ingestion dose to local population in Jalandhar district of Punjab, India. Indoor Built Environ 28:611–618
13. Balaram V, Rani A, Rathore DPS (2022) Uranium in groundwater in parts of India and world: a comprehensive review of sources, impact to the environment and human health, analytical techniques, and mitigation technologies. Geosyst Geoenviron 1:100043
14. Patel D, Pamidimukkala P, Chakraborty D (2024) Groundwater quality evaluation of Narmada district, Gujarat using principal component analysis. Groundw Sustain Dev 24:101050
15. Sar S, Diwan V, Biswas S, Singh S, Sahu M, Jindal M, Arora A (2017) Study of uranium level in groundwater of Balod District of Chhattisgarh state, India and assessment of health risk. Human Ecol Risk Assess Int J 24:1–8
16. Kumar A, Rout D, Narayanan U, Mishra M, Tripathi R, Singh J, Kumar S, Kushwaha Hs (2011) Geochemical modeling of uranium speciation in the subsurface aquatic environment of Punjab State in India. J Geol Min Res 3
17. Sahoo P, Virk H, Powell M, Kumar R, Pattanaik J, Salomão G, Mittal S, Chouhan L, Nandabalan Y, Tiwari R (2021) Meta-analysis of uranium contamination in groundwater of the alluvial plains of Punjab, Northwest India: status, health risk, and hydrogeochemical processes. Sci Total Environ 807:151753
18. Kumar A, Rout S, Narayanan U, Mishra MK, Tripathi R, Singh J, Kumar S, Kushwaha H (2011) Geochemical modelling of uranium speciation in the subsurface aquatic environment of Punjab State in India. J Geol Min Res 3:137–146
19. Sahoo PK, Virk HS, Powell MA, Kumar R, Pattanaik JK, Salomão GN, Mittal S, Chouhan L, Nandabalan YK, Tiwari RP (2022) Meta-analysis of uranium contamination in groundwater of the alluvial plains of Punjab, Northwest India: status, health risk, and hydrogeochemical processes. Sci Total Environ 807:151753
20. Soderholm L, Skanthakumar S, Wilson R (2011) Structural correspondence between uranyl chloride complexes in solution and their stability constants. J Phys Chem A 115:4959–4967
21. Endrizzi F, Rao L (2014) Chemical speciation of uranium(VI) in marine environments: complexation of calcium and magnesium ions with [(UO2)(CO3)3]4− and the effect on the extraction of uranium from seawater. Chem Eur J 20:14499–14506
22. Alam MS, Cheng T (2014) Uranium release from sediment to groundwater: influence of water chemistry and insights into release mechanisms. J Contam Hydrol 164:72–87
23. Li J, Szabó Z, Jonsson M (2022) Stability of Studtite in saline solution: identification of uranyl–peroxo–halo complex. Inorg Chem 61:8455–8466
24. Stanley DM, Wilkin RT (2019) Solution equilibria of uranyl minerals: role of the common groundwater ions calcium and carbonate. J Hazard Mater 377:315–320
25. Schneider AB, Velasquez I, Paul SAL, Krause CH, Koschinsky A (2022) Mixing and partially non-conservative behavior of molybdenum, uranium and vanadium along the salinity gradients of the Amazon and Pará estuaries and associated plume. Mar Chem 247:104182
26. Rahaman W, Singh SK, Raghav S (2010) Dissolved Mo and U in rivers and estuaries of India: implication to geochemistry of redox sensitive elements and their marine budgets. Chem Geol 278:160–172
27. Paikaray S, Chander S (2022) Geochemical variations in uranium and fluoride enriched saline groundwater around a semi-arid region of SW Punjab, India. Appl Geochem 136:105167

28. Pandit P, Saini A, Chidambaram S, Kumar V, Panda B, Ramanathan AL, Sahu N, Singh AK, Mehra R (2022) Tracing geochemical sources and health risk assessment of uranium in groundwater of arid zone of India. Sci Rep 12:2286
29. Srivastava M, Srivastava PK, Kumar D, Kumar A (2021) A systematic study of uranium in groundwater and its correlation with other water quality parameters. Water Supply 22:2478–2492
30. Bairwa AK, Gupta AK, Mishra V, Sahoo SK, Tiwar SN, Menaria T, Gupta K (2024) Seasonal variation and spatial distribution of uranium in sources of water in Tonk District of Rajasthan, India. Pollution 10:511–527
31. Mehta D (2012) Uranium in ground water, water-logging in Malwa region, scientific opinion & fact sheet, Unpublished Report. Available on: https://physics.puchd.ac.in/dmehta/uranium-facts-201207.pdf
32. Patel D, Pamidimukkala P, Chakraborty D, Yadav A (2022) Bharuch District, Gujarat, India: Factor analysis and geographical distribution of water quality characteristics. Environ Nanotechnol Monit Manag 18:100732
33. Jindal MK, Pandit SA, Karunakara N, Chandrashekara MS, Kumara S, Kumar V, Salim D, Srinivasan R (2024) High uranium dose from the groundwater in a granitic terrain in the eastern part of Karnataka, India. J Radioanal Nucl Chem 333:2905–2913

Organic Matter: Friend or Foe to Uranium Solubility

Rashmi Raghav, Kamal Kant Tiwari, and Pankaj Kandwal

Abstract Carbon-rich substances found in both natural and artificial habitats on land and water are referred to as organic matter. It is composed of organic constituents derived from the waste and remains of living things, including plants and animals. Due to its complex and dualistic behavior, organic matter plays an important role in the solubility of uranium in environmental systems, acting as both a friend and a foe to the solubility of uranium depending upon the conditions. By forming stable soluble complexes with uranium ions, organic matter can enhance uranium mobility, thus increasing uranium solubility in aqueous environments. The precipitation of uranium is prevented by generating compounds with humic and fulvic acids, which may be transported through groundwater systems more readily. Organic materials can play a key role in reducing uranium(VI) to uranium(IV), which tends to precipitate as uraninite, which is less soluble, resulting in a lower solubility of uranium. The dual effects of organic matter are affected by pH, redox conditions, the type and concentration of organic matter, and competing ions. To develop effective uranium remediation strategies and predict uranium mobility in natural and contaminated environments, it is essential to consider the specific interactions between organic materials and uranium. The complex relationship between organic matter and uranium solubility and the significance for environmental management and remediation effects are illustrated in this chapter.

Keywords Uranium(VI) · Organic matter · Uranium solubility

R. Raghav (✉) · K. K. Tiwari · P. Kandwal
Department of Chemistry, National Institute of Technology (NIT), Uttarakhand, Srinagar, Uttarakhand, India
e-mail: kamalkanttiwari@nituk.ac.in

M. K. Jindal, K. Patni (eds.), *Water Quality-Uranium Solubility Nexus: Unlocking the Secrets of Uranium Solubility*, Earth and Environmental Sciences Library, https://doi.org/10.1007/978-3-031-98909-4_9

1 Introduction

One common heavy metal found in the Earth's crust is uranium, which occurs naturally. It plays a large role in both the generation of energy and environmental issues [1]. Uranium is essential for producing nuclear power, which produces a significant amount of electricity worldwide, as it is a decisive element of nuclear fuel [2, 3]. Moreover, uranium is utilized in nuclear weapons, which has an impact on nonproliferation efforts and national security [4]. However, uranium poses threats to human health and the environment because of its chemical toxicity and radioactive characteristics, especially in areas where uranium mining, processing, or nuclear activities take place [5]. Rocks, soil, and water all naturally contain uranium. The minerals carnotite $K_2(UO_2)_2(VO_4)_2{\cdot}3H_2O$ and autunite ($Ca(UO_2)_2(PO)_2{\cdot}10\text{-}12H_2O$) are among the most common forms. Uraninite (UO_2) is its primary form. When these minerals weather, uranium is released into the environment and ends up in soils, surface waters, and groundwater [6]. The uranium level in the environment varies greatly; background levels usually range from 0.001 to 5 micrograms per liter (μg/L) in surface water and groundwater and from 0.3 to 11 parts per million (ppm) in soils. In certain places, the concentration of uranium has considerably increased due to human activity. Important human sources include the following:

- Nuclear Power Plants: During operation, some uranium may be discharged into the environment through leaks, accidents, or cooling water discharges. Uranium is the fuel used in nuclear reactors. There are additional long-term dangers of uranium contamination associated with the storage and disposal of radioactive waste and spent nuclear fuel.
- Phosphate fertilizers: Uranium is frequently present in phosphate rocks, which are used to make fertilizers. Phosphate fertilizer may cause uranium to accumulate in agricultural soils, which may then seep into groundwater systems.
- Uranium Mining and Milling: These processes produce significant amounts of waste materials containing uranium, such as tailings, which have the potential to leak uranium into the environment through erosion and leaching. Elevated quantities of uranium are frequently found in contaminated water from these locations.
- Production and Testing of Nuclear Weapons: Uranium and its decay byproducts have been released into the atmosphere and adjacent areas as a result of past nuclear weapon testing. Localized contamination has also been facilitated by locations that produce weapons and process uranium.

Numerous factors, including its chemical form, the existence of complexing agents (including carbonate, sulfate, and organic matter), pH, redox conditions, and the properties of the surrounding media, affect uranium mobility in the environment [7]. The most common form of uranium under oxidizing conditions is the hexavalent uranyl ion (UO_2^{2+}), which is soluble and may move through groundwater. The tetravalent ion (U^{4+}), which is less soluble and likely to precipitate as uraninite (UO_2) or other minerals, is the most prevalent form of uranium under reducing conditions [8, 9]. The depletion of soil, disturbance of microbial populations, and impairment of aquatic ecosystems are among the effects of uranium poisoning on the environment [10]. This limits the mobility of the element. Because of its radioactivity and chemical toxicity, uranium exposure can be harmful to one's health. The radioactive decay of uranium and its decay products can cause kidney damage and an increased risk of cancer when uranium-contaminated food or water is consumed [11].

1.1 Overview of Organic Matter

The intricate combination of carbon-based substances known as organic matter (OM) is derived from the detritus of plants, animals, and microorganisms [12]. It is essential to environmental chemistry because it affects nutrient availability, soil structure, water retention, and the fate of pollutants [13].

Types of Organic Matter
In general, organic matter may be divided into the following:

- Humic substances, including humic, fulvic, and humic acids, make up a significant portion of the organic matter. These are highly molecular weight, complex chemicals that are produced when plant and animal wastes breakdown. They play crucial roles in soil and water systems because of their interactions with contaminants, metals, and nutrients [14].
- Particulate organic matter (POM): Plant detritus, soil organic particles, and microbial biomass are examples of particulate organic matter (POM), which is an organic material that is not soluble in water. POM influences the way sediment layers and soils are physically structured.
- Dissolved organic matter (DOM): Organic compounds that dissolve in water are referred to as dissolved organic matter (DOM). These compounds might be simple molecules such as sugars and amino acids or more complicated ones such as humic substances. Because it influences both the chemistry of water and the behavior of contaminants, DOM is an important part of aquatic ecosystems [15].

Both natural and man-made sources provide organic matter. Microbial biomass, animal waste, plant litter, and root exudates are natural sources. Phytoplankton and aquatic plants are important sources of organic matter input in aquatic environments. The main sources of organic matter in soil are plant leftovers and microbial activities [16]. Anthropogenic sources include urban stormwater, industrial effluents, wastewater discharge, and agricultural runoff. The amount of organic matter in the environment can be increased by human activity, which frequently results in eutrophication and changes in water quality [17].

In the environment, organic matter affects chemical, physical, and biological processes in a variety of ways.

- Contaminant Adherence and Detoxification: Radionuclides, heavy metals, and organic contaminants can bind to organic matter, which decreases its mobility and bioavailability. Covalent bonding, adsorption, and complexation are some of the ways in which this binding might occur. For example, humic chemicals can influence the movement and destiny of metals such as copper, lead, and uranium in the environment by forming stable complexes with them [18, 19].

- Soil Fertility and Structure: By encouraging the development of soil aggregates, which increase porosity and water retention, organic matter strengthens the soil structure. Additionally, it increases the cation exchange capacity (CEC) of soils, which enables them to retain and gradually release important minerals such as calcium, magnesium, and potassium to plants.
- Nutrient cycling: Organic matter is an essential aspect of nutrient cycling, especially for carbon, nitrogen, and phosphorus. It provides nutrition and energy to microbes, which breakdown organic materials and release nutrients in accessible forms. Maintaining soil fertility and promoting plant development depend on this mechanism [20, 21].
- Buffering procedure: The presence of organic matter in soils and aquatic systems helps maintain stable conditions that are favorable for the growth of microorganisms and plants.
- Water Absorption and Erosion Management: Organic matter improves soil structure, which helps hold onto water and lessens runoff and soil erosion. This phenomenon is especially crucial in agricultural environments since soil erosion can result in topsoil loss and lower crop yields [22].

1.2 Chemistry of Uranium

Comprehending the chemical properties of uranium is indispensable for forecasting its behavior in the environment and creating effective plans for cleanup and management. For example, uranium can be immobile in polluted areas or removed more efficiently from groundwater by regulating redox conditions or by adding complexing agents [23]. In the environment, uranium mostly occurs in two oxidation states:

- Uranium(IV) (U^{4+}): U^{4+} is the tetravalent state of uranium under reducing circumstances. In general, insoluble uranium(IV) tends to precipitate as uraninite (UO_2) and coffinite ($USiO_4$) crystals. Compared with uranium(VI), this reduced form is less mobile and tends to accumulate in soils and residues [23, 24].
- Uranium(VI) (U^{6+}): Under oxidizing conditions, U^{6+} is the most prevalent and stable oxidation state of uranium. Uranyl ions (UO_2), which are highly soluble in water and may form several kinds of aqueous complexes, are the usual form of uranium(VI). The mobility of uranyl ions in aquatic environments is increased by their interaction with ligands such as phosphate, sulfate, and carbonate [23, 24].

Environmental conditions such as redox conditions, pH, the presence of complexing agents and the oxidation state are the factors affecting the movement and behavior of uranium in the environment. The behavior of uranium in natural waterways, soils, and residues is modulated by these features and impacts its transit, bioavailability, and likely threat to ecosystems and human health.

- Potential Redox: Redox conditions play a significant role in uranium separation detection. In oxidizing environments, uranium remains in the mobile U(VI) state; in reducing environments, it transforms into the less mobile U(IV) state and precipitates into crystalline particles.
- pH: Under acidic conditions, uranium is more likely to be found in a soluble form, remarkably as the uranyl ion. When the pH increases (alkaline conditions), uranium tends to form complexes with carbonate that either increase its mobility or cause carbonate minerals to precipitate [25].
- Adsorption: Organic matter, iron oxides, and clays are among the mineral surfaces on which uranium may adhere. The adsorbent material composition, pH, and redox potential affect the adsorption capacity. As a natural attenuation procedure, adsorption usually reduces uranium mobility by depositing uranium on solid phases.
- Presence of Complex Compounds: The solubility and mobility of uranium are influenced by the existence of ligands that may form complexes with it. Phosphate can cause uranium to become immobile, whereas carbonate ions increase its mobility. The organic ligands can either improve or reduce uranium mobility, depending on its composition and concentration. Organic ligands that can alter the movement of uranium include fulvic and humic acids. Depending on the environmental matrix, these complexing agents tend to bind with U(VI) and U(IV), which might influence the solubility and movement of uranium [26].
- Microbial Activities: Through redox processes, microorganisms can affect the mobility of uranium through changes in the oxidation state of uranium. Several bacteria can convert U(VI) to U(IV), which enhances the precipitation of insoluble U(IV) minerals and decreases the mobility of uranium [27].

To quantify the environmental influence of uranium and create remediation techniques for polluted sites, comprehending the solubility and mobility of uranium is important [28].

- Contamination in Soils and Residues: Adsorption and redox conditions regulate uranium mobility in soils and residues. To improve adsorption or stimulate the reduction of U(IV), remediation activities may require increasing the chemistry of the soil [23].
- Groundwater Contamination: Uranium may become mobile and endanger the quality of groundwater in areas with high carbonate contents and oxidizing environments. Altering the redox conditions or introducing substances that encourage uranium precipitation are two possible remediation possibilities.
- Surface water systems: pH, carbonate concentration, and organic matter percentage affect uranium mobility in surface water systems. One way to stop uranium migration is through control measures that encompass keeping an eye on and handling these elements [26].

The objective of this chapter is to investigate the two-faceted impacts of organic matter on uranium solubility and the importance of recognizing this connection for environmental management.

2 Organic Matter: Friend or Foe to Solubility of Uranium

The Earth's crust, water bodies, and even living things contain naturally occurring radioactive element uranium in various amounts. Because of its radioactive nature and harmful effects on health, its mobility and bioavailability in the atmosphere are important concerns. The interaction of uranium with organic matter is a substantial component determining its behavior in the environment. Decayed plants and animals constitute organic matter, which is found in aquatic systems, soils, and sediments. Its solubility and mobility can be affected by complexing with metal ions, such as uranium, due to the many functional groups it has [24].

The solubility of uranium can be influenced by organic matter in both positive and negative ways, which are reliant upon environmental factors and the molecular makeup of the organic molecules in question. In some situations, by generating stable soluble complexes, organic matter can improve the mobility of uranium in water systems, increasing uranium solubility. However, by accelerating sorption and precipitation processes or by encouraging reduction events, organic matter can also reduce the solubility of uranium, which changes uranium into less soluble forms [29]. Envisaging the behavior of uranium in polluted and natural settings involves an understanding of the dual nature of organic matter in terms of uranium solubility. This information is decisive for determining the hazards associated with uranium acquaintance in the environment and for making cleanup plans.

2.1 *Organic Matter as a Solubility Enhancer*

In the natural environment, uranium becomes more soluble because of complexation interactions between organic ligands and uranium ions, which increases the stability and water solubility of uranium ions. A new study confirmed the impact of organic matter on increasing uranium mobility, which has noteworthy consequences for environmental contamination, remediation efforts, and uranium movement [30].

- Impact on the Environment: Uranium behavior and mobility in the environment are greatly prejudiced by the interactions it has with organic matter. Soluble complexes can make uranium more mobile and available in areas rich in organic matter, such as wetlands and soils with high organic contents. This raises the possibility that uranium will find its way into the surface and groundwater, compromising ecosystems as well as human health [31].
- Complexation with Organic Ligands: Hydroxyl, carboxyl, and phenolic groups are examples of functional groups found in natural organic materials, such as humic and fulvic acids. These groups can interact with uranium ions to create

stable complexes. Because uranium is more soluble in its hexavalent form (U(VI)) than in its tetravalent state (U(IV)), these interactions aid in keeping uranium in this state. According to recent studies, this complexation can significantly improve uranium solubility and increase its mobility in groundwater and soils. For example, it has been demonstrated that humic materials in aquatic environments increase uranium solubility by creating effective complexes that prevent uranium from precipitating or sticking to mineral surfaces [32].

- Microbial Activity and Organic Ligands: By creating organic ligands that bind to uranium, microbial activities can tangentially increase the solubility of uranium. These bacteria use organic materials as a carbon source and use their biological processes to make organic acids and other ligands that can promote the solubility of uranium by forming complexes with it. Moreover, the redox state of uranium can be altered by microbial activity, which can affect how uranium interacts with organic materials and how it performs during complexation [33].

2.2 *Organic Matter as a Solubility Reducer*

The solubility and mobility of uranium in the environment can be significantly reduced by organic matter, mostly through sorption and reduction reactions. These methods and their influence on uranium behavior in various ecological situations have been illuminated by recent studies.

- Sorption mechanisms: Organic matter binds to uranium ions and causes them to adsorb onto soil and sediment particles, and organic matter contributes to a decrease in uranium solubility. According to recent research, organic materials such as humic substances improve uranium adsorption on mineral surfaces such as clays and iron oxides. When uranium binds to organic-rich sediments or peat, it forms strong complexes that drastically reduce the amount of uranium in water. Given that a high organic content can trap uranium and restrict its movement, these interactions are particularly crucial in wetlands and other environments rich in organic matter [34].
- Reduction reactions: Organic matter can help reduce uranium from its soluble U(VI) form to the less soluble U(IV) form in anaerobic environments. Microbial activity, in which microorganisms consume organic materials as electron donors, is frequently the driving force behind this process. Studies have shown that microorganisms such as Shewanella and Geo-bacter may use organic carbon to convert U(VI) to U(IV), which precipitates uraninite (UO_2) or other insoluble chemicals. The solubility and mobility of uranium are significantly reduced by this bioreduction, which aids in immobilizing it under subterranean conditions [35].

3 Reasons and Factors Affecting the Dual Nature of Organic Matter

Organic matter has a dual nature in that it can have opposing effects on environmental procedures, either facilitating or hindering specific interactions. Its behavior is dependent on a variety of factors, such as its chemical makeup, the conditions in its surroundings, and its microbial activity.

- Decomposition products and effects of temperature on decomposition: OM has a dual nature because decomposition products such as fulvic and humic acids have special physical chemical appearances that alter how they interact with metals and other pollutants [13]. As organic matter breaks down more quickly at higher temperatures, simpler organic molecules are released into the environment. These molecules interact differently with several elements of the environment, which may either enhance or impede processes such as sorption [36].
- Interactions with minerals: Metals and organic matter can combine to form complexes, depending on the kind of organic ligand and environmental factors such as pH, which may either make the metals more soluble or stabilize them in place [37]. When organic matter adheres to minerals, it can either enhance or inhibit the adsorption of hazardous substances. For example, humic compounds can help heavy metals adsorb onto surfaces made of clay or iron oxide, but they can also hinder the adsorption sites in certain circumstances [38].
- Organic matter concentration: OM can enhance contaminant sorption by binding to solid particles at low concentrations. On the other hand, at high concentrations, more soluble complexes can be produced, which increases the dispersion of contaminants in water [39].
- Organic matter composition: Molecules that make up organic matter include proteins, carbohydrates, and humic compounds. These elements bind with metals and nutrients in different ways, which have variable impacts on processes such as adsorption and mobility [40]. Organic compounds have functional groups such as phenolic and carboxyl groups that can chelate with metals to form stable complexes. Under some conditions, this interaction may decrease the solubility of the metal, whereas under other conditions, it may increase its mobility [13].
- Environmental Conditions and Microbial Activity: Environmental factors such as pH and ionic strength affect the interaction between hazardous substances and organic matter. OM binds to metals more strongly at low pH values, whereas organic–metal complexes may become more soluble at high pH values. Redox processes can be aided or hindered by organic matter. It behaves as an electron donor in anaerobic environments, resulting in the reduction of metals such as U(VI) to U(IV) ions, hence decreasing their mobility [41]. Organic matter may stabilize metals in aerobic environments, which might increase their solubility. Complex organic material is broken down by microorganisms into simpler molecules that have various interactions with nutrients and metals. While certain microorganisms may promote the transport of pollutants, others may use organic carbon to help reduce metals [42].

4 Remediation Techniques

Due to uranium's toxicity and radioactive properties, removing uranium from soil and water is critical. There are numerous remediation techniques that are used for uranium removal from soil as well as from water, primarily focusing on adsorption, chemical reduction, ion exchange, phytoremediation, and bioremediation.

- Adsorption: Because of its efficacy, affordability, and ease of handling, the adsorption technique is a popular process for removing uranium from contaminated soil and water. In this process, uranium ions (UO_2^{2+}) build up on the surface of a solid substance (adsorbent). This technique is well established and extensively used for both rural water treatment systems and industrial applications [43]. The key components of the adsorption techniques are as follows:
  - Activated Carbon: Due to its high surface area and porosity, activated carbon is effectively used for the removal of uranium.
  - Biochar has shown high affinity for uranium removal, which is a low-cost alternative method to activated carbon made from biomass.
  - Nanomaterials: Nanomaterials, carbon nanotubes and graphene oxide are used for uranium removal because of their high surface area and are often functionalized to increase their efficiency for uranium removal.
  - Functionalized materials: materials modified with various functional groups, such as amines, phosphates, sulfonic, carboxylic, and sulfonate groups, to increase their uranium adsorption efficiency.
  - Metal oxides, such as titanium oxide (TiO_2), iron oxide (Fe_3O_4), and zirconium oxide, have good efficiency for uranium removal.
  - Ion exchange resins: These synthetic polymers are designed to exchange ions attached to the resin itself with specific ions present in solution that are specifically used for selective uranium removal.
  - Bioadsorbents: Agricultural products such as rice husk, banana peels and natural materials such as bacteria, fungi and algae can bind uranium ions through numerous functional groups.
- Ion-Exchange: It is highly effective and widely used for the removal of uranium because of its selective nature in groundwater, industrial effluents, and contaminated water. The material contains charged functional groups that bind with uranium ions from water. It consists of a cation exchange resin, an anion exchange resin and chelating resins [44].
- Bioremediation: The use of living organisms, mainly microbes and plants, is a cost-effective and eco-friendly approach for removing contaminants from the environment. In regard to removing uranium from both soil and water, this approach makes use of the biological processes associated with specific microbes and plants that can modify, immobilize, or remove uranium [43]. The key mechanisms involved in uranium bioremediation are as follows:
  - Bioaccumulation: Some microbes can vigorously transport uranium into their cells, where it can accumulate in compartments within the cells or within the plasma membrane.

- Biotransformation: By using enzymatic activity, microbes convert U(VI) (soluble) to U(IV) (insoluble), resulting in uranium to precipitate as uraninite (UO_2) because uranium becomes less mobile and bioavailable in both soil and water. In anaerobic environments, several bacteria, such as Shewanella species and Geobacter, reduce uranium via uranium reduction.
- Biosorption: In this technique, the biomass gradually accumulates uranium, which is subsequently possible to extract. The functional groups such as carboxyl, phosphate and amine groups found in the cell walls of natural materials such as bacteria, fungi and algae can remove uranium from soil and water.
- Phytoremediation: Some hyperaccumulator plants, including *Brassica juncea*, *Hordeum vulgare*, Vicia faba, and Agrostis castellana, can carry uranium from contaminated soil and water to their shoots via roots. After the gathering process, these plants are processed to eliminate uranium. On a broad scale, this technique is very helpful for eliminating uranium [27].
- Bioprecipitation: By altering the pH or generating compounds such as phosphate, microorganisms can cause the precipitation of uranium, which leads to the formation of insoluble minerals such as uranyl phosphate.

5 Conclusion

OM can act as both a friend and a foe to the solubility of uranium, depending on the environmental context. It enhances uranium solubility by forming stable complexes with uranium ions, increasing its mobility under oxidizing conditions. This process can lead to greater uranium transport and potential environmental contamination. Conversely, organic matter can reduce uranium solubility through microbial reduction and adsorption processes, converting uranium to less soluble forms or binding it to organic-rich sediments. These interactions can immobilize uranium, limiting its spread and bioavailability. Thus, organic matter plays a dual role in influencing uranium behavior in the environment, with implications for both contamination risk and remediation efforts.

Acknowledgments The authors thank NIT Uttarakhand for supporting this work.

References

1. Emsley, J. (2001). Nature's building blocks: an A-Z guide to the elements Oxford University Press
2. Gagarinskii AY, Ignatev VV, Ponomarev-Stepnoi NN, Subbotin SA, Tsibulskii VF (2005) Role of nuclear power in world energy production in the 21st century. Atom Energy 99:759–769
3. Mudd GM (2008) Sustainability reporting and water resources: a preliminary assessment of embodied water and sustainable mining. Mine Water Environ 27(3):136–144

4. World Nuclear Association (2021) Nuclear power and the environment. Retrieved from www.world-nuclear.org
5. Cochran TB (2005) Nuclear weapons complex: overview of design, production, and stockpiling processes. Natural Resources Defense Council
6. Ganguly S, Bhan U (2023) Occurrences and mobility of uranium in soil profile due to groundwater-soil interaction. In: Hydrogeochemistry of aquatic ecosystems. Wiley, pp 181–198
7. Favaro DIT (2005) Natural radioactivity in phosphate rock, phosphogypsum and phosphate fertilizers in Brazil. J Radioanal Nucl Chem 264:445–448
8. Langmuir D (1997) Aqueous environmental geochemistry. Prentice Hall
9. Markich SJ (2002) Uranium speciation and bioavailability in aquatic systems: an overview. Sci World J 2(1):707–729
10. Ratnikov AN, Sviridenko DG, Popova GI, Sanzharova NI, Mikailova RA (2020) The behaviour of uranium in soils and the mechanisms of its accumulation by agricultural plants. In: Uranium in plants and the environment. Springer, Cham, pp 113–135
11. ATSDR (2013) Toxicological profile for uranium. Agency for Toxic Substances and Disease Registry
12. Stevenson FJ (1994) Humus chemistry: genesis, composition, reactions, 2nd edn. Wiley
13. Tipping E (2002) Cation binding by humic substances. Cambridge University Press
14. Lehmann J, Kleber M (2015) The contentious nature of soil organic matter. Nature 528(7580):60–68
15. McCarthy JF, Zachara JM (1989) Subsurface transport of contaminants. Environ Sci Technol 23(5):496–502
16. Wershaw RL, Aiken GR (1985) Molecular size and weight measurements of humic substances. In: Aiken GR, McKnight DM, Wershaw RL, MacCarthy P (eds) Humic substances in soil, sediment, and water. Wiley-Interscience, pp 469–481
17. Sposito G (2008) The chemistry of soils. Oxford University Press
18. Leenheer JA, Croue JP (2003) Characterizing aquatic dissolved organic matter. Environ Sci Technol 37(1):18A–26A
19. Neff JC, Asner GP (2001) Dissolved organic carbon in terrestrial ecosystems: synthesis and a model. Ecosystems 4(1):29–48
20. Schlesinger WH, Bernhardt ES (2013) Biogeochemistry: an analysis of global change, 3rd edn. Academic Press
21. Kalbitz K, Solinger S, Park JH, Michalzik B, Matzner E (2000) Controls on the dynamics of dissolved organic matter in soils: a review. Soil Sci 165(4):277–304
22. Emsley J (2001) Nature's building blocks: an A-Z guide to the elements. Oxford University Press
23. Abdelouas A (2006) Uranium mill tailings: geochemistry, mineralogy, and environmental impact. Elements 2(6):335–341
24. Dong W, Ball WP, Liu C (2005) Influence of calcite and dissolved calcium on uranium(VI) sorption to a Hanford subsurface sediment. Environ Sci Technol 39(20):7949–7955
25. Seder-Colomina M, Mangeret A, Stetten L, Merrot P, Diez O, Julien A, Morin G (2018) Carbonate facilitated mobilization of uranium from lacustrine sediments under anoxic conditions. Environ Sci Technol 52(17):9615–9624
26. Lovley DR, Phillips EJP, Gorby YA, Landa ER (1991) Microbial reduction of uranium. Nature 350(6317):413–416
27. Wu W-M, Carley J, Gentry T (2006) Pilot-scale in situ bioremediation of uranium in a highly contaminated aquifer. 1. Conditioning of a pH- and U(VI)-contaminated aquifer. Environ Sci Technol 40(12):3978–3985
28. Bone SE, Dynes JJ, Cliff J, Bargar JR (2017) Uranium (IV) adsorption by natural organic matter in anoxic sediments. Proc Natl Acad Sci 114(4):711–716
29. Westrop JP (2021) Influence of nitrate on biogeochemical uranium redox cycling and mobility in fluvial sediments (Doctoral dissertation, The University of Nebraska-Lincoln)
30. Li WC, Victor DM, Chakrabarti CL (1980) Effect of pH and uranium concentration on interaction of uranium (VI) and uranium (IV) with organic ligands in aqueous solutions. Anal Chem 52(3):520–523

31. Shuai W, Qin Z, Li J, Jia Y, Liang Y, Wang Y (2023) Uranium photo-precipitation coupled with fulvic acid oxidation under anoxic and toxic conditions. Chem Eng J 471:144554
32. Janot N, Lezama Pacheco JS, Pham DQ, O'Brien TM, Hausladen D, Noel V, Bargar JR (2016) Physico-chemical heterogeneity of organic-rich sediments in the Rifle aquifer, CO: impact on uranium biogeochemistry. Environ Sci Technol 50(1):46–53
33. Newsome L, Morris K, Lloyd JR (2014) The biogeochemistry and bioremediation of uranium and other priority radionuclides. Chem Geol 363:164–184
34. Suzuki Y, Suko T (2006) Geomicrobiological factors that control uranium mobility in the environment: update on recent advances in the bioremediation of uranium-contaminated sites. J Mineral Petrol Sci 101(6):299–307
35. Von Lutzow M, Kogel-Knabner I (2009) Temperature sensitivity of soil organic matter decomposition-what do we know? Biol Fertil Soils 46:1–15
36. Gustafsson JP, Pechová P, Berggren D (2003) Modelling metal binding to soils: the role of natural organic matter. Environ Sci Technol 37(12):2767–2774
37. Sparks DL (2003) Environmental soil chemistry. Academic Press
38. Lo KSL, Yang WF, Lin YC (1992) Effects of organic matter on the specific adsorption of heavy metals by soil. Toxicol Environ Chem 34(2–4):139–153
39. Tan KH (2003) Humic matter in soil and the environment. CRC Press
40. Zhang H, Liu W, Xiong Y, Li G, Cui J, Zhao C, Zhang L (2024) Effects of dissolved organic matter on distribution characteristics of heavy metals and their interactions with microorganisms in soil under long-term exogenous effects. Sci Total Environ 947:174565
41. Wang J, Guo X (2020) Adsorption kinetic models: physical meanings, applications, and solving methods. J Hazard Mater 390:122156
42. Singh A (2019) Ion exchange technologies for uranium recovery from water and wastewater. Water Res 165:114976
43. Lovley DR, Phillips EJP (1992) Reduction of uranium by Desulfovibrio desulfuricans. Appl Environ Microbiol 58(3):850–856
44. Ebbs SD, Brady DJ, Kochian LV (1998) Role of uranium speciation in the uptake and translocation of uranium by plants. J Exp Bot 49(328):1183–1190

Impact of Physicochemical Parameters on Uranium Mobility and Solubility in Water

Vikram Singh, Manoj Kumar Jindal, and R. S. Aswal

Abstract The occurrence of uranium content in water sources is presently a topic of concern at the global level owing to health-associated implications among inhabitants. In this chapter, an attempt has been made to understand the role of numerous physicochemical characteristics in its transportation, solubility, and complexation. These characteristics include redox potential, pH, and different complexing agents, such as carbonate, phosphate, and sulfate. Furthermore, uranium occurrence in groundwater sources in different parts of the country is also a focus. This chapter provides significant insights into the understanding of uranium behavior in different aqueous systems to aid in the adoption of effective remedial measures.

Keywords Physicochemical parameters · Solubility · Mobility · Complexing agents · Aqueous systems

1 Background

Humans have been exposed to radiation from natural and anthropogenic sources since ancient times. Radon, which is a byproduct of uranium, contributes approximately 55% of natural background radiation exposure. However, the concentration of uranium in a naturally occurring element with a heterogeneous distribution greatly depends on the geological material and location [1]. Cosmic, terrestrial,

V. Singh
Independent Researcher, Almora, Uttarakhand, India

M. K. Jindal
Center for Sustainable Drinking Water, Indian Institute of Technology Jodhpur, Jodhpur, Rajasthan, India

R. S. Aswal (✉)
Department of Environmental Sciences, H.N.B. Garhwal University, Tehri Garhwal, Uttarakhand, India

M. K. Jindal, K. Patni (eds.), *Water Quality-Uranium Solubility Nexus: Unlocking the Secrets of Uranium Solubility*, Earth and Environmental Sciences Library, https://doi.org/10.1007/978-3-031-98909-4_10

inhalation, and ingestion through food, drink, and air are the four primary categories of natural exposure [1]. Uranium exposure by drinking water and soil (via these four paths) can pose a health risk because of its chemical physiology rather than its radiological physiology [2–6]. Uranium is a primordial radioactive nuclide, it has a half-life of about several billion years (4.5 billion years). There is no documented evidence of the deterministic effects of uranium present in the environment. However, when it accumulates in the human body, it poses serious health hazards to humans as well as other living organisms because of its chemical and radiological impact [1]. Uranium toxicity has multiple functions in nature, such as the exposure path, particle solubility, contact duration, and elimination route [7].

Many studies have shown that excess uranium intake either through drinking water or via other routes is the cause of numerous varieties of health hazards, among which nephritis is the most common effect of high uranium exposure [7–9]. Therefore, investigating the uranium concentration in water across the globe is crucial. In addition to the upper limit (30 μg/L) of the World Health Organization (WHO) at the international level and the radiological threshold level (60 μg/L) of the Atomic Energy Regulatory Board (AERB) at the Indian national level for uranium content in drinking water, the threshold limit of the Bureau of Indian Standards (BIS) has also been established [10, 11].

In addition to temperature, uranium solubility in aqueous solutions is governed by various physicochemical characteristics, which critically influence its speciation, mobility, and environmental behavior. Therefore, studies and modeling of radionuclide speciation and other contaminants caused by anthropogenic and natural activities in aquatic environments have advantages in terms of physicochemical parameters. In this chapter, parameters such as the hydrogen ion concentration, ionic strength, and redox conditions, including the occurrence of complexing agents such as carbonate and organic ligands, are reviewed. The effects of competing ions, colloidal interactions, and microbial activity are also discussed, along with their impact on uranium transportation and precipitation. By focusing on these factors, this chapter also provides a detailed understanding of uranium behavior in aquatic systems, aiding in environmental risk assessments for uranium-contaminated water.

2 Redox Potential (Eh)

The redox potential (Eh), also referred to as the oxidation or reduction potential of a chemical species, is the propensity to either accept electrons (reduction) or donate electrons (oxidation) and quantifies the ability to oxidize and reduce them [12]. The redox interface forms between two zones, where groundwater invades and introduces oxidized U^{+6}, a hexavalent form of uranium that meets reducing circumstances in saturated sediments [13]. Thus, lowering the redox potential further reduces the chemical species while increasing the redox potential, resulting in oxidation. The solubility of uranium in aqueous solution strongly depends on redox conditions [14].

2.1 Effects of Eh on the Movement of Uranium

Uranium mobility is largely affected by Eh due to its influence on the uranium oxidation state, which further affects uranium solubility and interactions with ecological elements [15]. Many studies have revealed that U^{+4} and U^{+6} are predominantly found in reducing and oxidizing environments, respectively [16]. However, uranium exists in several oxidation states, such as +2, +3, +4, +5 and +6, but +4 and +6 are the most chemically relevant states [11, 17]. The oxidation state of uranium is crucial in determining uranium solubility in aqueous environments. The solubility of uranium is directly related to mobility, which means that greater solubility is associated with greater mobility [13, 15, 16, 18–20]. In aqueous environments, U^{+4} is quite insoluble and tends to solid, whereas U^{+6}, which is found near the Earth's surface, is generally more soluble than U^{+4}; U^{+4} forms complexes with various ligands, such as carbonate, sulfate, and hydroxide [17, 21]. According to Cumberland et al. [13], the uranium oxidation state is potentially influenced by both organic matter and microbial activity, which significantly facilitates its reduction from U^{+6} to U^{+4}. It is less soluble and more stable in the environment. Under strongly oxidizing conditions, uraninite (UO_2) may persist because of its potential to host the U^{+6} oxidation state both within its structure and at its edges [13, 22]. Two main uranium minerals are U^{+4} (uraninite and coffinite), which deteriorate and oxidize to yield uranyl compounds (U^{+6} compounds) [13]. Thus, uranium can undergo a metamorphosis from U^{+4} to the more soluble U^{+6}; however, this process moves very slowly naturally without the encouragement of additional catalysts [13] present in both organic and inorganic uranium compounds, which include carbonate oxide or hyperoxide, phosphate, rare earth elements, silicates and sulfates, and most of them have a +6 oxidation state. The significant hydrolysis potential of U^{+4} enables it to form bonds with the majority of inorganic ligands. Thus, the mobility and solubility of uranium increases with increasing redox potential. However, solubility may also be influenced by other factors, such as pH, microbial activity, ionic strength, complexation, etc., which will be discussed in a subsequent section.

2.2 Influence of pH on Uranium Solubility

The pH (concentration of hydrogen ions) is a scale from 0 to 14 that quantifies the acidic or alkaline behavior of a solution ($pH = 7$ indicates neutrality, $pH < 7$ indicates acidity, and $pH > 7$ indicates an alkaline or basic nature of the solution). The pH of a solution can have a significant effect on the chemical behavior of many elements, including uranium. pH is a crucial determinant of uranium speciation because of its dependency on solubility, sorption, complexation, and colloid formation [23]. The main factors influencing U^{+6} solubility in acidic environments (low pH, high [H]) are the formation of uranyl hydroxide complexes and the oxidation of U^{+4} into U^{+6}. The dominant species under moderately acidic conditions is the uranyl ion (UO_2^{2+}); however, with increasing pH, the formation of uranyl hydroxide complexes such as UO_2OH^+ and $(UO_2)_2(OH)_2^{2+}$ intensifies, which are still soluble but present in decreasing free uranyl ion concentrations. In the presence of oxidizing agents, which include dissolved oxygen (DO) and ferric ions, insoluble U^{+4} is oxidized and becomes more soluble U^{+6}, which increases overall uranium solubility. The primary species found in acidic groundwater are complexes of uranyl ions, uranyl fluoride, and uranyl phosphate. Uranus fluoride complexes are more common in anoxic ground fluids with pH values of 4 and lower, whereas at higher pH values, the Uranus hydroxy complex [24].

With increasing pH, uranyl ions (UO_2^{2+}) undergo a sequence of complexation reactions, predominantly with carbonate ions (CO_3^{2-}) [24]. Carbonate, which is abundant in natural waters, especially in alkaline environments, significantly increases uranium solubility due to the formation of uranyl carbonate complexes such as $UO_2(CO_3)_2^{2-}$ and $UO_2(CO_3)_3^{4-}$ [25]. However, at extremely high pH values (i.e., pH > 10), the free carbonate ion concentration may decrease due to the formation of hydroxide and bicarbonate, which could reduce the solubility, although this is inconsequential [21]. Thus, natural water typically includes complexed uranium such as $UO_2(CO_3)_2^{-2}$ and $UO_2(CO_3)_3^{4-}$. The effect of pH on uranium solubility exhibited a characteristic 'U' shaped curve. Complexation processes significantly increase the solubility of uranium in moderately acidic and alkaline environments [26, 27]. However, an area of relatively low or minimal solubility exists at nearly neutral pH because the free uranyl ion does not dominate, and complexation with carbonate at neutral pH is relatively low compared with that at higher pH. The behavior at which uranium exhibits reduced solubility at a certain pH and enhanced solubility at other pH values is termed amphoteric behavior.

3 Role of Complexing Agents in Uranium Speciation

Generally, uranium speciation is largely influenced by UO_2^{+2} cations (approximately pH 4–5) in the absence of significant complexing ligands such as carbonate, phosphate, sulfate, and vanadate. The majority of UO_2-OH complexes dominate near neutral pH, with $(UO_2)_3(OH)_7^-$ being the most prevalent species at pH 7 and higher, whereas in a carbonate-rich environment, $UO_2CO_3^{4-}$ is present at pH >9 [13]. Uranium can yield both di- and trioxides such as UO_2, UO_3, and U_3O_8. Research has revealed that both the U^{+4} and U^{+6} oxidation states may yield numerous pure uranium oxides, such UO_2(uraninite) and $(UO_2)O_2(OH)_{12}\cdot 12H_2O$ (Schoepite) [13, 28]. The solubility of uranium oxides is significantly influenced by the oxidation state of uranium, with $U^{+4}O_2$ being insoluble and U^{+6} hydroxides and oxides being more soluble [13]. Uranium speciation is anticipated to be dominated by carbonate and phosphate complexes in the majority of oxygen-rich environments without organic matter. The ability of carbonate (CO_3^{-2}) to solubilize UO_2^{+2} as carbonate complexes increases the possibility of uranium mobility under alkaline conditions, making them highly effective [13]. Elless and Lee highlighted the significant correlation between time and alkalinity and uranium concentration in groundwater by emphasizing the critical role of carbonate geochemistry [29]. In a strongly oxidizing environment, CO_3^{-2} ions may dissolve U^{+4} via an oxidation state transition and function as a powerful chelator, mobilizing U^{+6} even in the absence of oxygen [13, 30, 31]. Uranyl phosphates are more readily soluble than vanadates but are frequently less soluble than uranyl carbonates or silicates, notwithstanding their ability to compete for uranyl ions with CO_3^{-2} and VO_4^{-3} [13, 32]. Phosphate minerals that inhabit groundwater may additionally prevent uranium mobility. Due to the limited solubility of uranyl phosphates, an assortment of phosphate-based compounds have been built extensively to be employed as lixiviates for uranium capture. According to Vazquez et al., pH has a substantial effect on uranium phosphate solubility [33]. Although uranium-hydroxide precipitation predominates at pH >6, producing insoluble mixed-phase uranium-hydroxophosphate species, an insoluble uranium hydrogen phosphate (U-HPO_4) complex develops at pH <6. The uranium–polyphosphate complex exhibited enhanced solubility throughout the pH spectrum; precipitation occurred at pH<5, although its solubility increased at pH >6. The solubility of uranyl silicates is somewhat between that of uranyl carbonates or oxyhydrates and that of uranyl phosphates at neutral pH values [13]. Uranyl silicates, formed from uranyl ions or uranium oxides, are crucial minerals for modifying spent nuclear fuel under moist and oxidizing conditions. On the other hand, coffinite is derived from uraninite, while hexavalent uranyl silicate compounds such as soddyite, boltwoodite, and uranophane are also important minerals [34]. Compared with other anions, uranyl ions are less susceptible to complex formation with sulfates. Uranyl vanadates are significant because they are commonly identified paired in deposits, particularly in organic matter-rich ores. The degree of solubility of uranium vanadates fluctuates with uranium minerals. Vanadium is similar to uranium in that it is redox sensitive and has three oxidation states (+3, +4, and +5).

Uranium solubility in groundwater is influenced by a combination of geogenic, physicochemical, and environmental factors. One of the primary sources of uranium in groundwater is natural leaching from uranium-bearing rocks. The concentration of uranium varies among different rock types, with granite containing approximately 3.4 mg/kg uranium, limestone and dolomite containing 2.2 mg/kg uranium, and argillaceous shale containing approximately 3.7 mg/kg uranium. Sediments have a broader range of 1.4 to 53 mg/kg, whereas phosphate rocks can contain significantly higher concentrations ranging from 30 to 100 mg/kg [35]. The geological composition of an aquifer thus plays a crucial role in determining the background uranium levels in water. Uranium solubility in groundwater is governed by a complex interplay of physicochemical parameters, such as pH, redox potential, temperature, ionic composition, and climate-driven changes. These factors directly influence uranium mobility and transformation between soluble and insoluble forms. Groundwater often contains a variety of cations, such as copper, iron, and zinc, which can be beneficial to human health. However, toxic elements such as lead, mercury, arsenic and uranium present significant health risks. Uranium behavior is particularly complex because of its multiple oxidation states and tendency to form various aqueous complexes, especially those with carbonate, phosphate, and fluoride ions. For example, the carbonate concentration can significantly increase uranium mobility by forming stable uranyl carbonate complexes. Conversely, low pH and reducing conditions often promote uranium precipitation, limiting uranium mobility. Seasonal variation, groundwater exploitation, and contamination from anthropogenic sources further influence uranium chemistry. The mobility and retention of uranium in aquifers also depend on interactions with minerals, particularly calcium-rich host rocks, and on the seasonal hydrological cycle, which governs oxidation–reduction transitions in aquifers. The typical reaction mechanisms for uranium mobilization are presented in Fig. 1 [36]. Further studies have demonstrated the non uniform relationships between uranium

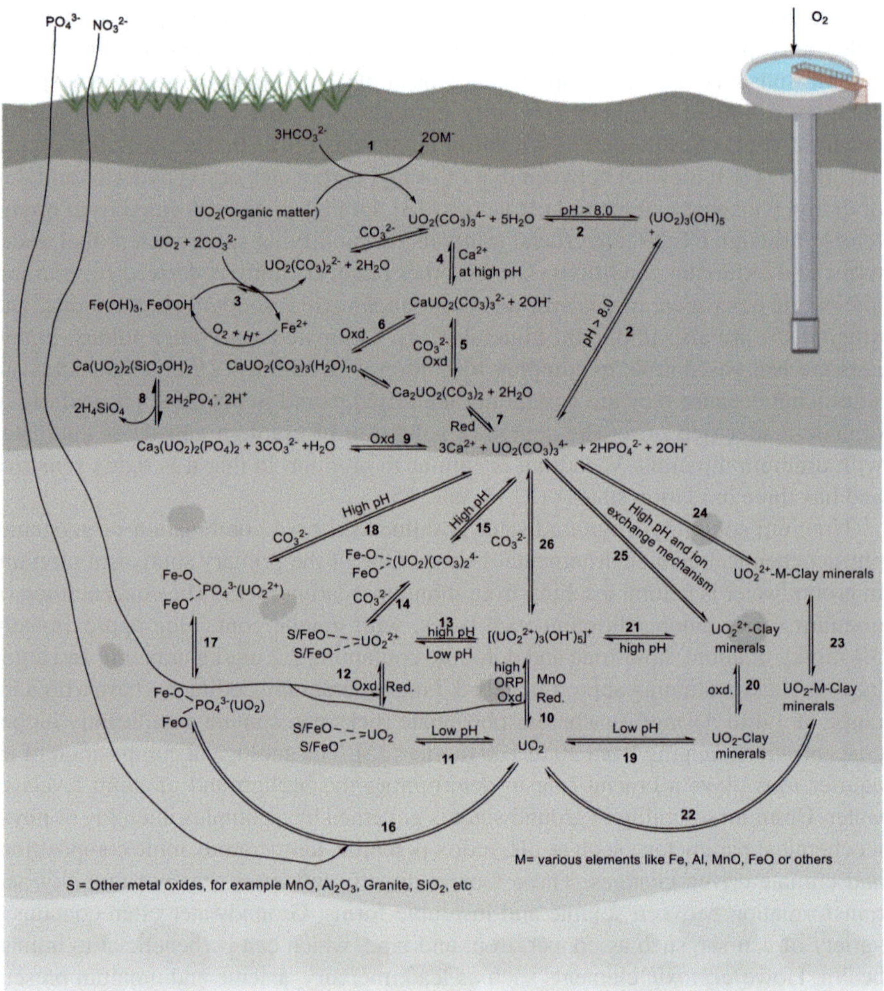

Fig. 1 Typical physicochemical reaction mechanisms for the mobilization of uranium in groundwater [36] (Reproduced from [36] with permission from Elsevier, publisher)

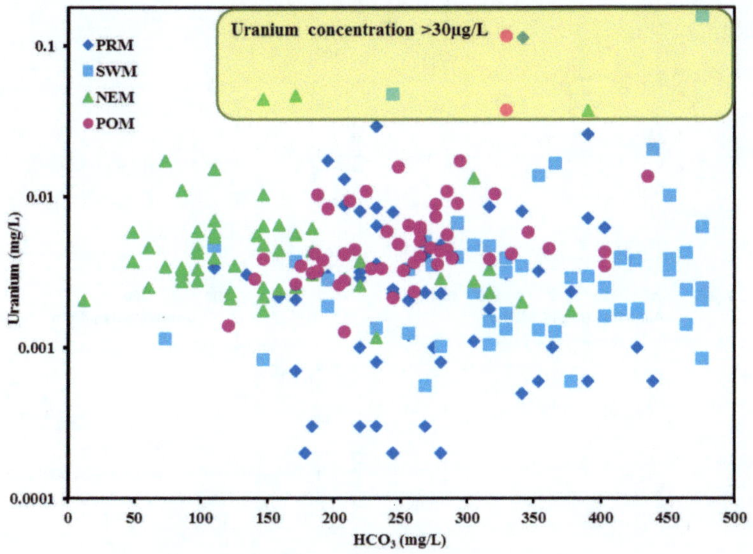

Fig. 2 Typical example of bicarbonate vs uranium during season sampling [37]

and various water quality parameters, such as bicarbonate, pH, alkalinity, conductivity, calcium, and strontium. Figures 2 and 3 illustrate such variability across sampling seasons and geochemical profiles [37, 38].

3.1 Uranium in Indian Groundwater: A Case Study of Regional Variability

Groundwater serves as the primary source of drinking water for most Indian populations. However, uranium contamination poses a serious concern, particularly in arid and semiarid regions, where alternative water sources are limited. In 2020, the Central Groundwater Board (CGWB) conducted a comprehensive survey involving 14,377 groundwater samples from various parts of India (Fig. 4). The uranium concentrations in these samples ranged up to 2876 µg/L, significantly exceeding the WHO (30 µg/L) and AERB guidelines at several locations [12]. The Ananthapur, Chittoor, Guntur, Cuddapah, East Godavari, Krishna, Kurnool, and Prakasam districts of Andhra Pradesh have uranium levels above the WHO recommended limit and might be sourced primarily from younger granitic intrusives [11]. Uranium is highly soluble in the calcium-rich minerals of host rocks, leading to leakage into ground water. Saikia et al. reported average uranium levels in the Nalbari district of Assam of 2.15 µg/L (ranging from 0.3–7.1 µg/L) and 2.75 µg/L (ranging from 0.6–10.3 µg/L) during the pre- and postmonsoon periods, respectively. The elevated uranium level during the postmonsoon season may be

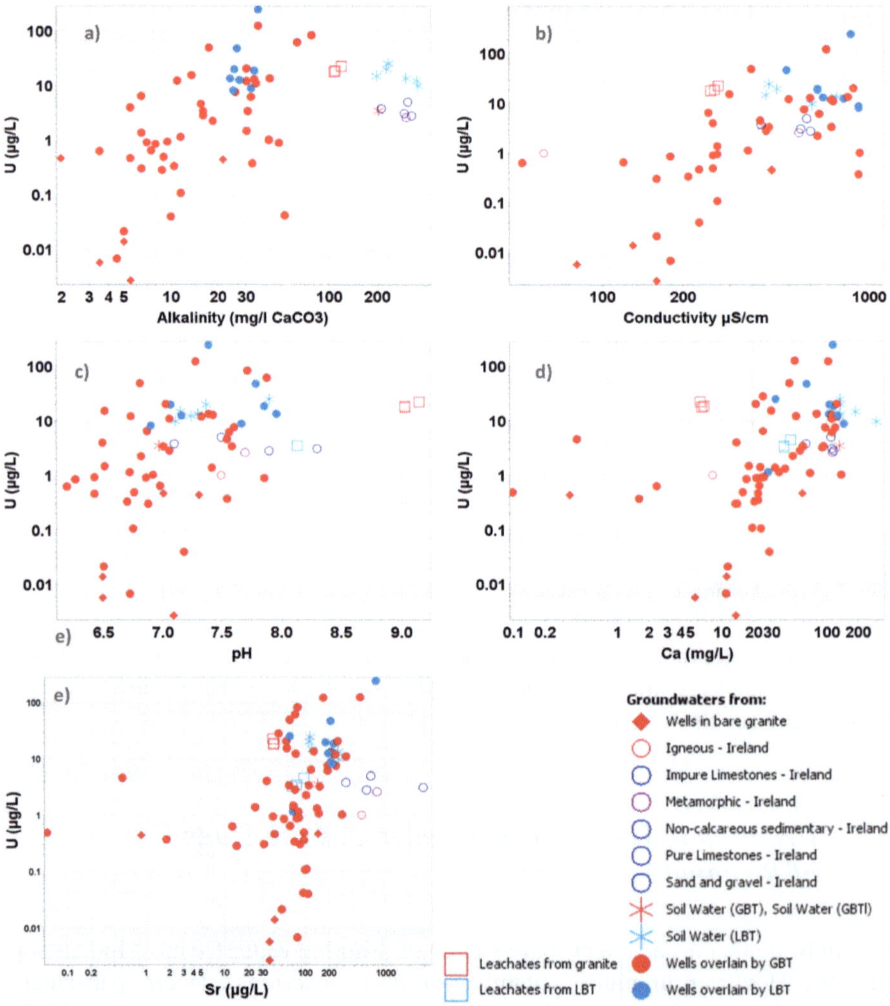

Fig. 3 Variation in uranium with multiple water quality parameters [38]

attributed primarily to the breakdown of uranium from soil sediments during the rainy season [39]. These observations underscore the spatial and temporal heterogeneity of uranium contamination in Indian groundwater systems. The contributing factors include geogenic sources such as uranium-rich host rocks, the depth and location of borewells, and hydrogeological interactions affecting uranium mobilization.

Table 1 lists the minimum and maximum uranium concentrations in groundwater for 23 Indian states acquired from various studies.

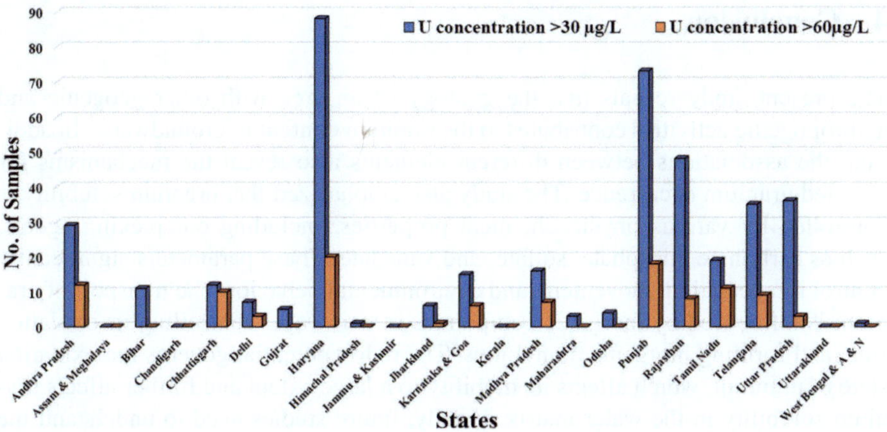

Fig. 4 Samples with uranium concentrations beyond the WHO and AERB limits (data extracted from the CGWB report 2020) [11]

Table 1 Uranium concentrations in groundwater sources for 23 Indian states

S.N.	States	Uranium Content (µg/L)	Within AERB Limit	Within WHO Limit	References
1	Andhra Pradesh	upto2876.0	No	No	[11, 40]
2	Assam & Meghalaya	upto10.7	Yes	Yes	[11, 39]
3	Bihar	0.1–80.5	No	No	[11, 41]
4	Chandigarh	upto15.6	Yes	Yes	[11]
5	Chhattisgarh	0.2–96.0	No	No	[42, 43]
6	Delhi	upto89.4	No	No	[11, 44]
7	Gujarat	upto57.3	Yes	No	[11, 45, 46]
8	Haryana	0.1–274.0	No	No	[47, 48]
9	Himachal Pradesh	0.2–90.5	No	No	[49–51]
10	Jammu & Kashmir	0.2–23.8	Yes	Yes	[52]
11	Jharkhand	0.1–69.9	No	No	[1, 11, 53]
12	Karnataka	0.1–8649.0	No	No	[54–57]
13	Kerala	0.1–12.5	Yes	Yes	[11, 58–60]
14	Madhya Pradesh	upto233.9	No	No	[11]
15	Maharashtra	upto48.0	Yes	No	[11, 61, 62]
16	Odisha	upto59.0	Yes	No	[11]
17	Punjab	0.4–1340	No	No	[10, 46, 51, 63–70]
18	Rajasthan	0.6–260.0	No	No	[71–74]
19	Tamil Nadu	0.2–302.0	No	No	[11, 75, 76]
20	Telangana	upto158.2	No	No	[11, 77]
21	Uttar Pradesh	upto189.0	No	No	[11, 78]
22	Uttarakhand	upto35.83	Yes	No	[11, 44, 79–81]
23	West Bengal	0.2–34.3	Yes	No	[11, 82]

4 Conclusion

The present study reveals that the geology of an area with other geogenic and anthropogenic activities contributes to the uranium content in groundwater. In addition, the associations between different elements also reveal the mechanisms for elevated uranium occurrence. The study also emphasized that uranium solubility is controlled by various physicochemical properties, including complexing agents, such as carbonate phosphate, sulfate, and vanadate. These parameters significantly control its speciation, movement, and environmental behavior. The transport of uranium is influenced by inorganic compounds in two ways: controlling its oxidation state and forming inorganic uranyl ions. The redox potential governs the oxidation state of uranium, which affects its mobility to a large extent and further affects uranium solubility in the water matrix. Finally, future studies need to understand the movement of uranium while considering hydrogeochemical factors that are responsible for the prevalence of high uranium content in aqueous solutions.

References

1. Srivastava M, Srivastava PK, Kumar D, Kumar A (2022) A systematic study of uranium in groundwater and its correlation with other water quality parameters. Water Supply 22(3):2478–2492. https://doi.org/10.2166/ws.2021.458
2. Zamora MLL, Zielinski JM, Moodie GB, Falcomer RAF, Hunt WC, Capello K (2009) Uranium in drinking water: renal effects of long-term ingestion by an aboriginal community. Arch Environ Occup Health 64(4):228–241. https://doi.org/10.1080/19338240903241267
3. Jenny-Burri J et al (2020) Uranium exposure of the Swiss population based on 24-hour urinary excretion. Swiss Med Wkly 150:w20207. https://doi.org/10.4414/smw.2020.20207
4. Okaneku J, Vearrier D, Mckeever R, Lasala G, Greenberg MI (2015) Urine uranium concentrations and renal function in residents of the United States - 2001 to 2010. Clin Toxicol. https://doi.org/10.3109/15563650.2015.1094704
5. Wade-Gueye NM, Delissen O, Gourmelon P, Aigueperse J, Dublineau I, Souidi M (2012) Chronic exposure to natural uranium via drinking water affects bone in growing rats. Biochim Biophys Acta Gen Subj 1820(7):1121–1127. https://doi.org/10.1016/j.bbagen.2012.04.019
6. Zhang W et al (2020) Association of adverse birth outcomes with prenatal uranium exposure: a population-based cohort study. Environ Int 135. https://doi.org/10.1016/j.envint.2019.105391
7. Domingo JL (2001) Reproductive and developmental toxicity of natural and depleted uranium: a review. [Online]. Available: www.elsevier.com/locate/reprotox
8. Mehra R, Kaur S (2020) Biokinetic modelling and risk assessment of uranium in humans. In: Uranium in plants and the environment. Springer, Cham, pp 217–241. https://doi.org/10.1007/978-3-030-14961-1_11
9. Brugge D, De Lemos JL, Oldmixon B (2005) Exposure pathways and health effects associated with chemical and radiological toxicity of natural uranium: a review. [Online]. Available: http://www.antenna.nl/wise/
10. Bajwa BS, Kumar S, Singh S, Sahoo SK, Tripathi RM (2017) Uranium and other heavy toxic elements distribution in the drinking water samples of SW-Punjab, India. J Radiat Res Appl Sci 10(1):13–19. https://doi.org/10.1016/j.jrras.2015.01.002
11. Sudhir D, Srivastava K (2021) Occurence of high uranium in shallow ground water in India. [Online]. Available: https://www.researchgate.net/publication/362860215

12. Mobilian C, Craft CB (2022) Wetland soils: physical and chemical properties and biogeochemical processes. In: Mehner T, Tockner K (eds) Encyclopedia of inland waters, 2nd edn. Elsevier, Oxford, pp 157–168. https://doi.org/10.1016/B978-0-12-819166-8.00049-9
13. Cumberland SA, Douglas G, Grice K, Moreau JW (2016) Uranium mobility in organic matter-rich sediments: a review of geological and geochemical processes. Earth Sci Rev. https://doi.org/10.1016/j.earscirev.2016.05.010
14. Bjørklund G et al (2020) Uranium in drinking water: a public health threat. Arch Toxicol. https://doi.org/10.1007/s00204-020-02676-8
15. Suzuki Y, Suko T (2006) Geomicrobiological factors that control uranium mobility in the environment: update on recent advances in the bioremediation of uranium-contaminated sites. J Miner Petrol Sci 101. https://doi.org/10.2465/jmps.jmps.060322
16. Rosendahl CD, Roebbert Y, Schippers A, Weyer S (2023) U mobilization and associated U isotope fractionation by sulfur-oxidizing bacteria. Front Microbiol 14. https://doi.org/10.3389/fmicb.2023.1190962
17. Denecke M, Bryan N, Kalmykov S, Morris K, Quinto F (2018) Sources and behaviour of actinide elements in the environment. In: Experimental and theoretical approaches to actinide chemistry: from fundamental systems to practical applications. Wiley, pp 378–444. https://doi.org/10.1002/9781119115557.ch8
18. Zhang M, Zhong R, Yu C, Cui H (2023) The immobility of uranium (U) in metamorphic fluids explained by the predominance of aqueous U(IV). Minerals 13(3). https://doi.org/10.3390/min13030427
19. Chardi KJ, Schenkeveld WDC, Kumar N, Giammar DE, Kraemer SM (2023) Effect of competing metals and humic substances on uranium mobilization from noncrystalline U(IV) induced by anthropogenic and biogenic ligands. Environ Sci Technol 57(42):16006–16015. https://doi.org/10.1021/acs.est.3c01705
20. Bali E, Audétat A, Keppler H (2011) The mobility of U and Th in subduction zone fluids: an indicator of oxygen fugacity and fluid salinity. Contrib Mineral Petrol 161(4):597–613. https://doi.org/10.1007/s00410-010-0552-9
21. Kertes AS, Guillaumont R (1985) Solubility of UO2. A comparative review. Nucl Chem Waste Manag 5(3):215–219. https://doi.org/10.1016/0191-815X(85)90080-4
22. Finch RJ, Ewing RC (1992) The corrosion of uraninite under oxidizing conditions. J Nucl Mater 190(C):133–156. https://doi.org/10.1016/0022-3115(92)90083-W
23. Mühr-Ebert EL, Wagner F, Walther C (2019) Speciation of uranium: compilation of a thermodynamic database and its experimental evaluation using different analytical techniques. Appl Geochem 100:213–222. https://doi.org/10.1016/j.apgeochem.2018.10.006
24. Vaaramaa K, Pulli S, Lehto J (2000) Effects of pH and uranium concentration on the removal of uranium from drinking water by ion exchange. Radiochim Acta 88:845–850
25. Akçay H (1998) Aqueous speciation and pH effect on the sorption behavior of uranium by montmorillonite. J Radioanal Nucl Chem 237:133
26. Saenen E et al (2013) Effects of pH on uranium uptake and oxidative stress responses induced in Arabidopsis thaliana. Environ Toxicol Chem 32(9):2125–2133. https://doi.org/10.1002/etc.2290
27. Ollila K, Olin M, Lipponen M (1996) Solubility and oxidation state of uranium under anoxic conditions (N 2 atmosphere). Radiochim Acta 74:9–14
28. Jang JH, Dempsey BA, Burgos WD (2006) Solubility of schoepite: comparison and selection of complexation constants for U(VI). Water Res 40(14):2738–2746. https://doi.org/10.1016/j.watres.2006.04.014
29. Elless MP, Lee SY (1998) Uranium solubility of carbonate-rich uranium-contaminated soils. Water Air Soil Pollut 107:147–162
30. Zhou P, Gu B (2005) Extraction of oxidized and reduced forms of uranium from contaminated soils: effects of carbonate concentration and pH. Environ Sci Technol 39(12):4435–4440. https://doi.org/10.1021/es0483443

31. D. Langmuir, "Uranium solution-mineral equilibria at low temperatures with applications to sedimentary ore deposits," 1978
32. Breit GN, Wanty RB (1991) Vanadium accumulation in carbonaceous rocks: a review of geochemical controls during deposition and diagenesis. Chem Geol 91:83–97
33. Vazquez GJ, Dodge CJ, Francis AJ (2007) Interactions of uranium with polyphosphate. Chemosphere 70(2):263–269. https://doi.org/10.1016/j.chemosphere.2007.06.019
34. Shvareva TY et al (2011) Thermodynamic characterization of boltwoodite and uranophane: enthalpy of formation and aqueous solubility. Geochim Cosmochim Acta 75(18):5269–5282. https://doi.org/10.1016/j.gca.2011.06.041
35. Central Ground Water Board (2024) Annual ground water quality report 2024. Accessed May 04, 2025. [Online]. Available: https://cgwb.gov.in/cgwbpnm/public/uploads/documents/17363272771910393216file.pdf
36. Kumar V et al (2023) Critical review on uranium and arsenic content and their chemical mobilization in groundwater: a case study of the Malwa region Punjab, India. Sci Total Environ. https://doi.org/10.1016/j.scitotenv.2023.163885
37. Chandrasekar T et al (2021) Potential interplay of Uranium with geochemical variables and mineral saturation states in groundwater. Appl Water Sci 11(4). https://doi.org/10.1007/s13201-021-01396-3
38. Papageorgiou F, McDermott F, Van Acken D (2022) Uranium in groundwaters: insights from the Leinster granite, SE Ireland. Appl Geochem 139. https://doi.org/10.1016/j.apgeochem.2022.105236
39. Saikia R, Chetia D, Bhattacharyya KG (2021) Estimation of uranium in groundwater and assessment of age-dependent radiation dose in Nalbari district of Assam, India. SN Appl Sci 3(1). https://doi.org/10.1007/s42452-020-04071-5
40. Kumar MP, Nagalakshmi K, Jayaraju N, Prasad TL, Lakshmanna B (2020) Deciphering water quality using WQI and GIS in Tummalapalle uranium mining area, Cuddapah Basin, India. Water Sci 34(1):65–74. https://doi.org/10.1080/11104929.2020.1765450
41. Richards LA, Kumar A, Shankar P, Gaurav A, Ghosh A, Polya DA (2020) Distribution and geochemical controls of arsenic and uranium in groundwater-derived drinking water in Bihar, India. Int J Environ Res Public Health 17(7). https://doi.org/10.3390/ijerph17072500
42. Diwan V, Sar SK, Biswas S, Dewangan R, Baghel T (2019) Uranium in ground water of Rajnandgaon District of Central India. J Radioanal Nucl Chem 321(1):293–302. https://doi.org/10.1007/s10967-019-06568-9
43. Sahu M, Sar SK, Dewangan R, Baghel T (2020) Health risk evaluation of uranium in groundwater of Bemetara district of Chhattisgarh state, India. Environ Dev Sustain 22(8):7619–7638. https://doi.org/10.1007/s10668-019-00539-6
44. Rohit M, Anamika K, Praveen M (2018) A comparative study of uranium concentration using two different analytical techniques and assessment of physicochemical parameters in groundwater. J Radiat Nucl Appl 3(3):149–156. https://doi.org/10.18576/jrna/030304
45. Balaram V, Rani A, Rathore DPS (2022) Review paper Uranium in groundwater in parts of India and world: a comprehensive review of sources, impact to the environment and human health, analytical techniques, and mitigation technologies. Geosyst Geoenviron 1:43. https://doi.org/10.1016/j.geogeo.2022.10
46. Coyte RM et al (2018) Large-scale uranium contamination of groundwater resources in India. Environ Sci Technol Lett 5(6):341–347. https://doi.org/10.1021/acs.estlett.8b00215
47. Chahal A et al (2019) Study of uranium in drinking water around the Sohna Fault Line in Haryana. J Geol Soc India 94(4):428–436. https://doi.org/10.1007/s12594-019-1332-4
48. Duggal V, Sharma S, Singh A (2021) Risk assessment of uranium in drinking water in Hisar district of Haryana, India. Water Sci Technol Water Supply 21(1):249–261. https://doi.org/10.2166/ws.2020.313
49. Bhardwaj S, Shukla DP, Halder A (2020) Spatial distribution of uranium and chemo-radiological assessment in Hamirpur district, Himachal Pradesh, India. J Radioanal Nucl Chem 324(2):467–480. https://doi.org/10.1007/s10967-020-07088-7

50. Kumari P et al (2021) Evaluation of uranium and other toxic heavy metals in drinking water of Chamba district, Himachal Pradesh, India for possible health hazards. Environ Earth Sci 80(7). https://doi.org/10.1007/s12665-021-09574-x
51. Rani A, Singh S, Duggal V, Balaram V (2013) Uraniumestimation in drinkingwater samples from some areas of Punjab and Himachal Pradesh, India using ICP-ms. Radiat Prot Dosim 157(1):146–151. https://doi.org/10.1093/rpd/nct112
52. Ajay K et al (2016) Quantification and assessment of health risk due to ingestion of uranium in groundwater of Jammu district, Jammu & Kashmir, India. J Radioanal Nucl Chem 310(2):793–804. https://doi.org/10.1007/s10967-016-4933-z
53. Patra AC et al (2013) Age-dependent dose and health risk due to intake of uranium in drinking water from Jaduguda, India. Radiat Prot Dosim 155(2):210–216. https://doi.org/10.1093/rpd/ncs328
54. Nagaiah N, Mathews G, Balakrishna KK, Rajanna A, Naregundi K (2013) Influence of physico-chemical parameters on the distribution of uranium in the ground water of Bangalore, India. Radiat Prot Environ 36(4):175. https://doi.org/10.4103/0972-0464.142389
55. Babu MNS, Somashekar RK, Kumar SA, Shivanna K, Krishnamurthy V, Eappen KP (2008) Concentration of uranium levels in groundwater. Int J Environ Sci Technol 5(2):263–266
56. Mathews G, Nagaiah N, Karthik Kumar MB, Ambika MR (2015) Radiological and chemical toxicity due to ingestion of uranium through drinking water in the environment of Bangalore, India. J Radiol Prot 35(2):447–455. https://doi.org/10.1088/0952-4746/35/2/447
57. Jindal MK et al (2024) High uranium dose from the groundwater in a granitic terrain in the eastern part of Karnataka, India. J Radioanal Nucl Chem 333(6):2905–2913. https://doi.org/10.1007/s10967-023-09053-6
58. Shalumon CS et al (2021) Analysis of uranium and other water quality parameters in drinking water sources of 5 districts of Kerala in southern India and potability estimation using water quality indexing method. HydroResearch 4:38–46. https://doi.org/10.1016/j.hydres.2021.04.003
59. Ben Byju S et al. (2012) Uranium in drinking water from the south coast districts of Kerala, India
60. Jose R, A S, S B, Pereira C, K V, John J (2011) Heterogeneity of uranium in water from south peninsular India. Int J Phys Fundam Res 1(1):11. https://doi.org/10.14331/ijfps.2011.330003. ISSN 2231-8186
61. Koliyar J, Patil P, Sahoo S, Patra AC (2021) Spatial distribution of uranium and its associated water quality parameters (drinking/ground water) in Raigad District of Maharashtra. In: Sustainable climate action and water management. Springer, Singapore, pp 125–130. https://doi.org/10.1007/978-981-15-8237-0_10
62. Kale AB, Bandela NN, Kulkarni JA (2018) Radiological and chemo-toxicological risk assessment of naturally occurred uranium in ground water from Aurangabad District of Maharashtra. https://doi.org/10.13140/RG.2.2.19688.83201
63. Saini K, Singh Bajwa B (2016) Uranium distribution study in the drinking water samples of SW Punjab, India. Pelagia Res Lib Adv Appl Sci Res 7(2):103–108. [Online]. Available: www.pelagiaresearchlibrary.com
64. Singh S, Rani A, Mahajan RK, Walia TPS (2003) Analysis of uranium and its correlation with some physico-chemical properties of drinking water samples from Amritsar, Punjab. J Environ Monit 5(6):917–921. https://doi.org/10.1039/b309493f
65. Sharma DA et al (2017) Distribution of uranium in groundwaters of Bathinda and Mansa districts of Punjab, India: inferences from an isotope hydrochemical study. J Radioanal Nucl Chem 313(3):625–633. https://doi.org/10.1007/s10967-017-5288-9
66. Singh H, Singh J, Singh S, Bajwa BS (2009) Uranium concentration in drinking water samples using the SSNTDs. Indian J Phys:1039–1044. https://doi.org/10.1007/s12648-009-0065-4
67. Sharma N, Singh J (2016) Radiological and chemical risk assessment due to high uranium contents observed in the ground waters of Mansa District (Malwa Region) of Punjab State, India: an area of high cancer incidence. Expo Health 8(4):513–525. https://doi.org/10.1007/s12403-016-0215-9

68. Rishi MS, Keesari T, Sharma DA, Pant D, Sinha UK (2017) Spatial trends in uranium distribution in groundwaters of Southwest Punjab, India - a hydrochemical perspective. J Radioanal Nucl Chem 311(3):1937–1945. https://doi.org/10.1007/s10967-017-5178-1
69. Sharma DA et al (2020) Distribution and correlation of radon and uranium and associated hydrogeochemical processes in alluvial aquifers of Northwest India. Environ Sci Pollut Res 27(31):38901–38915. https://doi.org/10.1007/s11356-020-10015-8
70. Singh KP, Kishore N, Tuli N, Loyal RS, Kaur M, Taak JK (2018) Uranium contamination of groundwater in southwest parts of Punjab State, India, with special reference to role of basement granite. In: Clean and sustainable groundwater in India. Springer, Singapore, pp 95–106. https://doi.org/10.1007/978-981-10-4552-3_7
71. Pandit P et al (2022) Tracing geochemical sources and health risk assessment of uranium in groundwater of arid zone of India. Sci Rep 12(1). https://doi.org/10.1038/s41598-022-05770-2
72. Mittal S, Rani A, Mehra R, Balaram V, Satyanarayanan M, Sawant SS (2017) Assessment of uranium in correlation with physico-chemical properties of drinking water of northern Rajasthan. J Geol Soc India 90(2):233–238. https://doi.org/10.1007/s12594-017-0704-x
73. Duggal V, Rani A, Balaram V (2016) Assessment of age-dependent radiation dose due to intake of uranium and thorium in drinking water from Sikar district, Rajasthan, India. Radiat Prot Dosim 171(2):257–261. https://doi.org/10.1093/rpd/ncw070
74. Rani A, Mehra R, Duggal V, Balaram V (2013) Analysis of uranium concentration in drinking water samples using ICPMS. Health Phys 104(3):251–255. https://doi.org/10.1097/HP.0b013e318279ba05
75. Adithya VS, Chidambaram S, Keesari T, Mohokar HV, Prasanna MV (2019) Occurrence of uranium in groundwater along the lithological contacts in Central Tamilnadu, India: an isotope hydrogeochemical perspective. Expo Health 11(4):277–290. https://doi.org/10.1007/s12403-017-0269-3
76. Raja V, Sahoo SK, Sreekumar K, Neelakantan MA (2021) High background radiation places and spatial distribution of uranium in groundwater of monazite placer deposit in Kanniyakumari district, Tamil Nadu, India. J Radioanal Nucl Chem 328(3):925–939. https://doi.org/10.1007/s10967-021-07727-7
77. Satyanarayanan M et al (2007) Assessment of groundwater quality in a structurally deformed granitic terrain in Hyderabad, India. Environ Monit Assess 131(1–3):117–127. https://doi.org/10.1007/s10661-006-9461-9
78. Prakash R, Bhartariya KG, Singh S, Singh K, Rajak M, Kaushik YB (2020) Uranium and its correlation with other Geogenic contaminants in ground water of Ganga Yamuna Doab, Fatehpur District, Uttar Pradesh, India. J Geol Soc India 95(4):359–365. https://doi.org/10.1007/s12594-020-1444-x
79. Patni K, Pande C, Pande AP, Tewari G, Joshi T (2020) Distribution of naturally occurring uranium and other heavy toxic elements in selected spring water samples of Pithoragarh District, Uttarakhand, India. SN Appl Sci 2(12). https://doi.org/10.1007/s42452-020-03947-w
80. Ramola RC, Singh S, Virk HS (1988) Uranium and radon estimation in some water samples from Himalayas. Int J Radiat Appl Instrument D Nucl Track Radiat Measur 15:791–793
81. Prasad M, Ranga V, Kumar GA, Ramola RC (2020) Radiological impact assessment of soil and groundwater of Himalayan regions in Uttarakhand, India. J Radioanal Nucl Chem 323(3):1269–1282. https://doi.org/10.1007/s10967-019-06827-9
82. Das A, Das SS, Chowdhury NR, Joardar M, Ghosh B, Roychowdhury T (2020) Quality and health risk evaluation for groundwater in Nadia district, West Bengal: an approach on its suitability for drinking and domestic purpose. Groundw Sustain Dev 10. https://doi.org/10.1016/j.gsd.2020.100351

An Overview on the Effect of Heavy Metals, Microorganisms, and Radioactive Elements on Uranium Solubility

Mohammad Hassan Ehrampoush, Hossein Arfaeinia, Farshad Hamidi, Danial Nayeri, Fatemeh Mortezazadeh, Mahdi Ahmadi Nasab, and Niloufar Borhani Yazdi

Abstract This chapter explores the complex influences of heavy metals (HMs), microorganisms, and other radioactive elements on uranium solubility in aqueous environments. The presence of HMs such as lead (Pb), cadmium (Cd), and arsenic (As) can significantly alter uranium speciation and solubility through competitive adsorption, complexation, and co-precipitation processes. The role of microorganisms is examined, with a focus on microbial oxidation and reduction of uranium, biofilm formation, and the impact of microbial metabolites on uranium mobility. This chapter also addresses the synergistic and antagonistic interactions between uranium and other radioactive elements, such as thorium and radium, in natural and contaminated water systems. In addition, integrated case studies provide useful insights into the multifaceted interactions in real-world scenarios, highlighting the challenges in predicting uranium behavior in the presence of multiple contaminants. Advanced analytical techniques, including spectroscopic and microscopic methods, are discussed for their utility in elucidating these interactions. The combined health and environmental risks posed by complex contaminant mixtures are assessed, with

M. H. Ehrampoush
Department of Environmental Health Engineering, School of Public Health, Shahid Sadoughi University of Medical Science, Yazd, Iran

H. Arfaeinia
Addiction and Lifestyle Research Center, Bushehr University of Medical Sciences, Bushehr, Iran

F. Hamidi · F. Mortezazadeh · M. A. Nasab · N. B. Yazdi (✉)
Department of Environmental Health Engineering, School of Public Health, Tehran University of Medical Sciences, Tehran, Iran

D. Nayeri
Department of Environmental Health Engineering, School of Public Health, Kermanshah University of Medical Sciences, Kermanshah, Iran

Student Research Committee, Kermanshah University of Medical Sciences, Kermanshah, Iran

M. K. Jindal, K. Patni (eds.), *Water Quality-Uranium Solubility Nexus: Unlocking the Secrets of Uranium Solubility*, Earth and Environmental Sciences Library, https://doi.org/10.1007/978-3-031-98909-4_11

an emphasis on developing robust risk assessment frameworks. Future research directions are proposed to address the emerging challenges in understanding and managing uranium solubility in the context of multi-contaminant environments.

Keywords Heavy metals · Microorganisms · Radioactive elements · Uranium solubility

1 Introduction

Uranium (U) ranks as the 49th most abundant element in the Earth's crust and is comparable in abundance to beryllium, tin, and arsenic. It is also fairly common in seawater, where it is ranked 26th, alongside arsenic and vanadium. This dense metal, with a density of 19.1 kg/m^3, is similar to gold [1]. U is a radioactive element that naturally occurs and is categorized as an actinide alongside thorium (Th) and synthetic elements like plutonium (Pu) in the periodic table. It has an atomic number and an atomic mass of 92 and 238, respectively. Known for its toxicity, U can sometimes be present in high concentrations in drinking water and groundwater used for irrigation. It is typically found only in water that contains oxygen [2, 3].

In environmental systems, uranium primarily exists in the +IV and + VI oxidation states. U(IV) is mostly found as the mineral uraninite (UO_2), which remains stable in reducing environments. Other forms of U(IV) can also be present in these conditions. Due to the low solubility of uraninite, remediation efforts often aim to convert U(VI) to U(IV) biologically. However, under oxidizing conditions, U(IV) can revert to the more soluble U(VI). Consequently, the presence of oxygen at many contaminated sites can reduce the long-term success of bioreduction-based cleanup approaches [4]. Uranium limits are set mainly based on chemical toxicity rather than radiological concerns. The European Union, United States Environmental Protection Agency, and World Health Organization established a maximum U concentration of 30 µg/L, which corresponds to about 0.78 Bq/L of radioactivity. Radiological calculations indicate that the WHO's annual dose limit of 0.1 mSv would be exceeded only if U levels exceeded 100 µg/L. For radioactivity, the WHO recommends limits of 1 Bq/L for ^{234}U and 10 Bq/L for ^{238}U in drinking water [1]. Uranium oxides (UO_2) can be found in U mining sites and metallic U production facilities, particularly as a byproduct of the use of depleted U ammunition in the military. When metallic U ignites at 500 °C, it forms UO_2 and U_2O_3. The proportion of metallic U that turns into UO_2 can vary widely, from as little as 2% to as much as 100%. U and its oxides dissolve in nitric, hydrochloric, and sulfuric acids, but they are only slightly soluble in water. When U oxide particles settle on the ground, they dissolve in water, move through the soil, and can be taken up by plants [5]. Uranium is relatively mobile near the surface because U(VI) complexes are stable in water.

However, it can be precipitated as U(IV) or as various U minerals, such as phosphates and silicates. The release of U into water depends on the solubility and dissolution rates of these minerals, which vary with pH and water composition. Therefore, determining the solubility of U compounds is key to evaluating their environmental movement [6]. U typically exists in two oxidation states: U^{4+} and U^{6+}. At room temperature, uranyl compounds (U^{6+}) are soluble and form stable species in water. Conversely, uranous compounds (U^{4+}) have very low solubilities in water under standard conditions (25 °C and 1 bar). Therefore, in aqueous systems at lower temperatures, U is predominantly found in the U(VI) form [7].

Understanding the factors that influence U solubility is crucial for assessing its mobility, distribution, and potential risks in various environments. This knowledge is essential for developing effective remediation strategies and mitigating the adverse effects associated with U contamination. U poses environmental and health risks due to its toxic and radioactive properties. It is found in high concentrations in certain rocks, like black shales and phosphorites, and can contaminate groundwater and soils [8]. As U mining and usage increase, so does environmental contamination. Remediation techniques include physical methods for small water areas, chemical methods for efficient U removal (although mostly experimental), and biological methods such as phytoremediation for larger, low-concentration sites. Combining these methods based on environmental conditions is essential for effective cleanup [9].

Birke et al. (2009) highlighted that U solubility and concentration are influenced by several factors. Geological formations, such as the Bunter and Keuper strata, are significant sources of U. Agricultural practices, especially the use of phosphate fertilizers, increase U levels in loess soils. Historical glacial activity has contributed to the redistribution of U, whereas old mining districts display high U concentrations due to past extraction. These factors collectively affect U levels in surface waters, illustrating their impact on U solubility and distribution [10].

Bacteria engage with U through various processes, such as biosorption on cell surfaces, precipitation, and redox reactions (oxidation and reduction) [11]. Biological reduction is commonly suggested as a technique for extracting U from groundwater through its reductive precipitation. Various microorganisms, such as iron- and sulfate-reducing bacteria and occasionally denitrifying bacteria, can reduce dissolved U [12].

On the other hand, Duff et al. (1999) revealed that microorganisms, particularly native algae (Chlorella) and bacteria, play crucial roles in the solubility of U. Their study investigated the effects of these algae and reducing agents on U redox chemistry in artificial lake sediments. The results revealed that biosorption by algae and bacteria is the dominant mechanism for U deposition in sediments. Additionally, U reduction to U(IV) was observed under certain conditions, including in the presence of algae and reducing agents. Overall, this study highlights that microbial activities can significantly influence U solubility [13].

2 Impact of HMs on Uranium Solubility

The solubility of U in water is influenced by several important factors. Redox conditions are vital, as U is soluble in the U(VI) state under oxidizing conditions but becomes insoluble as U(IV) in reducing environments. The pH of the water also impacts U solubility, with greater solubility in near-neutral to alkaline conditions. Additionally, the interactions of alkalinity and hardness with U ions affect its solubility. The behavior of U in solution is influenced by competing ligands, including carbonates, humic substances, and metal oxides. In particular, uranium–carbonate complexes, such as U–CO_3, play a major role in sustaining U mobility in aquatic systems [1]. Therefore, heavy metals can significantly impact U solubility through various interaction mechanisms. For instance, the presence of certain heavy metals might shift redox conditions or modify the pH, thus affecting U solubility [14–16]. Banzi et al. (2015) reported that elevated levels of specific heavy metals and changes in the TDS ratio, particularly in the dry season, can signal potential pollution caused by U mining [17]. Additionally, heavy metals can affect solids in water, such as sediments [18]. This effect can lead to a change in the absorption or excretion of U [19].

Research into the influence of heavy metals on U solubility has provided valuable insights into their interactions. The following examples illustrate various aspects of these effects. Research by Liu et al. (2020) indirectly indicates that heavy metals might influence U solubility, with this effect being analyzed through variations in the solubilities and behaviors of these metals under different pH levels of simulated rain. Their study explored how simulated rain at various pH levels affects the movement and transformation of manganese (Mn), strontium (Sr), lead (Pb), and U in U tailings. They employed semi-dynamic leaching tests and sequential extraction to analyze the leaching behavior and speciation of these metals. The findings showed that as the pH of the rain decreased, the leachability of Mn, Sr, Pb, and U increased. The metals initially release rapidly but then reach a relatively stable state. After leaching with simulated rain, the concentrations of metal-exchangeable and metal-carbonate forms in the tailings significantly decrease. The leaching of Mn and U is driven mainly by wash-off processes, whereas Sr and Pb leaching is controlled primarily by dissolution mechanisms [20]. Additionally, Li et al. (2019) investigated the use of Nymphaea tetragona Georgi (N. tetragona) for the remediation of water contaminated with U and uranium-accompanied heavy metals (UAHMs). This study explored how these heavy metals influence the enrichment of U in plants. These results indicate that the simultaneous presence of U and UAHMs tends to decrease the amount of U accumulated in the whole plant. Under co-contamination conditions, manganese (Mn) and mercury (Hg) can increase the enrichment of U from water in N. tetragona, whereas lead (Pb) and arsenic (As) generally inhibit it. Overall, the impact of heavy metals on U enrichment is examined in a direct manner and through complex interactions [21]. Sitte et al. (2010) investigated the effects of sulfate-reducing bacteria (SRB) on metal mobility in groundwater-affected soils from a former U mining area in Ronneburg, Germany.

They reported that SRB activity, measured via the $^{35}SO_4^{2-}$ radiotracer method, was limited to reduced soil horizons, where heavy metals were mostly retained in solid phases, and pore water concentrations were low. While SRB activity led to reduced concentrations of soluble metals such as nickel, cobalt, and zinc, it unexpectedly increased the solubility of U, which reached 1407 nM in carbon-amended soils. This increase in U solubility suggests that U behavior in contaminated soils can be unpredictable, potentially leading to downstream contamination of surrounding ecosystems [22].

3 Role of Microorganisms

3.1 Microbial Reduction and Oxidation: Mechanisms

Because uranium accumulation can pose a significant risk to ecosystems and humans, strict control, monitoring, and remediation and protection measures are necessary. There are several different physical and chemical techniques, some of which are more sophisticated than others, but each has its own restrictions and disadvantages, such as high cost, high complexity, and lengthy processing times [23]. As a result, simpler and more environmentally friendly alternatives—such as biologically based techniques—are needed. In uranium-contaminated sites, microorganisms are frequently present and can affect the mobility, toxicity, and distribution of uranium [24]. Various well-known processes, such as biosorption, bioaccumulation, biomineralization and redox transformations, are responsible for interactions with uranium (Fig. 1). In turn, uranium exerts permanent pressure on the prevailing microbial population, disrupting microbial communities and processes [11].

The release of metallic pollutants can be controlled by a range of microorganisms as a result of their small size and high surface area, which is the main reason

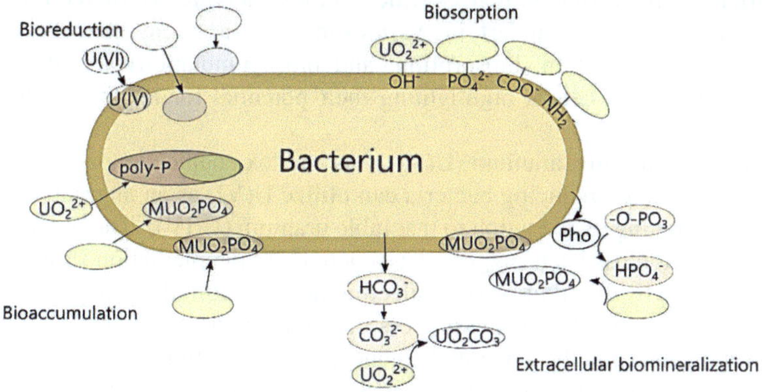

Fig. 1 Bacterial interactions with uranium [11]

for the occupation of a large portion of the reactive site to interact with dissolved molecules and nanominerals that can be released into the environment [25, 26]. A key process for uranium immobilization in aquatic sediments and the creation of some uranium ores is the reduction of the soluble, oxidized form of uranium, U(VI), to insoluble U(IV). U(VI) reduction is typically thought of as a biological reaction involving organic compounds, sulfur dioxide, or hydrogen molecules as reductants [27]. Research is currently being conducted at various Department of Energy field sites on the stimulation of native dissimilatory metal-reducing bacteria (DMRB) for uranium remediation. Evaluating the success of a single tactic (e.g., changes in the aqueous geochemistry detected in monitoring wells) is usually the basis for ethanol addition. Owing to the complicated range of redox reactions that may occur in these subsurface settings, interpreting these data is frequently difficult. For example, adding an electron donor will assist in reducing conditions but has several electron acceptors (e.g., G. either promotes or obstructs U(VI) reduction, such as nitrate, Mn(III/IV) oxides, Fe(III) oxides, or sulfate). Similarly, U(VI) reduction and U(IV) reoxidation can also be influenced by the concentration and flux of electron donor addition [28]. In an anoxic environment, U(VI) can precipitate from groundwater by reducing to sparingly soluble U(IV) minerals. The remediation of uranium-contaminated sites in situ may benefit from the bacterially mediated reduction of U(VI) to uraninite [29]. Generally, uranium mobility has been regulated throughout Earth's history by redox reactions between soluble hexavalent uranium [U(VI)] and insoluble tetravalent uranium [U(IV)]. Reduced conditions lead to the reduction of U(VI) to U(IV). Following that, uranium precipitates as minerals, frequently uraninite (UO_2). In the presence of oxidizing agents such as oxygen, Fe(III), and Mn(IV), the oxidative dissolution of U(IV) minerals occurs quickly. These are the largest current global sinks for dissolved uranium, formed as a result of the reduction in U(VI) in organic-rich sediments that gave rise to economically significant uranium ore deposits. In natural organic-rich sediments, U(VI) reduction is important, but the essential pathways for this process are still unknown [30]. There is still some debate regarding the reduction of electron acceptors such as U(VI), Fe(III), nitrate, and sulfate in accordance with their redox potential. If not, more information needs to be provided about what influences which of these ions are prioritized for electron transfer. The mechanisms underlying various microbe–uranium interactions (bioreduction, biomineralization, biosorption, and bioaccumulation) are illustrated in Fig. 2 and discussed below, highlighting their potential for facilitating long-term uranium removal [31].

At circumneutral pH, uranium (U(VI)) has a redox couple comparable to that of iron (Fe(III)). Fe(III)-reducing bacteria can utilize U(VI) as an alternative electron acceptor, facilitating its reduction to insoluble uranium (U(IV)). The process of uranium bioreduction has been suggested as a potential bioremediation strategy, which can be enhanced by the addition of an electron donor to stimulate enzymatic activity [31, 32]. Biomineralization is the process through which metals precipitate in conjunction with microbially generated ligands, such as sulfide or phosphate, or as carbonates or hydroxides, in response to localized alkaline conditions at the cell

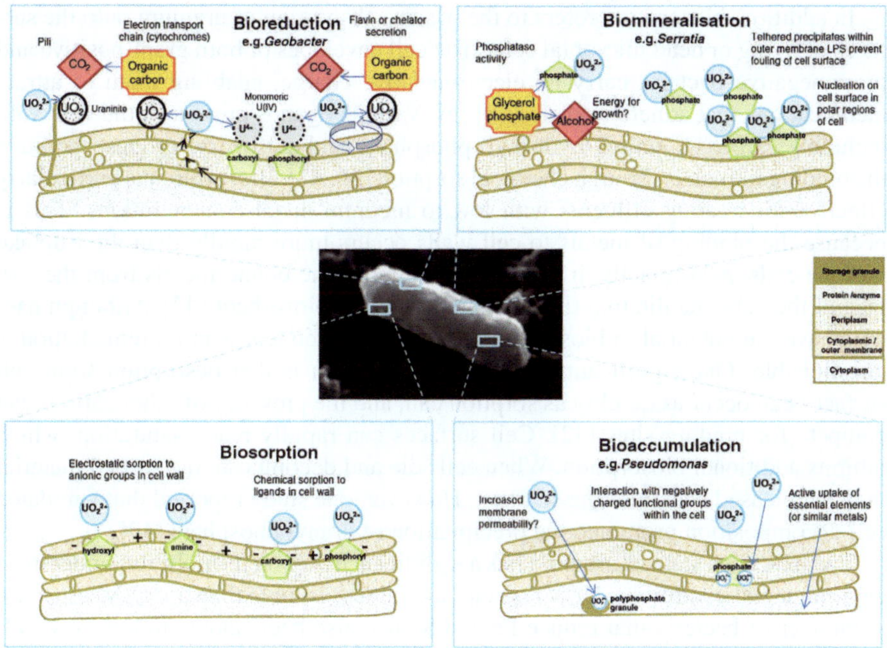

Fig. 2 Schematic illustrating the mechanisms of microbe–uranium interactions [32]

surface. The biomineralization of uranium has been demonstrated via the use of a species of Citrobacter, which has since been reclassified as a species of Serratia [33, 34]. When glycerol phosphate is provided, the cell phosphatase activity cleaves the organic phosphate, releasing inorganic phosphate that precipitates with U(VI) to form extracellular hydrogen uranyl phosphate minerals [HUO_2PO_4]. This phenomenon has also been demonstrated via an environmental isolate from the US DOE Oak Ridge site, as well as a Pseudomonas species supplied with a tributyl phosphate donor. Observations of microbial cells completely enveloped in uranium phosphate minerals in uraniferous soils suggest that bacterial biomineralization occurs naturally within this system [35–37]. A potential issue with biomineralization is that the rapid precipitation of metals on the cell surface could theoretically form a barrier to cellular metabolism, although this phenomenon has not been directly observed. A recent review emphasized the contradiction between certain studies that propose biomineralization as a mechanism for resistance to toxicity and others that assume it is harmful to cells [38, 39]. Microbial cells can accumulate a wide variety of metal ions through “bioaccumulation.” In the case of certain metals, unintentional uptake may occur because these transported metals resemble essential elements required for cellular function, leading to their active absorption into the cell. Although uranium has no recognized biological role, it has been proposed that its uptake by cells may occur due to increased membrane permeability, which can be induced by uranium toxicity, for example [40].

In addition, biosorption refers to the passive absorption of uranium onto the surfaces of living or dead microbial cells. The cell envelopes of both gram-positive and gram-negative bacteria carry an electronegative charge, enabling them to attract metal cations that adhere to their surfaces. Various ligands present in the cell wall, including carboxyl, amine, hydroxyl, phosphate, and sulfhydryl groups, facilitate the binding of metals through chemical sorption [38, 41]. Biosorption is particularly effective for treating effluents with low to medium metal concentrations. This is because the binding of metals to cell walls occurs more rapidly than their uptake into the cells. Additionally, it is simpler to remove the bound metals from the surface of the cells, facilitating the regeneration of the biosorbent [42]. Although bacteria have the potential to biosorb uranium, their effectiveness in bioremediation is questionable. One significant issue with biosorption is that desorption from cell surfaces can occur as quickly as sorption can, and the presence of other cations can compete for binding sites [42]. Cell surfaces can rapidly reach saturation, which inhibits additional biosorption. When cells die and decompose, the sorbed material may be released back into the solution. However, one study reported that simulated cell decomposition promoted the precipitation of uranyl phosphate [43].

The oxidation state of uranium has a significant effect on its mobility; under most environmental conditions, U(IV) species are far less soluble than are U(VI) species. A number of bacteria that reduce Fe and S have also been shown to reduce U(VI) enzymatically and indirectly [44]. Numerous prokaryotic species are able to reduce U(VI) to U(IV) [45]. U(VI) can be reduced to insoluble U(IV) minerals by the most prevalent Fe(III)-reducing bacteria by using it as an alternate electron acceptor. Sulfate-reducing bacteria (SRB) can also convert U(VI) to U(IV) [46]. Fe(III) reducers are associated with *Geobacter uraniireducens* and *Geobacter daltonii*, which reduce uranium, whereas members of e are known to be sulfate reducers. *G. Desulfotomaculum*, *Desulfovibrio*, and *Desulfobacterium* genera [47]. Because nitrate and other metal ions, which also promote anaerobic respiration, are common electron acceptors in natural environments, identifying the specific microbes causing U(VI) reduction is challenging. Nonetheless, the most prevalent microorganisms in the environment are frequently credited with contributing to U(VI) reduction [48]. U(VI)-reducing bacteria are the basis for biochemistry. PCR amplification of the 16S rRNA gene, high-throughput sequencing and phospholipid fatty acid (PLFA) analysis revealed that U(VI)-reducing bacteria belong to three main phyla, Proteobacteria and Firmicutes, and are associated with actinomists [49, 50]. In addition, Proteobacteria, which contains many U(VI)-reducing genera and plays a pivotal role in curtailing the toxicity and bioavailability of uranium, is the main phylum commonly found in U(VI)-contaminated sites. For example, Shewanella from γ-Proteobacteria, Desulfovibrio and Geobacter from δ-Proteobacteria, which can survive in oligotrophic environments, function under low water activity conditions, adapt to U(VI) reduction metabolism, exhibit metal resistance, and play a critical role in uranium immobilization and remediation, are commonly detected in radionuclide-contaminated sediments [51–53].

4 Influence of Other Radioactive Elements

Uranium often coexists with other radioactive elements in geological formations, particularly with elements from its decay chain, such as radium, thorium, and polonium. The presence of these cooccurring elements influences the geochemical behavior of uranium, notably its solubility, mobility, and bioavailability in the environment. This section examines the nature of the interactions among uranium and these elements, focusing on synergistic and antagonistic effects and their implications for uranium mobility and solubility in various environmental settings.

4.1 Co-occurrence with Uranium

Recent studies examining the impact of simultaneous exposure to various contaminants—referred to as the "mixture effect"—have highlighted that such combined exposures can lead to heightened toxicity [54]. Drinking water is a key pathway through which people are exposed to various chemicals each day. As the global reliance on groundwater as a source of drinking water continues to grow, geogenic contaminants present in groundwater are becoming a more significant concern for human health. These naturally occurring contaminants, released through interactions between aquifer rocks and groundwater, are especially susceptible to co-occurrence, posing potential combined exposure risks [55]. Uranium is highly soluble in oxidizing groundwater as hexavalent uranium (U(VI)), where it commonly forms carbonate complexes (Fig. 3) [56]. The occurrence of sorption/desorption processes, speciation, and mineral dissolution may also influence the occurrence of uranium in groundwater. Even under oxidizing conditions, uranium can adsorb to mineral surfaces, particularly near neutral pH levels, in the absence of strong complexing agents. However, strong complexing agents, such as carbonate or organic ligands, can significantly reduce uranium adsorption [56, 57]. Groundwater containing uranium is most frequently found in oxic conditions, particularly in granitic and granitic-derived sedimentary aquifers, as well as in arid to semiarid closed-basin systems [58, 59]. In contrast, reduced tetravalent uranium (U(IV)) has low solubility, which results in lower uranium concentrations in suboxic groundwater.

Uranium is rarely found in isolation and is commonly associated with elements such as radium and thorium in both groundwater and sedimentary deposits [56, 59]. These elements, which are part of the uranium and thorium decay series, can significantly influence uranium solubility because of their similar chemical properties and their capacity to participate in complexation and competitive sorption [60]. For example, radium, as a decay product of uranium, often accumulates in uranium-rich sites and shares similar geochemical behaviors, influencing uranium solubility by affecting pH and competitive ion binding [60].

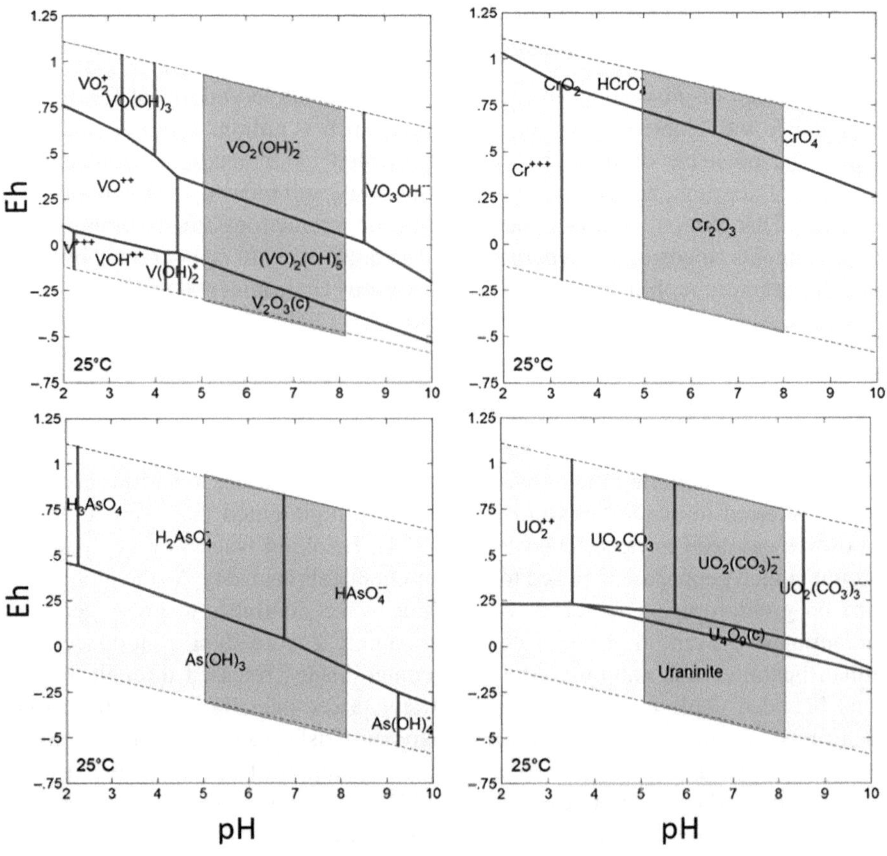

Fig. 3 Simplified Eh-pH diagrams for V, Cr, As, and U, calculated with element concentrations of 10^{-6} M and HCO^{-3} activities of 10^{-2} [56]

Thorium, another naturally occurring radioactive element, frequently coexists with uranium in mineral deposits. Thorium and uranium often compete for the same ligands in soil and water systems, especially anionic ligands such as sulfate and carbonate, affecting the environmental distribution of uranium [61]. Under acidic conditions, these coexisting radionuclides can form soluble complexes that increase uranium bioavailability and mobility in groundwater, contributing to environmental contamination concerns [60, 61].

4.2 *Synergistic and Antagonistic Effects*

The interactions among uranium and other radioactive elements can produce either synergistic or antagonistic effects, depending on factors such as chemical concentration, pH, and redox conditions in the environment.

Synergistic effects occur when the copresence of another radioactive element increases uranium solubility and mobility. Radium ions, for example, enhance

Table 1 Summary of Interactions between Uranium and Cooccurring Radioactive Elements

Co-occurring Element	Typical Environment	Interaction Type	Mechanism	Effect on Uranium Solubility
Thorium	Sediment, groundwater	Antagonistic	Competes for sorption sites	Decreases uranium solubility
Radium	Mining sites, groundwater	Synergistic/ Antagonistic	Alters redox, competes for sorption	Variable; may enhance or reduce
Polonium	Reducing environments	Synergistic	Stabilizes uranium complexes	Increases bioavailability in specific conditions

uranium solubility by influencing ion exchange reactions and competitive sorption. This effect has been observed in environments where radium and uranium compete for mineral surface sites, which can increase the mobility of both elements in groundwater [62]. Furthermore, the strong affinity of thorium for anions can promote uranium desorption, which has been shown to increase uranium mobility and bioavailability under certain environmental conditions [63]. The influence of polonium, although less frequently studied, can also affect uranium solubility, especially in reducing environments where polonium interactions with uranium ions lead to the stabilization of uranium species [63]. This increased stability of uranium complexes under specific redox conditions can increase uranium bioavailability and environmental persistence, especially in aquatic systems [62].

Conversely, antagonistic interactions among uranium and other radioactive elements can limit uranium solubility or restrict its mobility. Thorium, for example, competes with uranium for adsorption sites on mineral surfaces, which can reduce uranium solubility and immobilize it in soil and sediment matrices [60, 62]. Under certain pH and ionic strength conditions, radium ions may also contribute to uranium immobilization by forming coprecipitated minerals or insoluble compounds, thus reducing uranium mobility [60].

Another notable effect is observed when sulfate ions are present alongside uranium and radium; this combination can lead to the precipitation of insoluble uranium sulfate compounds, which significantly reduces uranium solubility and transport [60]. Table 1 shows a summary of the interactions between uranium and cooccurring radioactive elements.

5 Integrated Case Studies

5.1 Hanford Site, Washington State, USA

The Hanford site is one of the most contaminated nuclear sites in the U.S., where uranium was processed during the Manhattan Project. Heavy metals such as chromium and lead are present alongside uranium. Microbial communities adapted to these conditions influence uranium solubility. Research has indicated that microbial reduction of uranium(VI) to uranium(IV) leads to decreased solubility and increased precipitation of uraninite, demonstrating bioremediation potential [64].

5.2 The Chernobyl Exclusion Zone, Ukraine

Following the Chernobyl disaster, significant amounts of radioactive elements, including uranium, were released into the environment. The presence of heavy metals and radionuclides influences microbial community structures and their metabolic pathways. Studies have shown that certain bacteria can sequester uranium and other heavy metals, affecting their mobility and solubility in contaminated soils [65].

5.3 Lessons Learned from Field Studies

Field studies have shown that microorganisms play crucial roles in the biogeochemical cycling of uranium and its solubility in contaminated environments. In the case of the Hanford site, the presence of specific microbial strains that can reduce uranium facilitated uranium immobilization, emphasizing the potential for bioremediation techniques [66].

The interactions among heavy metals, radionuclides, and environmental conditions are complex and site specific. In the Chernobyl exclusion zone, varying soil types and organic matter contents influence the behavior of uranium and its solubility, highlighting the need for site-specific assessments in remediation efforts [67].

6 Analytical Methods and Techniques

6.1 Mass Spectrometry (MS)

A powerful analytical technique used to measure the mass–charge ratio of ions. Identifying and quantifying heavy metals, organic contaminants, and isotopes of uranium in environmental samples. High sensitivity, the ability to analyze complex mixtures, and the ability to provide structural information [68, 69] exist.

6.2 Gas Chromatography–Mass Spectrometry (GC–MS)

Gas chromatography can be used to separate volatile compounds, and mass spectrometry can be used for identification. This method is useful for analyzing organic pollutants and their interactions with heavy metals in environmental matrices. High resolution and specificity enable the detection of trace-level contaminants [70, 71].

6.3 *Inductively Coupled Plasma–Mass Spectrometry (ICP–MS)*

A technique for detecting metals and several nonmetals at trace levels. Heavy metals and radionuclides such as uranium in soil and water samples were quantified. Extremely low detection limits and the ability to analyze multiple elements simultaneously exist [72, 73].

6.4 *X-Ray Fluorescence (XRF)*

An analytical technique used to determine the elemental composition of materials. Nondestructive analysis of heavy metal contamination in soils and sediments. Rapid analysis, minimal sample preparation, and the ability to analyze solid samples directly [74, 75].

6.5 *Nuclear Magnetic Resonance (NMR) Spectroscopy*

A technique that exploits the magnetic properties of certain nuclei to provide information about molecular structure. The interactions of contaminants with organic matter and microbial metabolites should be studied. Provides detailed structural information and can be used for both qualitative and quantitative analysis [76, 77].

6.6 *Advances in Analytical Chemistry*

Development of smaller, portable analytical devices that can be used in the field. Handheld the XRF analyzers and portable GC–MS systems. This method allows for onsite analysis, reducing the need for extensive sample transport and enabling real-time monitoring of contamination [78]. Innovations in spectroscopic methods such as Raman spectroscopy and infrared spectroscopy. This method is used for characterizing contaminants, understanding interactions at the molecular level, and monitoring changes in soil chemistry. Nondestructive analysis, rapid results, and the ability to analyze samples without extensive preparation [79, 80]. Techniques that allow for the rapid analysis of multiple samples simultaneously. This method is useful in environmental monitoring programs for assessing large numbers of samples for contaminants. This increases efficiency and reduces costs in environmental assessments [81]. Integration of data analytics and machine learning algorithms to interpret complex datasets generated via analytical techniques. To predict contaminant behavior, understand microbial interactions, and optimize remediation strategies. The ability to make sense of large datasets can lead to more informed decision-making [82]. The development of environmentally friendly analytical methods that minimize waste and reduce hazardous substances. Use of alternative solvents, energy-efficient processes, and biodegradable materials in sample preparation and analysis. Sustainability in analytical chemistry practices should be promoted while maintaining accuracy and reliability [77].

7 Health and Environmental Risks

The simultaneous presence of multiple contaminants in the environment poses significant health and environmental risks because of their potential to interact in synergistic, antagonistic, or additive ways. This phenomenon, referred to as the "mixture effect," can amplify or mitigate the toxicity of individual contaminants, resulting in unpredictable outcomes for ecosystems and human health. For example, heavy metals such as arsenic and lead, when cooccurring with organic pollutants such as polychlorinated biphenyls (PCBs), have been observed to exhibit enhanced toxicological impacts due to their combined oxidative stress mechanisms and interference with cellular pathways [83]. Geogenic contaminants, including uranium and radium, often cooccur in groundwater systems, compounding the challenges in understanding their combined toxicological impacts. These interactions can alter their bioavailability and mobility, leading to greater or lesser uptake by organisms depending on the geochemical environment. These combined effects necessitate more complex risk assessments to accurately evaluate potential health outcomes (Table 2) [84]. Assessing the risks associated with multiple contaminants requires integrated methodologies that account for their interactions. Traditional risk assessment models, which typically evaluate contaminants in isolation, may underestimate or overestimate risks when applied to mixtures. Modern approaches incorporate advanced toxicological and computational tools, such as:

Table 2 Examples of Combined Effects and Associated Risks

Contaminants	Interaction Type	Environmental Impact	Health Risks
Uranium and Radium	Synergistic	Increased mobility in oxidizing conditions	Elevated risks of kidney and bone cancer
Lead and PCBs	Additive	Enhanced bioaccumulation in aquatic organisms	Neurological damage and developmental issue
Arsenic and Fluoride	Antagonistic	Reduced arsenic toxicity in high fluoride levels	Variable effects on bone mineralization
Mercury and Cadmium	Synergistic	Amplified oxidative stress in soils	Increased renal and hepatic toxicity

1. **Cumulative risk assessment (CRA):** A framework that evaluates the combined health risks posed by multiple contaminants, considering factors such as dose addition and chemical interactions.
2. **High-throughput screening (HTS):** This method enables the rapid identification of potential toxic effects of contaminant mixtures through in vitro and computational modeling techniques.
3. **Quantitative structure–activity relationship (QSAR):** Models that predict the toxicity of chemical mixtures on the basis of their molecular structures and interactions.
4. **Probabilistic risk assessment (PRA):** PRA incorporates variability and uncertainty in exposure and toxicological data to provide more robust risk estimates [85, 86].

8 Future Research Directions

To date, uranium has been identified as a trigger for toxicity through mechanisms such as oxidative stress, genetic damage, protein dysfunction, inflammation, and metabolic disorders, leading to cellular damage and cell death. However, our understanding of the health risks and underlying toxicological mechanisms associated with uranium contamination, as well as its chemotoxicity, remains limited, indicating that many areas require further investigation [87]. Undoubtedly, as research has investigated the toxicity and epidemiology of uranium in greater depth, numerous new effects have been uncovered. The previously mentioned estrogenic and neurological effects are particularly novel. Even within the most established area of toxicity—nephrotoxicity—recent findings are enhancing our understanding, including new molecular and mechanistic evidence. Currently, both the established health concerns and other risks associated with uranium exposure clearly warrant further investigation, as they appear plausible on the basis of the evidence available to date. Uranium possesses both radiological and chemical properties [88].

Over the past few decades, significant efforts have been made to remediate radioactively contaminated sites worldwide. Numerous countries have launched programs to assess and clean up radionuclide-contaminated locations resulting from various industrial activities. As a result, extensive experience has been accumulated in applying control and treatment technologies to different types of contaminated sites. A wide array of proven remediation technologies, along with their combinations and adaptations, have been developed to address specific radionuclide contamination conditions. However, many of these techniques and technologies may be prohibitively expensive for large-scale implementation or insufficient to address the complexity and severity of contamination issues at particular sites. Importantly, there is no one-size-fits-all remediation plan. Furthermore, considerable effort is needed to organize remediation activities effectively, optimize the use of available resources and leverage advancements in technology research and development. International organizations, such as the International Atomic Energy Agency and the U.S. Environmental Protection Agency, have made substantial contributions by providing guidelines and recommendations for various remediation approaches on the basis of research studies, practical experiences, case studies, and lessons learned.

9 Conclusion

Various factors affect the solubility and stability of uranium in aqueous environments, among which heavy metals, microorganisms, and other radioactive elements play significant roles. The present study aimed to provide a brief overview of the effects of several microbial mechanisms of uranium solubility, and it was confirmed that biotransformation and oxidation–reduction are important for uranium solubility. In addition, under favorable conditions, microorganisms can modify the chemical speciation, solubility, and sorption characteristics of uranium; thus, their concentration in solution and environmental mobility and bioavailability can be increased or decreased. In another section of this chapter, comprehensive case studies on intricate interactions in practical situations are reviewed, emphasizing the difficulties in forecasting uranium behavior amidst numerous contaminants. However, after this chapter is written, some recommendations are suggested for future research as follows.

- The details of sophisticated analytical methods, including spectroscopic and microscopic techniques for determining the effects of interactions among heavy metals, microorganisms, and other radioactive materials on uranium solubility, should be investigated further.
- Research on how bacterial cells can interact with uranium and then biotransformation mechanisms

References

1. Smedley PL, Kinniburgh DG (2023) Uranium in natural waters and the environment: distribution, speciation and impact. Appl Geochem 148:105534
2. Bjørklund G et al (2020) Uranium in drinking water: a public health threat. Arch Toxicol 94:1551–1560
3. Abolli S et al (2024) Understanding uranium distribution: a systematic review and meta-analysis in the context of drinking water resources. Resul Eng:102152
4. Mehta VS et al (2014) Effect of co-solutes on the products and solubility of uranium (VI) precipitated with phosphate. Chem Geol 364:66–75
5. Anicin I et al (2005) Investigation of the uranium solubility and absorption. Phys Scr 2005(T118):39
6. Zavodska L et al. (2008) Environmental chemistry of uranium. HV ISSN, pp 1418–7108
7. van Hartesveldt NF (2020) Uranium solubility in high temperature, reduced systems. Mississippi State University
8. Alloway BJ (2013) Uranium. In: Alloway BJ (ed) Heavy metals in soils: trace metals and metalloids in soils and their bioavailability. Springer Netherlands, Dordrecht, pp 565–577

9. Li J, Zhang Y (2012) Remediation technology for the uranium contaminated environment: a review. Procedia Environ Sci 13:1609–1615
10. Birke M, Rauch U, Lorenz H (2009) Uranium in stream and mineral water of the Federal Republic of Germany. Environ Geochem Health 31(6):693–706
11. Merroun ML, Selenska-Pobell S (2008) Bacterial interactions with uranium: an environmental perspective. J Contam Hydrol 102(3–4):285–295
12. Galitskaya I, Putilina V, Yuganova T (2017) The role of microorganisms in uranium behavior in the water–rock system. Water Resour 44:892–902
13. Duff MC et al (1999) Factors influencing uranium reduction and solubility in evaporation pond sediments. Biogeochemistry 45:95–114
14. Casas I et al (1998) The role of pe, pH, and carbonate on the solubility of UO2 and uraninite under nominally reducing conditions. Geochim Cosmochim Acta 62(13):2223–2231
15. Bots P, Behrends T (2008) Uranium mobility in subsurface aqueous systems: the influence of redox conditions. Mineral Mag 72(1):381–384
16. Chuan M, Shu G, Liu J (1996) Solubility of heavy metals in a contaminated soil: effects of redox potential and pH. Water Air Soil Pollut 90:543–556
17. Banzi FP, Msaki PK, Mohammed NK (2015) Assessment of heavy metal concentration in water around the proposed Mkuju River Uranium Project in Tanzania. Tanzania J Sci 41(1):8–18
18. Linnik PM, Zubenko IB (2000) Role of bottom sediments in the secondary pollution of aquatic environments by heavy-metal compounds. Lakes Reserv Res Manag 5(1):11–21
19. Crawford SE, Lofts S, Liber K (2017) The role of sediment properties and solution pH in the adsorption of uranium(VI) to freshwater sediments. Environ Pollut 220:873–881
20. Liu B et al (2020) Transport and transformation of uranium and heavy metals from uranium tailings under simulated rain at different pH. Environ Chem Lett 18(2):495–503
21. Li C, Wang M, Luo X (2019) Uptake of uranium from aqueous solution by Nymphaea tetragona Georgi: the effect of the accompanying heavy metals. Appl Radiat Isot 150:157–163
22. Sitte J et al (2010) Microbial links between sulfate reduction and metal retention in uranium- and heavy metal-contaminated soil. Appl Environ Microbiol 76(10):3143–3152
23. Godheja J et al (2016) Xenobiotic compounds present in soil and water: a review on remediation strategies. J Environ Anal Toxicol 6(392):2161–0525
24. Choudhary S, Sar P (2015) Interaction of uranium (VI) with bacteria: potential applications in bioremediation of U contaminated oxic environments. Rev Environ Sci Biotechnol 14:347–355
25. Francis AJ et al (2004) Uranium association with halophilic and non-halophilic bacteria and archaea. Radiochim Acta 92(8):481–488
26. Merroun M et al (2006) Interaction mechanisms of bacterial strains isolated from extreme habitats with uranium. Radiochim Acta 94(9–11):723–729
27. Lovley DR et al (1991) Microbial reduction of uranium. Nature 350(6317):413–416
28. Wan J (2009) Influences of organic carbon supply rate on uranium bioreduction in initially oxidizing, contaminated sediment
29. Zhang G et al (2011) Microbial reduction of chlorite and uranium followed by air oxidation. Chem Geol 283(3–4):242–250
30. Suzuki Y et al (2005) Direct microbial reduction and subsequent preservation of uranium in natural near-surface sediment. Appl Environ Microbiol 71(4):1790–1797
31. Lovley DR et al (1993) Reduction of uranium by cytochrome c3 of Desulfovibrio vulgaris. Appl Environ Microbiol 59(11):3572–3576
32. Newsome L, Morris K, Lloyd JR (2014) The biogeochemistry and bioremediation of uranium and other priority radionuclides. Chem Geol 363:164–184
33. Macaskie LE et al (1992) Uranium bioaccumulation by a Citrobacter sp. as a result of enzymically mediated growth of polycrystalline HUO2PO4. Science 257(5071):782–784
34. Pattanapipitpaisal P et al (2002) Reduction of Cr (VI) and bioaccumulation of chromium by gram positive and gram negative microorganisms not previously exposed to Cr-stress. Environ Technol 23(7):731–745

35. Beazley MJ et al (2007) Uranium biomineralization as a result of bacterial phosphatase activity: insights from bacterial isolates from a contaminated subsurface. Environ Sci Technol 41(16):5701–5707
36. Thomas RA, Macaskie L (1996) Biodegradation of tributyl phosphate by naturally occurring microbial isolates and coupling to the removal of uranium from aqueous solution. Environ Sci Technol 30(7):2371–2375
37. Mondani L et al (2011) Influence of uranium on bacterial communities: a comparison of natural uranium-rich soils with controls. PLoS One 6(10):e25771
38. Lloyd JR, Macaskie LE (2000) Bioremediation of radionuclide-containing wastewaters. In: Environmental microbe-metal interactions. Wiley, pp 277–327
39. Benzerara K et al (2011) Significance, mechanisms and environmental implications of microbial biomineralization. Comptes Rendus Géosci 343(2–3):160–167
40. Suzuki Y, Banfield JF (1999) Geomicrobiology of uranium. In: Reviews in mineralogy, vol 38, pp 393–432
41. Beveridge T, Murray R (1980) Sites of metal deposition in the cell wall of Bacillus subtilis. J Bacteriol 141(2):876–887
42. Schiewer S, Volesky B (2000) Biosorption processes for heavy metal removal. In: Environmental microbe-metal interactions. Wiley, pp 329–362
43. Knopp R et al (2003) Laser spectroscopic studies of interactions of UVI with bacterial phosphate species. Chem Eur J 9(12):2812–2818
44. Wan J et al (2005) Reoxidation of bioreduced uranium under reducing conditions. Environ Sci Technol 39(16):6162–6169
45. Suzuki Y, Suko T (2006) Geomicrobiological factors that control uranium mobility in the environment: update on recent advances in the bioremediation of uranium-contaminated sites. J Mineral Petrol Sci 101(6):299–307
46. Wufuer R et al (2017) Uranium bioreduction and biomineralization. Adv Appl Microbiol 101:137–168
47. Akob DM et al (2012) Gene expression correlates with process rates quantified for sulfate-and Fe (III)-reducing bacteria in U (VI)-contaminated sediments. Front Microbiol 3:280
48. Williams KH et al (2011) Acetate availability and its influence on sustainable bioremediation of uranium-contaminated groundwater. Geomicrobiol J 28(5–6):519–539
49. Anderson RT et al (2003) Stimulating the in situ activity of Geobacter species to remove uranium from the groundwater of a uranium-contaminated aquifer. Appl Environ Microbiol 69(10):5884–5891
50. Zeng T et al (2020) Microbial characteristic and bacterial community assessment of sediment sludge upon uranium exposure. Environ Pollut 261:114176
51. Merroun ML et al (2011) Bio-precipitation of uranium by two bacterial isolates recovered from extreme environments as estimated by potentiometric titration, TEM and X-ray absorption spectroscopic analyses. J Hazard Mater 197:1–10
52. Kolhe N, Zinjarde S, Acharya C (2018) Responses exhibited by various microbial groups relevant to uranium exposure. Biotechnol Adv 36(7):1828–1846
53. You W et al (2021) Uranium bioremediation with U (VI)-reducing bacteria. Sci Total Environ 798:149107
54. Evans RM et al (2016) Should the scope of human mixture risk assessment span legislative/regulatory silos for chemicals? Sci Total Environ 543:757–764
55. Kortenkamp A (2014) Low dose mixture effects of endocrine disrupters and their implications for regulatory thresholds in chemical risk assessment. Curr Opin Pharmacol 19:105–111
56. Coyte RM, Vengosh A (2020) Factors controlling the risks of co-occurrence of the redox-sensitive elements of arsenic, chromium, vanadium, and uranium in groundwater from the eastern United States. Environ Sci Technol 54(7):4367–4375
57. Dong W, Brooks SC (2006) Determination of the formation constants of ternary complexes of uranyl and carbonate with alkaline earth metals (Mg2+, Ca2+, Sr2+, and Ba2+) using anion exchange method. Environ Sci Technol 40(15):4689–4695

58. Alam MS, Cheng T (2014) Uranium release from sediment to groundwater: influence of water chemistry and insights into release mechanisms. J Contam Hydrol 164:72–87
59. Kohler M et al (2004) Methods for estimating adsorbed uranium(VI) and distribution coefficients of contaminated sediments. Environ Sci Technol 38(1):240–247
60. Balaram V, Rani A, Rathore DPS (2022) Uranium in groundwater in parts of India and world: a comprehensive review of sources, impact to the environment and human health, analytical techniques, and mitigation technologies. Geosyst Geoenviron 1(2):100043
61. Atwood DA (2013) Radionuclides in the Environment. Wiley
62. Chen L et al (2019) Uranium occurrence state in the Tarangaole area of the Ordos Basin, China: implications for enrichment and mineralization. Ore Geol Rev 115:103034
63. Xia Y, Mou C, Wu H (2023) Uranium occurrence state and its implication for sandstone-type uranium mineralization within the Hanbazhai area of the Longchuanjiang Basin, China. Minerals 13(8):1037
64. He Y et al (2019) Bioremediation of Cr (VI) contaminated groundwater by Geobacter sulfurreducens: environmental factors and electron transfer flow studies. Chemosphere 221:793–801
65. de Menezes AB et al (2024) Relationships between radiation, wildfire and the soil microbial communities in the Chornobyl exclusion zone. Sci Total Environ 950:175381
66. Lovley DR, Phillips EJ (1992) Reduction of uranium by Desulfovibrio desulfuricans. Appl Environ Microbiol 58(3):850–856
67. Barescut J et al (2009) Isotopic composition of uranium in soil and ground water samples collected around 30 km Chernobyl exclusion zone. Radioprotection 44(5):785–790
68. Nier KA, Yergey AL, Gale PJ (2015) The Encyclopedia of Mass Spectrometry: Volume 9: Historical Perspectives, Part A: The Development of Mass Spectrometry. Newnes
69. Fenn JB et al (1989) Electrospray ionization for mass spectrometry of large biomolecules. Science 246(4926):64–71
70. Karasek FW, Clement RE (2012) Basic gas chromatography-mass spectrometry: principles and techniques. Elsevier
71. Chace DH (2001) Mass spectrometry in the clinical laboratory. Chem Rev 101(2):445–478
72. Beauchemin D (2008) Inductively coupled plasma mass spectrometry. Anal Chem 80(12):4455–4486
73. Krachler M (2007) Environmental applications of single collector high resolution ICP-MS. J Environ Monit 9(8):790–804
74. Sitko R, Zawisza B (2012) Quantification in X-ray fluorescence spectrometry. In: X-ray spectroscopy. InTech, pp 137–162
75. Van Grieken R, Markowicz A (2001) Handbook of X-ray Spectrometry. CRC press
76. Keeler J (2010) Understanding NMR spectroscopy. Wiley
77. Clark JH, Macquarrie DJ (2008) Handbook of green chemistry and technology. Wiley
78. Trinh TND et al (2024) Recent advances in portable devices for environmental monitoring applications. Biomicrofluidics 18(5):051501
79. Smith BC (2011) Fundamentals of Fourier transform infrared spectroscopy. CRC Press
80. Ong TT, Blanch EW, Jones OA (2020) Surface enhanced Raman spectroscopy in environmental analysis, monitoring and assessment. Sci Total Environ 720:137601
81. Radojevic M, Bashkin VN (1999) Practical environmental analysis. Royal society of chemistry
82. Zhong S et al (2021) Machine learning: new ideas and tools in environmental science and engineering. Environ Sci Technol 55(19):12741–12754
83. U.S. Department of Health and Human Services (1999) Agency for toxic substances and disease registry-ATSDR. Agency for Toxic Substances and Disease Registry (US), Atlanta, GA
84. Wu X et al (2016) A review of toxicity and mechanisms of individual and mixtures of heavy metals in the environment. Environ Sci Pollut Res 23:8244–8259
85. National Research Council (1994) Science and judgment in risk assessment. National Academy Press, Washington, DC
86. Escrivá L et al (2020) A validated search filter for the identification of endocrine disruptors based on the ECHA/EFSA guidance recommendations. Environ Int 142:105828

87. Ma M et al (2020) Emerging health risks and underlying toxicological mechanisms of uranium contamination: lessons from the past two decades. Environ Int 145:106107
88. Brugge D, Buchner V (2011) Health effects of uranium: new research findings. Rev Environ Health 26:231

Unveiling the Chemistry Behind Uranium Solubility and Interfering Agents

C. G. Poojitha, P. Manohar, B. S. Rohini, Bhupender Singh, and A. Pyngrope

Abstract This chapter explores the in-depth chemical process of the complex primordial radionuclide uranium (U), which is correlated with its solubility and the impact of interfering agents. The chemistry of uranium depends on the oxidation state, predominantly uranium (+4 and +6), which monitors the behavior of uranium in the environment. Uranium solubility poses potential health risks when it is transported into the environment and contaminated. The U (+6) oxidation state is soluble and leaches into natural sources, such as groundwater, soil, sediment and the food chain, by posing risks to drinking water and plantations, causing toxic effects on wildlife and humans. Organic compounds such as humic, citric and fulvic acids, in addition to synthetic organic pollutants and other ligands, increase uranium solubility and mobility in natural sources. Analogously, inorganic compounds such as carbonate, bicarbonate, chloride, sulfates, and iron and manganese oxides contribute to the mobility and solubility of uranium. Environmental factors, including microbial activities, redox conditions, and physicochemical parameters (pH, salinity, temperature and electrical conductivity), influence interfering agent interactions with the chemistry of uranium. By analyzing the complex interactions between the solubility of uranium and an interfering agent, this chapter reveals uranium chemistry and generates precautions for managing contamination in natural and industrial contexts.

C. G. Poojitha (✉) · B. S. Rohini
Department of Physics and Electronics, FET, JAIN (Deemed to be) University, Bengaluru, Karnataka, India
e-mail: poojitha.cg@jainuniversity.ac.in

P. Manohar
Department of Physics, School of Engineering, Dayananda Sagar University, Bengaluru, Karnataka, India

B. Singh
Department of Applied Sciences and Humanities, IIMT College of Engineering, Noida, Uttar Pradesh, India

A. Pyngrope
Department of Physics, Captain Williamson Sangma State University, Meghalaya, Meghalaya, India

M. K. Jindal, K. Patni (eds.), *Water Quality-Uranium Solubility Nexus: Unlocking the Secrets of Uranium Solubility*, Earth and Environmental Sciences Library, https://doi.org/10.1007/978-3-031-98909-4_12

Keywords Uranium solubility · Redox chemistry · Complexing ligands · Environmental mobility

1 Introduction: Uranium Solubility—A Nexus of Science, Environment, and Technology

Uranium is a naturally occurring actinide found in trace amounts in the Earth's crust, and it presents a fascinating duality, being significant in both Earth sciences and technology. Although it is naturally rare, human activities have concentrated on its presence, highlighting the importance of understanding how it behaves. Understanding the solubility of uranium may be useful, but it is a vital physicochemical property. The chemistry and solubility of uranium impact its convenient applications (nuclear reactors and weapons and its environmental risks), predominantly owing to the chemotoxicity and radiotoxicity associated with its long-lived isotopes, U-238 and U-235. The geochemical behavior of uranium in groundwater represents a major fraction of its impact on the environment, as a resource and for nuclear waste management. Hence, an in-depth understanding of its movement in groundwater, which is altered by physicochemical parameters (pH, redox conditions, complexing agents, and microbial activity), as discussed in previous chapters, is important. From determining the ongoing security of radioactive waste regions and purifying areas of contamination to examining ore formation and analyzing ecological and health concerns, this understanding is essential for a variety of applications. In this chapter, we delve into uranium solubility complexities, highlight the chemical and environmental parameters that influence uranium leaching in water, and correlate the results with studies of sustainable uranium management.

Uranium, a rare earth element, is known for its ability to produce energy globally [1]. In the present world, we frequently use nuclear power as a low-carbon energy alternative, which is particularly relevant in the framework of global climate change extenuation. The high energy density of uranium fuel, wherein a miniscule mass equates to the energy output of extensive quantities of fossil fuels, highlights its exceptional robust potential [1]. However, the debate surrounding nuclear energy is complicated, all-inclusive, enduring erudite deliberation concerning the prolonged administration of radioactive waste, reactor safety protocols, and the risks associated with nuclear production [2]. In addition to its vigorous applications, uranium has considerable geopolitical significance. As a strategically vital resource, assured access to uranium deposits is of dominant national interest for states engaged in nuclear energy programs or possessing nuclear arsenals. The geographically heterogeneous distribution of uranium ore bodies engenders complex international trade dynamics and geopolitical interdependencies [3].

Uranium is just as complicated when you look at it from the point of view of the world. As a naturally formed part of the Earth's crust, uranium is found at background levels of approximately 2–4 parts per million [4]. It is found in many different types of geological layers and is more common in certain types of minerals and rocks. Uranium is naturally present in the environment, but human activities have clearly changed how uranium is distributed in the environment. Uranium enters the environment through mining and milling, a variety of industrial processes, and nuclear accidents [5]. Uranium is spread in the environment by substances that do not seem dangerous, such as phosphate-based fertilizers made from phosphate rock that contains uranium [6]. Therefore, uranium can be found in both underground and surface water systems worldwide. It is usually found at low levels, but these levels are rising in places where people are doing more farming and industrial activities. Therefore, its environmental importance is closely linked to its natural movement, which is largely controlled by how easily it dissolves, as we will see below. In regard to human health, the fact that uranium is binary is still important. Uranium is known to be both radioactive and chemically poisonous [7], but it is also used in certain medical and industrial settings under controlled conditions. Uranium is an alpha-emitting radioactivity, so when it builds up inside of things, it can be radiotoxic. It can damage DNA and increase the risk of developing cancer over time. As a heavy metal, uranium is also chemically harmful, most notably through nephrotoxicity, which makes the kidneys work less well [8]. People are mostly exposed when they drink or breathe in dirty drinking water or particulate matter. This shows how important it is to keep a close eye on the environment and take proactive steps to protect people [9].

Nevertheless, to fully comprehend the various effects of uranium, one must carefully look at a feature that seems basic but is actually very complex: its ability to dissolve in water. The idea of "uranium solubility" seems simple at first glance, but it is complicated and situation dependent. Solubility is not an intrinsic, unchangeable property such as invariant atomic properties; instead, it changes greatly depending on the conditions in the world [10]. For uranium to dissolve or precipitate, a group of physical factors work together to control the process. The level of acidity or alkalinity (pH), the oxidation–reduction potential (Eh), the presence of complex-forming ligands, and the general ionic strength of the water medium affect solubility [11]. The "solubility puzzle" is already very hard, and it is even harder when you think about how many things affect each other. Changes in pH, for example, cause large changes in the chemistry of water with uranium and the solubility patterns that follow [12]. Redox conditions, which determine whether hexavalent uranium(VI), which dissolves more easily, or tetravalent uranium(IV), which dissolves less easily, are the most important factors [13]. Additionally, natural water systems are complicated mixtures that contain many different kinds of dissolved chemicals, such as carbonates, phosphates, and organic acids. These chemicals can change how uranium dissolves by complexing or precipitating it [14]. Adding to this difficulty is the fact that uranium's solubility is inherently contradictory. Uranium that dissolves more easily is actively sought after in some technological uses, such as uranium ore leaching methods [15]. On the other hand, reducing the solubility of uranium is very important for long-term nuclear waste management and for cleaning the environment so that contaminants do not move around and the containment remains intact [16]. Because of how these complex physical factors interact with each other, the idea of "interfering agents" becomes very important. These things, which can be specific chemical species or the general state of the world, work as "solubility modulators," which means that they either make it easier for uranium to dissolve or make it harder for it to dissolve. For example, carbonate anions are known to improve solubility by creating uranyl carbonate complexes that are very easy to dissolve [17]. On the other hand, phosphate anions can effectively lower solubility by forming uranyl phosphate phases that are only slightly soluble [18]. The main goal of this chapter is to systematically explain these "interfering agents," including their chemical names, how they work, and how important they are in different environmental settings.

This solubility paradox strongly affects regulations for protecting the environment, methods for managing resources, and people's health around the world. Effective ways to clean uranium-contaminated areas depend on changing the solubility to make it easier for uranium to be sucked up or removed from affected materials [19]. To improve the efficiency of uranium extraction in mining, it is important to have thorough knowledge of agents that improve the degree of dissolution [20]. To ensure that nuclear waste stays in approved repositories for a long time, the solubility of uranium must be carefully controlled so that radioactive materials do not leak into the geosphere [21]. It is also important to protect drinking water resources, which requires the use of strong monitoring and intervention methods on the basis of a full knowledge of uranium solubility and the factors that affect uranium regulation [22].

2 Uranium Chemistry

The electron configuration of uranium, while complex, can be simplified to [Rn] $5f^3$ $6d^1$ $7s^2$, highlighting the involvement of the 5f, 6d, and 7 s orbitals in its chemical behavior [23]. The actinide series is characterized by the gradual filling of the 5f electron shell, and for uranium, the occupancy and energetic proximity of the 5f, 6d, and 7 s orbitals, coupled with relativistic effects, contribute to its rich redox chemistry and diverse coordination behavior. Relativistic effects, which are particularly significant for heavy elements such as uranium, influence orbital energies and radial extensions, modifying bonding characteristics and spectroscopic properties.

A defining feature of uranium, inherent to all actinides, is radioactivity. Uranium isotopes possess unstable nuclei that undergo radioactive decay processes, emitting energy and particles. This nuclear instability underlies uranium's role in nuclear power and related applications. Furthermore, uranium exhibits a range of oxidation states, which is a consequence of its accessible 5f, 6d, and 7 s electrons. While oxidation states from +2 to +6 are formally possible, the most environmentally relevant and chemically significant states are +3, +4, +5, and +6. In aqueous systems, the hexavalent state, U(VI), and the tetravalent state, U(IV), are of paramount importance because of their prevalence and distinct solubility characteristics. Under oxidizing conditions, such as those found in oxygenated surface waters, uranium predominantly exists in its hexavalent form, typically as a uranyl ion (UO_2^{2+}). Conversely, reducing environments, such as anoxic groundwater or saturated sediments, favor the tetravalent state, U(IV), which generally exhibits lower aqueous solubility and tends to precipitate as solid phases, notably uranium dioxide ($UO_2(s)$), also known as uraninite in its mineral form [24]. Uranium(V), as dioxouranium(V) or uranyl(V) (UO_2^{+}), is less stable and prone to disproportionation, forming U(IV) and U(VI) species [25]. The thermodynamic stability of these oxidation states is critically dependent on the oxidation–reduction potential (Eh) and pH of the surrounding environment, as depicted in Pourbaix diagrams [26].

In aqueous solutions, uranium (VI) is predominantly present as the uranyl ion $[UO_2]^{2+}$, a linear dislocation with strong, covalent U=O bonds formed through the overlap of uranium 5f and oxygen 2p orbitals, as elucidated by molecular orbital theory [27]. Spectroscopic analyses, including UV–Vis, Raman, and infrared spectroscopy, confirmed the linear O=U=O geometry and characteristic electronic transitions. The uranyl ion is extensively solvated in water, forming a primary hydration shell and readily engaging in coordination chemistry with a variety of anionic ligands in aqueous media. The equatorial coordination sphere around the uranyl ion, perpendicular to the linear O=U=O axis, typically accommodates 4–6 ligands, coordinating through oxygen donor atoms originating from water molecules, hydroxide ions, carbonates, phosphates, and other relevant species [28]. The amount and shape of the coordination pairs, as well as the type of ligands that are used, have a large effect on how stable and mobile uranyl complexes are in terms of thermodynamics and kinetics. Computer-based chemistry techniques, such as density functional theory (DFT) and molecular dynamics (MD) models, help us understand how uranyl species dissolve and form complexes [29].

The most common type of U(VI) in oxygenated water is the uranyl ion (UO_2^{2+}). It strongly attracts carbonate ions, which creates a number of stable and soluble complexes. The most common of these complexes are $[UO_2(CO_3)_2]^{2-}$ and $[UO_2(CO_3)_3]^{4-}$, which are very important for making uranium move around more easily in different types of water. Carbonate makes it easier for uranium to move over long distances by encasing it in these soluble complexes. This could have an effect on both surface and groundwater supplies. How stable these complexes are and where they are found depend strongly on the pH of the solution. Conditions that are more alkaline help them form. This depends on pH because, as the pH increases, bicarbonate (HCO^{3-}), which is the main form of inorganic carbon in most natural liquids, is deprotonated more. This makes more carbonate ions available to combine with uranyl ions. Therefore, in alkaline settings, more uranium is found as soluble uranyl carbonate complexes, which makes it easier to dissolve and move around. Uranium oxides, especially uranium dioxide (UO_2, uraninite) and Tri-uranium octoxide ($U_3O_8(s)$), are the main rocks that contain uranium. Uranium dioxide has a fluorite-like cubic structure and is thermodynamically stable when reducing. It is not easily dissolved in water. Tri-urium octoxide, which possesses a more complex orthorhombic structure, represents a higher oxidation state of uranium oxide and has limited solubility. Uranium hydroxides, including $UO_2(OH)_2$ and hydrated forms of UO_3, can precipitate from aqueous solutions under circumneutral to alkaline pH conditions. These hydroxide phases also generally exhibit low aqueous solubility. In contrast to oxides and hydroxides, uranyl carbonate compounds, which include species such as $UO_2CO_3(aq)$ and anionic complexes such as $[UO_2(CO_3)_3]^{4-}$, are characterized by significantly increased aqueous solubility. The formation of uranyl carbonate complexes is particularly pertinent in natural waters rich in carbonate alkalinity, leading to increased uranium mobility in such systems [30]. The thermodynamic solubility of uranium compounds is fundamentally determined by the balance between the lattice energy, which stabilizes the solid phase, and the hydration enthalpy, which favors dissolution, as well as the overall Gibbs free energy of formation of the solid phase under specific environmental conditions [31].

3 Solubility of Uranium: Basic Principles

Solubility, a fundamental physicochemical property, describes the extent to which a substance, or solute, dissolves in a solvent, typically water, in environmental contexts. More precisely, solubility is defined as the concentration of a solute in a saturated solution at equilibrium [32]. This equilibrium state signifies a dynamic condition where the rate of dissolution of the solid phase is equal to the rate of precipitation from the solution phase; it is not a static cessation of the dissolution process. Thermodynamically, solubility is governed by the change in Gibbs free energy ($\Delta G_{dissolution}$) associated with the dissolution process. Spontaneous dissolution occurs when $\Delta G_{dissolution}$ is negative, indicating a thermodynamically favorable transition from the solid to the aqueous phase. In ideal solutions, solubility is primarily determined by the solubility product constant (K_{sp}).

The solubility product (K_{sp}) is an equilibrium constant that quantifies the solubility of sparingly soluble ionic compounds. For a generic solid compound M_aX_b that dissolves according to equilibrium.

$$M_aX_b \leftrightarrows aM^{n+} + bX^{m-}$$

The solubility product expression is defined as:

$$K_{sp} = \left[M^{n+}\right]^a \left[X^{m-}\right]^b$$

where $[M^{n+}]^a$ and $[X^{m-}]^b$ represent the equilibrium molar concentrations (more accurately, activities in non-ideal solutions) of the cation and anion, respectively, in a saturated solution. a and b are the stoichiometric coefficients from the balanced dissociation equation. A higher K_{sp} value indicates a greater extent of dissolution and thus a greater solubility of the compound. Conversely, a lower K_{sp} value signifies lower solubility. The K_{sp} principle allows for the prediction of precipitation: if the ion activity product (IAP), which is calculated from the instantaneous concentrations of ions in a solution, exceeds the K_{sp} value for a given solid phase, the solution is supersaturated with respect to that solid, and precipitation is thermodynamically favored to reduce the ion concentrations until equilibrium is re-established at the solubility limit defined by K_{sp}. However, it is crucial to recognize that K_{sp} values, typically determined for pure water systems, may deviate in complex environmental matrices because of factors such as ionic strength effects and competitive complexation reactions. In nonideal solutions, particularly at higher ionic strengths, activities, rather than concentrations, should be rigorously employed in thermodynamic calculations and solubility product expressions to account for interionic interactions and deviations from ideal solution behavior. Activity coefficients, as described by the Debye–Hückel theory and its extensions, are used to relate concentrations to activities [33].

Several environmental parameters exert significant control over uranium solubility in aqueous systems. While temperature and pressure can theoretically influence solubility, their effects are often less pronounced under typical ambient environmental conditions than other factors. For uranium, the dominant factors modulating solubility are pH, the oxidation–reduction potential (Eh), and the presence of complexing ligands. pH, a measure of solution acidity or alkalinity, profoundly affects uranium speciation and solubility. In acidic pH ranges, uranium solubility behavior differs significantly from that in alkaline conditions, impacting the dominant aqueous uranium species, and the solubility of uranium-bearing solid-phase redox conditions, quantified by the Eh parameter, is paramount in determining the oxidation state of uranium, thereby controlling its solubility. Oxidizing conditions generally favor the more soluble U(VI) form, whereas reducing conditions promote the less soluble U(IV) state and the precipitation of U(IV) solids [34]. The presence of various ligands in aqueous solution represents another critical determinant of uranium solubility. Ligands capable of forming stable, soluble complexes with uranyl ions, such as carbonate, citrate, or EDTA, can significantly increase uranium solubility by shifting the equilibrium toward the dissolved phase. Conversely, ligands that form sparingly soluble uranium compounds, such as phosphate or sulfide, can effectively reduce uranium solubility by promoting precipitation, and quantitative treatment of these factors necessitates the consideration of relevant equilibrium constants, including acidity constants for pH effects, redox potentials and Nernst equations for redox effects, and complex formation constants (stability constants) for ligand interactions. By using thermodynamic databases and computer programs, geochemical speciation modeling is a strong way to guess how uranium will dissolve and what kind of uranium it will be based on these connected environmental factors [35].

The most important mineral that contains uranium is uraninite (UO_2), which is often referred to as pitchblende when it is found in massive form. It is the key constituent of several uranium ores and a vital mineral in nuclear waste. Other important uranium minerals include:

Coffinite: $U(SiO_4)$: A uranium silicate often found in hydrothermal vein deposits.
Carnotite: $K_2(UO_2)_2(VO_4)_2 \cdot nH_2O$ (where n = 1–3): A secondary uranium mineral often found in sedimentary rocks, particularly sandstones.

Autunite: $Ca(UO_2)_2(PO_4)_2 \cdot nH_2O$ (where $n = 10–12$): A secondary uranium mineral commonly found in granitic rocks and hydrothermal veins.

Torbernite: $Cu(UO_2)_2(PO_4)_2 \cdot nH_2O$ (where $n = 8–12$): A secondary uranium mineral similar to autunite, often found in copper-bearing uranium deposits.

The solubility of uranium minerals strongly depends on both Eh and pH. Uraninite, for example, is much less soluble under reducing conditions (low Eh) than under oxidizing conditions. Increasing pH generally increases the solubility of U(VI) minerals due to the formation of uranyl carbonate complexes. However, the formation of less soluble U(VI) hydroxide species can decrease the solubility at very high pH values.

Several factors influence the solubility of uranium minerals, including redox potential (Eh) reducing conditions, which favor the less soluble U(IV) form, whereas oxidizing conditions lead to the more soluble U(VI) form. The pH of the solution plays a crucial role in determining the dominant uranium species and their solubility. Higher temperatures generally increase the solubility of most minerals, including uranium minerals. Increasing the ionic strength can decrease the solubility of uranium minerals due to the "common ion effect" and changes in the activity coefficients. The presence of other ions in the solution, such as calcium, can affect the solubility of uranium minerals by competing for complexing sites or forming new solid phases. pH is a master variable in controlling uranium solubility. Under near-neutral to alkaline conditions, uranyl carbonate complexes dominate, leading to increased solubility. At lower pH values, the formation of these complexes is less favored, and their solubility may decrease. However, very low pH can also increase U(VI) solubility due to the formation of other U(VI) species. Redox conditions are paramount. Under oxidizing conditions, U(VI) is dominant and relatively mobile. Reducing conditions favor the formation of less soluble U(IV), often uraninite. The presence of reducing agents such as organic matter, sulfide minerals, or certain microbial communities can significantly impact uranium solubility by promoting U(VI) to U(IV) reduction.

4 Uranium Solubility

Having established the fundamental principles governing uranium solubility, it is now pertinent to examine specific environmental factors and chemical species that function as "interfering agents," promoting the dissolution and enhancing the aqueous mobility of uranium. This section details key agents that enhance uranium solubility, including carbonate, selected inorganic ligands, organic ligands, pH effects, and redox conditions, elucidating their mechanisms of action and significance in natural and industrial systems.

4.1 Carbonate as a Solubility-Enhancing Agent

Carbonate ions $\left(CO_3^{2-}\right)$ represent a preeminent class of interfering agents that significantly increase uranium solubility, particularly in oxidizing and circumneutral to alkaline pH conditions, which are commonly encountered in natural waters [36]. The primary mechanism for this enhancement is the formation of stable, highly soluble uranyl carbonate complexes. In aqueous solutions containing uranyl UO_2^{2+} and carbonate, complexation reactions occur, yielding anionic species such as $[UO_2(CO_3)_2]^{2-}$ and $[UO_2(CO_3)_3]^{4-}$. The simplified reaction for the formation of the tris-carbonato complex is as follows:

$$UO_2^{2+}(aqueous) + 3CO_3^{2-}(aqueous) \rightleftharpoons \left[UO_2(CO_3)_3\right]^{4-}(aqueous)$$

This equation illustrates that a uranyl ion (UO_2^{2+}) in aqueous solution reacts with three carbonate ions (CO_3^{2-}) in aqueous solution to form a uranyl tris-carbonate complex, represented as $[UO_2(CO_3)_3]^{4-}$ in aqueous form, characterized by a 4- negative charge. The formation of these anionic complexes is thermodynamically favored due to their stability and high solubility, effectively shifting the uranium dissolution equilibrium toward the aqueous phase and increasing overall uranium solubility [37]. This effect is especially pronounced in carbonate-rich waters, contributing significantly to uranium mobility in various aquatic environments.

4.2 Influence of Other Inorganic Ligands

In addition to carbonate, other inorganic ligands can also contribute to uranium solubility enhancement, albeit generally to a lesser extent. Sulfate (SO_4^{2-}), fluoride (F^-), and chloride (Cl^-) are examples of such ligands capable of forming uranyl complexes. For instance, sulfate ions can form uranyl sulfate complexes through reactions such as the following:

$$UO_2^{2+}(aq) + SO_4^{2-}(aq) \rightleftharpoons \left[UO_2(SO_4)\right](aq)$$

This reaction results in the formation of a uranyl sulfate complex when a uranyl ion reacts with a sulfate ion in aqueous solution. Similarly, fluoride ions can form stable uranyl fluoride complexes, particularly at elevated fluoride concentrations and lower pH values, as shown by:

$$UO_2^{2+}(aq) + F^-(aq) \rightleftharpoons \left[UO_2F\right]^+(aq)$$

indicating the formation of a uranyl fluoride complex from uranyl and fluoride ions in solution. Chloride ions, while exhibiting weaker complexation than carbonate, sulfate, or fluoride, can also form uranyl chloride complexes, such as the following:

$$UO_2^{2+}(aq) + Cl^-(aq) \rightleftharpoons [UO_2Cl]^+(aq)$$

demonstrated the formation of a uranyl chloride complex from uranyl and chloride ions in aqueous media. While these inorganic ligands can increase uranium solubility through complexation, carbonate generally remains the most environmentally significant inorganic ligand in many natural water systems [38].

4.3 Organic Ligands and Uranium Solubility

Organic ligands, which encompass naturally occurring humic substances and synthetic chelating agents, also play crucial roles in modulating uranium solubility. Humic substances, complex macromolecules found in soils and aquatic environments, increase uranium solubility by complexing with uranyl ions through various functional groups [39]. Similarly, anthropogenic organic ligands, such as EDTA and citric acid, can form very stable uranyl complexes, leading to a significant increase in uranium solubility due to the chelation effect and strong ligand–metal interactions [40].

4.4 Effects of pH on Increasing Uranium Solubility

pH indirectly but profoundly influences uranium solubility enhancement, primarily by controlling the speciation and availability of carbonate and other pH-dependent ligands [41]. As the pH increases toward alkaline conditions, the concentration of carbonate ions (CO_3^{2-}) increases, favoring the formation of highly soluble uranyl carbonate complexes, as detailed in Sect. 4.1, and consequently enhancing uranium solubility. This pH-dependent carbonate speciation is a key driver for the observed correlation between alkaline pH and increased uranium solubility in many natural systems.

4.5 Redox Conditions and Uranium Solubility Enhancement

Oxidizing conditions, which are indicative of oxygenated environments, promote uranium solubility by favoring the U(VI) oxidation state, which is inherently more soluble than U(IV) [42]. When coupled with complexing ligands such as carbonate, oxidizing conditions further enhance uranium solubility. Oxidation maintains uranium in the soluble U(VI) form, while ligands such as carbonate stabilize uranium in the dissolved phase through complexation, resulting in synergistic solubility enhancement under oxidizing and carbonate-rich conditions.

5 Interfering Agents that Reduce Uranium Solubility

In contrast to agents that increase uranium solubility, several physicochemical factors and chemical species act to decrease uranium solubility in aqueous environments, promoting immobilization and precipitation. This section details key "interfering agents" that reduce uranium solubility, including phosphate precipitation, sulfide precipitation, hydroxide precipitation, adsorption and sorption processes, bioreduction and biomineralization, and the overarching role of reducing conditions.

5.1 *Phosphate-Induced Uranium Precipitation*

Phosphate ions (PO_4^{3-}) are potent agents for reducing uranium solubility by inducing the precipitation of highly insoluble uranyl phosphate minerals such as autunite and metaautunite [43]. The reaction of uranyl ions (UO_2^{2+}) with phosphate and calcium ions results in the formation of autunites, as represented by the simplified precipitation reaction:

$$2UO_2^{2+}(aqueous) + PO_4^{3-}(aqueous) + Ca^{2+}(aqueous) + n\,H_2O(liquid) \rightleftharpoons Ca(UO_2)_2(PO_4)_2 \cdot nH_2O(solid)$$

This equation demonstrates that uranyl ions, phosphate ions, calcium ions, and water react to form solid autunite $Ca(UO_2)_2(PO_4)_2 \cdot nH_2O$ (calcium uranyl phosphate hydrate), which precipitates out of solution due to its low solubility. The formation of these uranyl phosphate minerals effectively removes uranium from the aqueous phase, significantly reducing its solubility and mobility. Phosphate-induced precipitation is therefore a key strategy in uranium remediation and waste management technologies [44].

5.2 *Sulfide-Mediated Uranium Reduction and Precipitation*

Sulfide (S^{2-}), which is prevalent in reducing anaerobic environments, reduces and precipitates uranium, thereby decreasing its solubility. Sulfide promotes the reduction of soluble U(VI) to the less soluble U(IV), followed by the precipitation of uranium sulfide minerals [45]. A simplified reaction for sulfide-induced uranium reduction and precipitation can be depicted as:

$$2UO_2^{2+}(aq) + 3S^{2-}(aq) + 4H_2O(l) \rightarrow 2UO_2(s) + 3S(s) + 8H^+(aq)$$

This equation represents the reaction of uranyl ions with sulfide ions and water, resulting in the formation of solid uranium dioxide (UO_2) and solid elemental sulfur (S), along with the generation of hydrogen ions. Alternatively, and more directly focusing on uranium(IV) sulfide precipitation, the process can be simplified to:

$$U^{4+}(aq) + S^{2-}(aq) \rightarrow US(s)$$

representing the precipitation of a uranium sulfide solid (US) from U(IV) and sulfide ions. Biogenic sulfide, produced by sulfate-reducing bacteria under anaerobic conditions, is a significant driver of this process in natural environments, leading to uranium immobilization in reducing sediments and aquifers [46].

5.3 *Hydroxide Precipitation at Circumneutral to Alkaline pH*

The hydroxide ion OH^- can effectively reduce uranium solubility through the precipitation of uranium hydroxide solids, especially in the circumneutral to alkaline pH range [47]. The precipitation reaction can be represented as:

$$UO_2^{2+}(aq) + 2OH^-(aq) \rightarrow UO_2(OH)_2(s)$$

This equation illustrates the reaction between uranyl ions and hydroxide ions to form solid uranium hydroxide [$UO_2(OH)_2$], which precipitates from aqueous solution. This mechanism is particularly relevant in near-neutral to alkaline pH environments where sufficient hydroxide ions are available to drive precipitation.

5.4 *Adsorption and Sorption Processes*

Adsorption and sorption onto environmental solid phases, such as iron oxides, clay minerals, and natural organic matter, are crucial processes that reduce uranium mobility and effective solubility [48]. These solid materials bind uranyl ions from solution onto their surfaces, effectively removing uranium from the dissolved phase. While a single, simple equation does not fully capture the complexity of adsorption, it is a surface-driven phenomenon that significantly controls uranium distribution in many environments.

5.5 *Bioreduction and Biomineralization of Uranium*

Microbial activity, particularly bioreduction by certain bacteria, is a significant mechanism for reducing uranium solubility. Microorganisms can reduce soluble U(VI) to the less soluble U(IV), leading to the immobilization of uranium, often as uranium dioxide (UO_2) [49]. A simplified bioreduction reaction in which organic matter is used as an electron donor can be depicted as:

$$2UO_2^{2+}(aq)+CH_2O(aq)+4H^+(aq)\rightarrow 2UO_2(s)+CO_2(g)+3H_2O(l)$$

This equation represents the microbial reduction of uranyl ions using organic matter (simplified as CH_2O) as an electron donor under acidic conditions, resulting in the precipitation of solid uranium dioxide, carbon dioxide gas, and water. Additionally, some microorganisms can promote uranium biomineralization, forming specific low-solubility uranium minerals such as uranyl phosphates, further contributing to uranium immobilization [50].

5.6 *Overarching Role of Reducing Conditions*

Reducing conditions are a master variable controlling the reduction in uranium solubility. Low redox potential (Eh) environments favor a less soluble U(IV) oxidation state, promote sulfide-induced precipitation, and facilitate microbial bioreduction, all of which significantly decrease uranium solubility and mobility. Understanding and maintaining reducing conditions is thus a key strategy in natural attenuation and engineered remediation approaches for uranium contamination [51].

6 Techniques for Studying Uranium Solubility and Speciation

A comprehensive understanding of uranium solubility and speciation in complex environmental and industrial systems necessitates the application of a range of sophisticated analytical techniques. These techniques allow for the quantitative determination of total uranium concentrations, as well as the identification and quantification of different uranium species present in aqueous and solid phases. This section details several key analytical techniques commonly employed in uranium solubility and speciation studies, including spectrophotometry, inductively coupled plasma–mass spectrometry (ICP–MS), geochemical modeling, and X-ray absorption spectroscopy (XAS).

6.1 Spectrophotometric Methods for Uranium Quantification

Spectrophotometry, particularly colorimetric methods, offers a relatively accessible and cost-effective approach for the quantitative determination of uranium concentrations in aqueous solutions. These methods are based on the principle of spectrophotometry, wherein the absorbance of light by a colored solution is measured at a specific wavelength. The intensity of the color, and hence the absorbance, is directly proportional to the concentration of the analyte, following the Beer–Lambert Law. For uranium analysis, specific color-forming reagents that react with uranyl ions (UO_2^{2+}) to produce colored complexes in solution are employed.

One widely used reagent is arsenazo III, which forms a blue–purple complex with uranyl ions in acidic solutions, resulting in a strong absorbance peak in the visible region of the spectrum (typically approximately 600–660 nm) [52]. The intensity of the blue–purple color, measured spectrophotometrically, is directly proportional to the uranyl concentration over a specific concentration range. Other colorimetric reagents for uranium include dibenzoylmethane (DBM) and 2-(5-bromo-2-pyridylazo)-5-diethylaminophenol (Bromo-PADAP), each offering different sensitivities, selectivities, and optimal pH ranges [53].

Spectrophotometric methods are relatively simple to implement, require readily available instrumentation, and are suitable for routine uranium concentration measurements in water samples. However, these methods often suffer from limitations in sensitivity compared with more advanced techniques such as ICP–MS and may be prone to interference from other metal ions or colored compounds present in complex environmental matrices. Sample preparation procedures, such as pH adjustment and reagent addition, are crucial for accurate and reliable spectrophotometric analysis. Spectrophotometry is particularly useful for rapid screening of uranium concentrations and for applications where high sensitivity is not paramount.

6.2 Inductively Coupled Plasma–Mass Spectrometry (ICP–MS) for Total Uranium Measurement

Inductively coupled plasma–mass spectrometry (ICP–MS) is a highly sensitive and versatile analytical technique widely employed for the determination of the concentrations of trace elements, including uranium, in a variety of sample matrices [54]. ICP–MS combines the high-temperature ionization capabilities of an inductively coupled plasma (ICP) source with the mass separation and detection capabilities of a mass spectrometer.

In ICP-MS, the liquid sample is introduced into an argon plasma, which is a high-temperature (approximately 6000–10,000 K) ionized gas. In the plasma, uranium atoms in the sample are ionized, mostly resulting in the formation of singly charged positive ions (U^+). The ions are subsequently removed from the plasma and analyzed via a mass spectrometer, which segregates ions according to their mass–charge ratio (m/z). A detector, commonly an electron multiplier or Faraday cup, quantifies the concentration of ions at each m/z value. Measurements for uranium are often conducted at m/z values associated with the predominant uranium isotopes, namely, 238-U and 235-U. Isotope dilution ICP-MS, a highly accurate and exact variant of the technique, utilizes isotopically enriched uranium spikes to mitigate matrix effects and experimental drift, hence improving the accuracy of uranium quantification [55].

ICP-MS provides remarkable sensitivity for uranium analysis, with detection limits typically ranging from subparts per billion (μg/L) to parts per trillion (ng/L), rendering it appropriate for quantifying trace uranium levels in environmental water, geological, and biological samples. It additionally offers isotopic data, which might be instrumental for source tracking and isotopic fingerprinting analyses. Compared with other techniques, ICP-MS results in minimal matrix interference; however, meticulous matrix matching and internal standards remain crucial for precise quantification, especially in intricate matrices. Sample preparation for ICP-MS generally involves acid digestion or dilution of liquid samples to guarantee thorough dissolution and compatibility with the instrument. ICP-MS is an effective method for quantifying total uranium concentrations; nevertheless, it does not yield direct insights into uranium speciation.

6.3 Geochemical Modeling for Uranium Speciation and Solubility Prediction

Geochemical modeling, which uses thermodynamic principles and computational algorithms, represents a powerful in silico approach for predicting uranium speciation and solubility under varying environmental conditions [56]. Geochemical models employ thermodynamic databases that contain equilibrium constants for a wide range of chemical reactions, including acid—base reactions, complexation reactions, redox reactions, and mineral dissolution/precipitation reactions involving uranium and relevant environmental components (e.g., water, ligands, minerals).

Geochemical modeling software, such as MINTEQA2, PHREEQC, and Visual MINTEQ, allows users to input solution composition parameters (e.g., pH, Eh, total concentrations of uranium and major ions, and ligand concentrations) and then calculates the equilibrium distribution of uranium species (speciation) and predicts saturation indices with respect to various uranium-bearing solid phases, thereby estimating uranium solubility [57]. These models perform equilibrium calculations on the basis of the principle of Gibbs free energy minimization to determine the thermodynamically most stable speciation and phase assemblage under the specified conditions. Speciation diagrams, solubility diagrams, and saturation indices generated by geochemical models provide valuable insights into the factors controlling uranium solubility and mobility in complex systems.

Geochemical modeling is a valuable tool for understanding and predicting uranium behavior, particularly in scenarios where experimental speciation measurements are challenging or costly to obtain. However, geochemical models are based on thermodynamic equilibrium assumptions and rely on the completeness and accuracy of thermodynamic databases. Kinetic limitations, non-equilibrium conditions, and uncertainties in thermodynamic data can introduce limitations and uncertainties in model predictions. Model validation via experimental data is crucial for assessing the reliability of geochemical modeling predictions. Geochemical modeling complements experimental techniques by providing a framework for interpreting experimental results, exploring hypothetical scenarios, and extrapolating laboratory findings to field-scale environmental conditions.

6.4 X-Ray Absorption Spectroscopy (XAS) for Uranium Speciation Determination

X-ray absorption spectroscopy (XAS) is a synchrotron-based spectroscopic technique that provides direct, element-specific information on the oxidation state, coordination environment, and local atomic structure of uranium in a variety of solid, liquid, and gaseous samples [58]. XAS measures the absorption of X-rays by a sample as a function of X-ray energy. The X-ray absorption spectrum exhibits characteristic features near the absorption edge of the element of interest (for uranium, the L_3 absorption edge is approximately 17.17 keV).

X-ray absorption near edge structure (XANES) spectroscopy, the near-edge region of the XAS spectrum, is sensitive to the oxidation state and electronic structure of the absorbing atom. The position and shape of the XANES edge for uranium are oxidation state dependent, allowing for the determination of the oxidation state of uranium in samples, even in complex mixtures [59]. Extended X-ray absorption fine structure (EXAFS) spectroscopy, the region of the XAS spectrum above the absorption edge, provides information on the local atomic environment around the absorbing atom, including the coordination number, interatomic distances, and types of neighboring atoms. By analyzing the EXAFS oscillations, the coordination environment of uranium, such as the number and type of ligands directly coordinated to uranium (e.g., U-O, U-C, and U-P distances and coordination numbers), can be determined [60].

XAS is a powerful technique for uranium speciation studies because it is element specific, requires minimal sample preparation, and can be applied to a wide range of sample types, including solids, suspensions, and solutions, under in situ conditions.

X-ray absorption spectroscopy (XAS) is especially advantageous for investigating uranium speciation in intricate environmental matrices, including polluted soils, sediments, and biofilms, where conventional chemical separation techniques may disrupt the initial speciation. XAS offers direct structural insights into uranium speciation, enhancing macroscopic solubility assessments and geochemical modeling forecasts. XAS measurements generally necessitate access to synchrotron radiation facilities and are comparatively more intricate and costly than various other analytical methods. The analysis and interpretation of XAS spectra necessitate specialized skills and software.

7 Environmental and Industrial Implications of Uranium Solubility

The solubility of uranium, which is influenced by complex physicochemical interactions and interfering substances, as previously outlined, has significant ramifications in several environmental and industrial contexts. A comprehensive understanding of uranium solubility is essential for addressing environmental contamination issues, enhancing uranium resource management, and guaranteeing the long-term safety of nuclear waste disposal. This section will examine the environmental and industrial ramifications of uranium solubility, emphasizing environmental contamination, uranium extraction and processing, and nuclear waste management.

7.1 Environmental Contamination and Uranium Solubility

Uranium contamination in water and soil presents considerable environmental and human health hazards, principally due to its radiotoxicity and chemotoxicity [61]. The mobility and bioavailability of uranium in the environment are fundamentally connected to its solubility. Diverse sources of anthropogenic uranium contamination include uranium mining and milling operations, nuclear fuel cycle activities, the application of phosphate fertilizers derived from uranium-bearing phosphate rock, and emissions from industrial processes and historical nuclear weapon testing [62]. Upon release into the environment, uranium can be disseminated through multiple channels, including surface runoff, groundwater movement, and atmospheric deposition. The degree and velocity of uranium transport are fundamentally governed by its solubility under existing environmental circumstances.

In aqueous systems, higher uranium solubility translates to increased mobility and wider dispersal, potentially leading to the contamination of surface water and groundwater resources used for drinking water and irrigation. Uranium in drinking water is a significant human exposure pathway, with both radiotoxic and chemotoxic health effects. Radiologically, uranium is an alpha-emitting radionuclide, and internal exposure increases the risk of radiation-induced health effects, including carcinogenesis [63]. Chemically, uranium is a heavy metal that is nephrotoxic and impacts kidney function [64]. In soils, uranium solubility influences its vertical migration, plant uptake, and potential transfer into the food chain. Understanding and predicting uranium solubility in specific environmental compartments is therefore essential for assessing contamination risk, delineating contaminated plumes, and designing effective remediation strategies. Remediation approaches often target the manipulation of uranium solubility to either enhance its removal (e.g., through solubility-enhanced leaching followed by extraction) or promote its immobilization (e.g., through phosphate-induced precipitation or bioreduction) [65].

7.2 Uranium Mining and Processing: Solubility as a Key Operational Parameter

Uranium solubility is a central parameter in uranium mining and ore processing operations. The economic extraction of uranium from ores fundamentally relies on processes that increase uranium solubility to facilitate its leaching and subsequent recovery from ore matrices [66]. Leaching, the process of selectively dissolving uranium from ore minerals, is typically achieved via chemical leaching agents that promote uranium oxidation and/or complexation, thereby increasing its solubility in the leaching solution. Common leaching agents include alkaline carbonate/bicarbonate solutions and acidic solutions (e.g., sulfuric acid), which are chosen on the basis of the mineralogy of the ore and environment [67].

In alkaline leaching, carbonate or bicarbonate ions complex with uranyl ions, forming highly soluble uranyl carbonate complexes (as detailed in Sect. 4.1), thereby enhancing uranium dissolution from uranium-bearing minerals. Acid leaching, which typically employs sulfuric acid, solubilizes uranium through a combination of oxidation (if U(IV) minerals are present), acid dissolution of gangue minerals, and the formation of uranyl sulfate complexes. The efficiency of uranium leaching is directly dependent on the solubility of uranium under the applied leaching conditions, and optimization of the leaching parameters (e.g., the leaching agent concentration, pH, redox potential, temperature, and leaching time) is crucial for maximizing uranium recovery and minimizing reagent consumption and environmental impact [68]. After leaching, uranium is recovered from the pregnant leach solution through various separation and concentration techniques, such as solvent extraction, ion exchange, or chemical precipitation, to produce uranium concentrates for nuclear fuel fabrication. Conversely, in the context of mine closure and the environmental management of mining sites, minimizing uranium solubility in mine tailings and waste rock is critical to prevent the release of uranium into the surrounding environment.

7.3 Nuclear Waste Management and Long-Term Uranium Solubility Control

In the context of long-term geological disposal of spent nuclear fuel and high-level radioactive waste, controlling uranium solubility is of paramount importance for ensuring the long-term safety and integrity of deep geological repositories [69]. Spent nuclear fuel contains significant quantities of uranium, along with other radionuclides, and the potential release of these radionuclides into the geosphere over extended timescales (thousands to millions of years) is a primary concern in repository design and performance assessment. The multibarrier concept for geological repositories relies on a combination of engineered and natural barriers to isolate nuclear waste and minimize radionuclide migration. Controlling uranium solubility is a key aspect of barrier systems.

In the reducing geochemical conditions expected in deep geological repositories in suitable geological formations (e.g., deep clay formations, crystalline rock), uranium solubility is anticipated to be intrinsically low owing to the predominance of the less soluble U(IV) oxidation state and the potential for precipitation of U(IV) solid phases, particularly uranium dioxide, which is a major component of the spent nuclear fuel matrix [Reference: Uranium Solubility in Geologic Repository Conditions]. However, perturbations to the repository environment, such as radiolysis-induced oxidation, groundwater inflow with oxidizing potential, or the presence of complexing ligands from waste materials or geological formations, could increase uranium solubility and radionuclide release [70]. Therefore, repository design and performance assessment include detailed analyses of uranium solubility under anticipated repository conditions, considering the long-term evolution of geochemical parameters, the potential for solubility-enhancing or solubility-reducing processes, and the retardation of uranium migration through geochemical interactions with repository materials and the host rock [71]. Engineered barriers, such as waste forms with low uranium leachability, corrosion-resistant waste canisters, and backfill materials with a sorption capacity for uranium, are designed to further limit uranium release and migration, complementing the natural geochemical barrier provided by geological formations. Long-term monitoring and research are essential to validate the predicted low uranium solubility and containment performance of geological repositories over extended timescales.

In summary, uranium solubility is a critical physicochemical characteristic with significant consequences for environmental safeguarding, resource management, and the secure handling of nuclear substances. A thorough comprehension of the determinants of uranium solubility, along with the capacity to predict and modify its solubility, is crucial for tackling uranium contamination, enhancing uranium extraction methods, and guaranteeing the enduring safety of nuclear waste disposal, thus fostering environmental sustainability and responsible technological advancement in the nuclear sector.

8 Conclusion: The Complex Interplay of Chemistry in Uranium Solubility

In our study, we extensively delved into the complicated phenomenon of uranium solubility, which is controlled by multifaceted interactions between chemical and environmental conditions. This chapter emphasizes the crucial role of chemistry in interpreting the natural distribution and extraction of uranium, encompassing the entire process from the fundamentals of uranium chemistry and the importance of solubility to the impacts of interfering agents and the range of analytical techniques used for its study.

Uranium is a miscellaneous and complicated chemical that signifies radiological and geochemical importance and is mostly determined by its electronic configuration, oxidation state, and distinct coordination adeptness. Uranium solubility in aqueous systems is a dynamic equilibrium that is highly sensitive to a variety of environmental parameters rather than a static feature. Uranium solubility is not merely an element that "dissolves" or "does not dissolve," but rather ranges throughout a continuum that is precisely regulated by the chemical conditions that are impart.

The main substances that can significantly increase uranium solubility have been studied. These include carbonate ions, which, in many natural waters, especially those with a pH between circumneutral and alkaline and noticeable alkalinity, greatly increase uranium mobility by forming highly soluble uranyl carbonate complexes.

Inorganic ligands such as sulfate, fluoride, and chloride, as well as both natural and anthropogenic organic ligands, facilitate improved solubility via complexation mechanisms, albeit generally to a lesser degree. Conversely, several "interfering agents" markedly reduce uranium solubility, promoting immobilization and attenuation. Precipitation reactions involving phosphate, sulfide, and hydroxide, as well as adsorption/sorption processes and microbial bioreduction, are critical mechanisms for diminishing uranium solubility and controlling uranium environmental transport, especially in reducing environments or under particular geochemical conditions.

The examination of uranium solubility and speciation requires the utilization of many analytical methods, each offering distinct and complimentary perspectives. Spectrophotometry and ICP-MS provide reliable techniques for measuring total uranium amounts. Geochemical modeling offers a robust in silico method for forecasting uranium speciation and solubility under diverse conditions, whereas X-ray absorption spectroscopy (XAS) furnishes direct, element-specific structural insights into uranium species inside intricate matrices. The combined use of these approaches is crucial for obtaining comprehensive knowledge of uranium behavior in intricate environmental and industrial systems.

The ramifications of uranium solubility are extensive, affecting environmental pollution, uranium resource exploitation in mining and processing, and the enduring management of nuclear waste. The solubility of uranium directly influences its environmental mobility, bioavailability, and associated health hazards. In uranium mining, solubility is a vital operational characteristic that can be adjusted to increase extraction efficiency. In nuclear waste disposal, reducing long-term uranium solubility is a critical safety goal, necessitating a comprehensive understanding of geochemical controls and the application of multibarrier containment solutions.

In conclusion, the solubility of uranium is not a straightforward intrinsic feature but rather a complicated emergent behavior resulting from the intricate interaction of its underlying chemistry with a dynamic and varied chemical environment. A thorough understanding of the intricate chemistry influencing uranium solubility is essential not only for academic inquiry but also for tackling urgent environmental issues, promoting sustainable resource management, and protecting both environmental and human health in the nuclear era. Ongoing investigations into the fundamental and applied dimensions of uranium solubility are essential for addressing the intricacies of the Anthropocene and securing a responsible, sustainable future.

References

1. Reyners P (1996) The convention on nuclear safety of 1994. Rev Eur Commun Int Environ Law 5(3):231–238. https://doi.org/10.1111/j.1467-9388.1996.tb00280.x
2. Cowan R (1990) Nuclear power reactors: a study in technological lock-in. J Econ Hist 50(3):541–567. https://doi.org/10.1017/s0022050700037153
3. Peiffert C, Nguyen-Trung C, Cuney M (1996) Uranium in granitic magmas: Part 2. Experimental determination of uranium solubility and fluid-melt partition coefficients in the uranium oxide-haplogranite-H_2O-NaX (X = Cl, F) system at 770°C, 2 kbar. Geochim Cosmochim Acta 60(9):1515–1529. https://doi.org/10.1016/0016-7037(96)00039-7
4. DeBoom MJ (2022) Radioactive strategies: geopolitical rivalries, African Agency, and the longue durée of nuclear infrastructures in Namibia. Policy Press eBooks, pp 137–152. https://doi.org/10.1332/policypress/9781529220773.003.0010
5. Langmuir D (1997). Aqueous environmental geochemistry. https://is.muni.cz/publication/681897/cs/Aqueous-environmental-geochemistry/Langmuir
6. Bear J (1981) Groundwater hydrology. J Hydrol 53(3–4):371–373. https://doi.org/10.1016/0022-1694(81)90015-9
7. Filippelli GM (2011) Phosphate rock formation and marine phosphorus geochemistry: the deep time perspective. Chemosphere 84(6):759–766. https://doi.org/10.1016/j.chemosphere.2011.02.019
8. Sam Keith MS et al (2002) Toxicological profile for uranium. CRC Press eBooks. https://doi.org/10.1201/9781420061888_ch157
9. Pyle G, Swanson S, Lehmkuhl D (2001) Toxicity of uranium mine-receiving waters to caged fathead minnows, Pimephales promelas. Ecotoxicol Environ Saf 48(2):202–214. https://doi.org/10.1006/eesa.2000.2016
10. Stumm W, Morgan JJ (1996) Aquatic chemistry: chemical equilibria and rates in natural waters. Choice Rev Online 33(11):33–6312. https://doi.org/10.5860/choice.33-6312
11. Riedel T, Kübeck C (2017) Uranium in groundwater—a synopsis based on a large hydrogeochemical data set. Water Res 129:29–38. https://doi.org/10.1016/j.watres.2017.11.001
12. Van Berk W, Fu Y (2016) Redox roll-front mobilization of geogenic uranium by nitrate input into aquifers: risks for groundwater resources. Environ Sci Technol 51(1):337–345. https://doi.org/10.1021/acs.est.6b01569
13. Maher K, Bargar JR, Brown GE (2012) Environmental speciation of actinides. Inorg Chem 52(7):3510–3532. https://doi.org/10.1021/ic301686d

14. Von Gunten HR, Beneš P (1995) Speciation of radionuclides in the environment. Radiochim Acta 69(1):1–30. https://doi.org/10.1524/ract.1995.69.1.1
15. Grenthe I (1992) Chemical thermodynamics of uranium. Elsevier Science and Technology
16. Morss L, Edelstein NM, Fuger J (2010b) The chemistry of the actinide and transactinide elements (set Vol.1-6). Springer Science & Business Media
17. Jarvis KE, Gray AL, Houk RS (2003) Handbook of inductively coupled plasma mass spectrometry. Twayne Publishers
18. Brookins DG (2012) Geochemical aspects of radioactive waste disposal. Springer Science & Business Media
19. Merkel BJ, Hasche-Berger A (2006) Uranium in the environment: mining impact and consequences. Springer Science & Business Media
20. Ojovan MI, Hyatt NC (2020) Materials for nuclear waste immobilization. MDPI
21. Brodie EL, DeSantis TZ, Joyner DC, Baek SM, Larsen JT, Andersen GL, Hazen TC, Richardson PM, Herman DJ, Tokunaga TK, Wan JM, Firestone MK (2002) Application of a high-density oligonucleotide microarray approach to study bacterial population dynamics during uranium reduction and Reoxidation. Appl Environ Microbiol 72(9):6288–6298. https://doi.org/10.1128/aem.00246-06
22. Merritt RC (1971) The extractive metallurgy of uranium. http://ci.nii.ac.jp/ncid/BA87394543
23. Kerr RA (1979) Geologic disposal of nuclear wastes: salt's lead is challenged. Science 204(4393):603–606. https://doi.org/10.1126/science.204.4393.603
24. Aly MM, Hamza MF (2013) A review: studies on uranium removal using different techniques. Overview. J Dispers Sci Technol 34(2):182–213. https://doi.org/10.1080/01932691.2012.657954
25. Parsons JG, Aldrich MV, Gardea-Torresdey JL (2002) Environmental and biological applications oExtended X-ray absorption fine structure (Exafs) and X-ray absorption near edge structure (Xanes) spectroscopies. Appl Spectrosc Rev 37(2):187–222. https://doi.org/10.1081/asr-120006044
26. Formenti P, Schütz L, Balkanski Y, Desboeufs K, Ebert M, Kandler K, Petzold A, Scheuvens D, Weinbruch S, Zhang D (2011) Recent progress in understanding physical and chemical properties of African and Asian mineral dust. Atmos Chem Phys 11(16):8231–8256. https://doi.org/10.5194/acp-11-8231-2011
27. Wieland E, Macé N, Dähn R, Kunz D, Tits J (2010) Macro- and micro-scale studies on U(VI) immobilization in hardened cement paste. J Radioanal Nucl Chem 286(3):793–800. https://doi.org/10.1007/s10967-010-0742-y
28. Aqeel M, Jamil M, Yusoff I (2014) Soil contamination, risk assessment and remediation. InTech eBooks. https://doi.org/10.5772/57287
29. Lambert D, Gilbert RO (1988) Statistical methods for environmental pollution monitoring. J Am Stat Assoc 83(404):1226. https://doi.org/10.2307/2290180
30. OECD (2015) Introduction of thorium in the nuclear fuel cycle, Nuclear Science. OECD. https://doi.org/10.1787/9789264241732-en
31. Laurent O, Gomolka M, Haylock R, Blanchardon E, Giussani A, Atkinson W, Baatout S, Bingham D, Cardis E, Hall J, Tomasek L, Ancelet S, Badie C, Bethel G, Bertho J, Bouet S, Bull R, Vathaire CC, Cockerill R et al (2016) Concerted Uranium Research in Europe (CURE): toward a collaborative project integrating dosimetry, epidemiology and radiobiology to study the effects of occupational uranium exposure. J Radiol Prot 36(2):319–345. https://doi.org/10.1088/0952-4746/36/2/319
32. Kalderis D, Juhasz AL, Boopathy R, Comfort S (2011) Soils contaminated with explosives: environmental fate and evaluation of state-of-the-art remediation processes (IUPAC technical report). Pure Appl Chem 83(7):1407–1484. https://doi.org/10.1351/pac-rep-10-01-05
33. Hecht G (2023) The hollow rand. In: Residual governance: how south africa foretells planetary futures. Duke University Press eBooks, Durham, NC, pp 46–83. https://doi.org/10.1215/9781478027263-003

34. Rawlings DE (2005) Characteristics and adaptability of iron- and sulfur-oxidizing microorganisms used for the recovery of metals from minerals and their concentrates. Microb Cell Factories 4(1). https://doi.org/10.1186/1475-2859-4-13
35. Li C, Liu Q, Liu L, Li Y, Zhao X, Liu Z (2024) Simulation of the natural attenuation of groundwater contaminated by acid in situ leaching at a uranium mine. J Radioanal Nucl Chem 333(10):4907–4917. https://doi.org/10.1007/s10967-024-09624-1
36. O'Gorman NG, Von Michaelis NH, Olson NGJ (2004) Novel in-situ metal and mineral extraction technology. https://doi.org/10.2172/835781
37. OECD (2015) Radioactive waste management and constructing memory for future generations. In: Radioactive waste management. OECD. https://doi.org/10.1787/9789264249868-en
38. Çevirim-Papaioannou N, Yalçıntaş E, Gaona X, Dardenne K, Altmaier M, Geckeis H (2018) Redox chemistry of uranium in reducing, dilute to concentrated NaCl solutions. Appl Geochem 98:286–300. https://doi.org/10.1016/j.apgeochem.2018.07.006
39. OECD (1999) Geological disposal of radioactive waste. In: Radioactive waste management. OECD. https://doi.org/10.1787/9789246180547-en
40. Witherspoon P (1991) Geological problems in radioactive waste isolation—a world-wide review. https://doi.org/10.2172/138336
41. Cox DP, Singer DA (1986) Mineral deposit models. https://doi.org/10.3133/b1693
42. Bryan CR, Siegel MD (2003) Environmental geochemistry of radioactive contamination. https://doi.org/10.2172/915153
43. Baker RJ (2013) Uranium minerals and their relevance to long term storage of nuclear fuels. Coord Chem Rev 266–267:123–136. https://doi.org/10.1016/j.ccr.2013.10.004
44. Berto S, Crea F, Daniele PG, Gianguzza A, Pettignano A, Sammartano S (2011) Advances in the investigation of dioxouranium(VI) complexes of interest for natural fluids. Coord Chem Rev 256(1–2):63–81. https://doi.org/10.1016/j.ccr.2011.08.015
45. Liddle ST (2015) The renaissance of non-aqueous uranium chemistry. Angew Chem Int Ed 54(30):8604–8641. https://doi.org/10.1002/anie.201412168
46. Kertes AS, Guillaumont R (1985) Solubility of UO_2. A comparative review. Nucl Chem Waste Manag 5(3):215–219. https://doi.org/10.1016/0191-815X(85)90080-4
47. Schreckenbach G, Hay PJ, Martin RL (1999) Density functional calculations on actinide compounds: survey of recent progress and application to [UO2X4]2? (X=F, Cl, OH) and AnF6 (An=U, Np, Pu). J Comput Chem 20(1):70–90. https://doi.org/10.1002/(sici)1096-987x(19990115)20:1
48. Bryan ND, Abrahamsen L, Evans N, Warwick P, Buckau G, Weng L, Van Riemsdijk WH (2011) The effects of humic substances on the transport of radionuclides: recent improvements in the prediction of behaviour and the understanding of mechanisms. Appl Geochem 27(2):378–389. https://doi.org/10.1016/j.apgeochem.2011.09.008
49. Bone SE, Dynes JJ, Cliff J, Bargar JR (2017) Uranium(IV) adsorption by natural organic matter in anoxic sediments. Proc Natl Acad Sci 114(4):711–716. https://doi.org/10.1073/pnas.1611918114
50. Saito T, Nagasaki S, Tanaka S, Koopal LK (2004) Application of the NICA-Donnan model for proton, copper and uranyl binding to humic acid. Radiochim Acta 92(9–11):567–574. https://doi.org/10.1524/ract.92.9.567.54984
51. Zachara JM, Brown CF, Christensen JN, Davis JA, Dresel PE, Liu C, Kelly SD, McKinley JP, Serne RJ, Um W (2007) A site wide perspective on uranium geochemistry at the Hanford site. https://doi.org/10.2172/920206
52. Haas JR, Northup A (2004) Effects of aqueous complexation on reductive precipitation of uranium by Shewanella putrefaciens. Geochem Trans 5(3). https://doi.org/10.1186/1467-4866-5-41
53. Brannock WW (1953) Contributions to geochemistry, 1949. https://doi.org/10.3133/b992
54. Froideval A, Del Nero M, Barillon R, Hommet J, Mignot G (2003) pH dependence of uranyl retention in a quartz/solution system: an XPS study. J Colloid Interface Sci 266(2):221–235. https://doi.org/10.1016/s0021-9797(03)00528-9

55. Goulet RR, Fortin C, Spry DJ (2011) Uranium. In: Fish physiology, pp 391–428. https://doi.org/10.1016/s1546-5098(11)31030-8
56. Smedley P, Kinniburgh D (2022) Uranium in natural waters and the environment: distribution, speciation and impact. Appl Geochem 148:105534. https://doi.org/10.1016/j.apgeochem.2022.105534
57. Philipp T, Azzam SSA, Rossberg A, Huittinen N, Schmeide K, Stumpf T (2019) U(VI) sorption on ca-bentonite at (hyper)alkaline conditions—spectroscopic investigations of retention mechanisms. Sci Total Environ 676:469–481. https://doi.org/10.1016/j.scitotenv.2019.04.274
58. Bailey E, Mosselmans J, Schofield P (2004) Uranyl-citrate speciation in acidic aqueous solutions—an XAS study between 25 and 200 °C. Chem Geol 216(1–2):1–16. https://doi.org/10.1016/j.chemgeo.2004.10.011
59. Banala UK, Das NPI, Toleti SR (2020) Microbial interactions with uranium: towards an effective bioremediation approach. Environ Technol Innov 21:101254. https://doi.org/10.1016/j.eti.2020.101254
60. Majumder EL, Wall JD (2017) Uranium bio-transformations: chemical or biological processes? Open J Inorgan Chem 7(2):28–60. https://doi.org/10.4236/ojic.2017.72003
61. Beazley MJ, Martinez RJ, Sobecky PA, Webb SM, Taillefert M (2007) Uranium biomineralization as a result of bacterial phosphatase activity: insights from bacterial isolates from a contaminated subsurface. Environ Sci Technol 41(16):5701–5707. https://doi.org/10.1021/es070567g
62. Khan MH, Warwick P, Evans N (2005) Spectrophotometric determination of uranium with arsenazo-III in perchloric acid. Chemosphere 63(7):1165–1169. https://doi.org/10.1016/j.chemosphere.2005.09.060
63. Lee JH, Wang Z, Liu J, Lu Y (2008) Highly sensitive and selective colorimetric sensors for uranyl (UO_2^{2+}): development and comparison of labeled and label-free DNAZyme-gold nanoparticle systems. J Am Chem Soc 130(43):14217–14226. https://doi.org/10.1021/ja803607z
64. Heumann KG (2004) Isotope-dilution ICP?MS for trace element determination and speciation: from a reference method to a routine method? Anal Bioanal Chem 378(2):318–329. https://doi.org/10.1007/s00216-003-2325-z
65. Thomas R (2003) Practical guide to ICP-MS: a tutorial for beginners. CRC Press
66. Albarède F (1995) Introduction to geochemical modeling. Cambridge University Press
67. WHO (1993) Guidelines for drinking-water quality. World Health Organization
68. Chapman DV (1996) Water quality assessments: a guide to the use of biota, sediments and water in environmental monitoring, 2nd edn. CRC Press
69. Caruso JA, Crews H, Heumann KG (2005) Handbook of elemental speciation II: species in the environment, food, medicine and occupational health. Wiley
70. Iaea. (2020) Occupational radiation protection in the uranium mining and processing industry. International Atomic Energy Agency
71. Choppin G, Liljenzin J, Rydberg J (2002) Radiochemistry and nuclear chemistry. Butterworth-Heinemann

Uranium Solubility in Various Water Systems

Exploring Uranium Solubility in Freshwaters (Groundwater & Surface Water)

Sai Kumar Punna, Suvarshitha Pusuluru, Karrun Velmurugan, Selvarajan Ethiraj, and Melvin Samuel

Abstract Uranium is a naturally occurring element with complex geochemical behavior that is influenced by mineralogical, redox, and isotopic processes across surface and groundwater systems. This chapter critically reviews the occurrence, mobilization, and isotopic signatures of uranium in freshwater environments, emphasizing the environmental and health risks associated with uranium and fluoride contamination in drinking water sources. This study examines the geogenic and anthropogenic drivers of uranium release, alongside microbial and abiotic mechanisms that govern uranium reduction and oxidation. Particular attention has been given to recent advances in bioremediation strategies, nanotechnology applications, and managed aquifer recharge for mitigating uranium and fluoride exposure. By integrating geochemical, microbial, and hydrogeological perspectives, this chapter highlights the urgent need for sustainable groundwater management and outlines future directions for minimizing contaminant risks to ecosystems and human health.

S. K. Punna
Department of Materials Science and Engineering, University of Houston, Houston, TX, USA

S. Pusuluru
Department of Engineering Data Science, University of Houston, Houston, TX, USA

K. Velmurugan
Department of Mechanical Engineering, Indian Institute of Technology – Kharagpur, Kharagpur, West Bengal, India

S. Ethiraj
Department of Genetic Engineering, College of Engineering and Technology, School of Bioengineering, SRM Institute of Science and Technology, Kattankulathur, Tamil Nadu, India

M. Samuel (✉)
Department of Bioengineering, Saveetha School of Engineering, Saveetha Institute of Medical and Technical, Chennai, Tamil Nadu, India

Department of Civil, Construction and Environmental Engineering, Marquette University, Milwaukee, WI, USA

M. K. Jindal, K. Patni (eds.), *Water Quality-Uranium Solubility Nexus: Unlocking the Secrets of Uranium Solubility*, Earth and Environmental Sciences Library, https://doi.org/10.1007/978-3-031-98909-4_13

Keywords Uranium mobilization · Groundwater contamination · Isotopic fractionation · Microbial bioremediation · Fluoride health risks

1 Introduction to Uranium in Aquatic Systems

Uranium, a heavy metal and naturally radioactive element, is pervasive in the Earth's crust and aquatic environments due to natural geological processes and increasingly intensified anthropogenic activities. Its behavior in surface water and aquifers is highly sensitive to environmental conditions, notably redox state, mineralogy, and water chemistry. Although global averages of uranium concentrations in river systems remain low, localized enrichments in uranium-rich geological settings, agricultural catchments, and limestone aquifers underscore the variability and complexity of its environmental distribution [1–3]. Simultaneously, the isotopic fractionation of uranium, particularly between ^{234}U and ^{238}U, provides valuable tools for tracing groundwater flow paths, redox transitions, and weathering dynamics. Rising concerns over groundwater contamination, primarily through the dual threats of uranium and fluoride, have underscored the need for comprehensive assessments of groundwater quality and public health risks. Urban growth fosters economic growth and leads to social development; on the other hand, it leads to significant environmental risks. Rapid urbanization in developing countries speeds up environmental deprivation, with groundwater resulting in the brunt of these impacts [4].

Over the years, the basin has experienced a significant increase in population growth, industrial accomplishments, and infrastructure advancements, all of which have reinforced pressure on the region's water sources. Groundwater, an essential resource in this region, is widely applied for drinking, irrigation, and industrialized goals. However, urbanization has resulted in improved groundwater extraction and improved contaminant consequences. Urban sources of toxic waste, including domestic sewage, industrial discharge, and surface runoff, introduce diverse pollutants into groundwater, comprising nitrates, heavy metals and pathogens [1–3]. Groundwater itself has high salinity and contains heavy metals, radioactive elements, and major and minor ions that can cause severe health hazards to humans. Owing to human activity, groundwater becomes susceptible to contamination, and the quality of the water is important. Notably, the inorganic element uranium (U) has contaminated several water resources through both geogenic and anthropogenic activities. The uranium is present in rocky substances and in certain regions of the ore body, migrates through the water table and contaminates the groundwater. Contamination can also occur via groundwater exploitation, such as the use of excess toxic fertilizer and chemicals [5].

In contrast, smaller catchments often exhibit greater fluctuations in uranium levels. In most rivers, uranium concentrations remain below 4 μg/L, although naturally elevated levels-exceeding 10 μg/L can be observed in under 1% of basins, typically those that drain uranium-rich geological formations [6–8]. While much remains unknown about natural uranium variability in small streams, recent interest has turned toward understanding uranium dynamics in agricultural catchments [9]. Groundwater systems, however, have much broader concentration ranges, ranging from 0.03 μg/L to as high as 120 μg/L [10]. Uranium mobility in these subsurface environments is heavily influenced by redox conditions: it tends to be more soluble under oxidizing conditions and less so in reducing environments. Alkaline aquifers, such as those dominated by limestone, often promote uranium mobility through the formation of stable uranyl–carbonate complexes [11].

The principal processes responsible for the release of uranium into groundwater include the weathering of natural mineral sources and desorption from mineral surfaces [12]. Both natural and anthropogenic activities, such as the use of nitrogen-based fertilizers, can significantly increase uranium solubility and transport, thereby deteriorating groundwater quality [13, 14]. Recharge events in which oxidizing waters are introduced into shallow aquifers also increase uranium mobility [15]. Selenium (Se), another trace element with similar redox-sensitive mobility, often cooccurs with uranium, especially in roll-front deposits where dissolved elements migrate from oxidized to reduced subsurface zones. Organic-rich deposits, including detrital sediments and peat and lignite layers, are important sinks for uranium accumulation [16–18]. Pyrite- and organic matter-enriched argillaceous sediments in karst cavities have also been identified as co-concentrators of uranium and selenium [19]. It has been reported that uranium is present in the Earth's crust, is distributed throughout the entire crust in trace quantities (2.7 mg/L), and it is known as a rocky formation. This uranium is classified into three isotopes, viz., ^{238}U, ^{235}U, and ^{234}U, in which ^{238}U and ^{234}U are generally found in groundwater [20, 21]. To prevent uranium contamination, in-depth research is needed to identify the occurrence and deposits of uranium. Several developing nations utilize groundwater and surface water for daily needs, and in certain areas, uranium contamination is high. A recent study revealed that the uranium concentration has exceeded the limit of the WHO standard for drinking water in both groundwater and surface water. Long-term exposure to uranium can lead to various health hazards related to carcinogenic as well as noncarcinogenic diseases, such as bone disorders, neurotoxicity, nephrotoxicity, and pulmonary disease [16, 22–25].

In general, three uranium isotopes are detected in rocks: ^{238}U (half-life: 4.47. 10^9 y), ^{235}U (7.04×10^8 y), ^{234}U (2.46×10^5 y) and ^{234}U (2.46×10^5 y). ^{238}U and ^{235}U are primordial radionuclides present on Earth since their formation, whereas ^{234}U has a radiogenic origin and is produced through the radioactive disintegration of ^{238}U. The radiogenic fraction yields preferential leaching of 234 U relative to 238 U during chemical weathering. Consequently, surface and groundwaters generally display an enrichment of 234 U with respect to 238 U, with corresponding (234 U/238 U) activity ratios (ARs) above 1. Waters depleted in 234 U indicate that the weathered solid surface is also depleted in ^{234}U. In soil leachates, ^{234}U deficit indicates that surface or subsurface soils have already experienced 234 U loss by leaching, as has been noted in soil or river catchments. The riverine value of AR is an average of 1.17. On a worldwide scale, observations of ^{234}U deficits regarding ^{238}U (AR below 1) in freshwater are scarce: a single example is given by the Strengbach watershed

(France), where a slight ^{234}U deficit has been occasionally documented in streams and springs [26]. Reports of AR values less than 1 in surface waters are rare worldwide. One documented case is the Strengbach watershed in France, where values as low as 0.966 in a stream and 0.819 in a spring were observed [21, 27]. Groundwater, however, has a broader AR spectrum, ranging from 0.9 to as high as 12. Isolated AR values have been recorded at or below 0.5, typically in hydrogeological settings characterized by high permeability, rapid groundwater movement, and active dissolution [28].

2 Uranium Isotopic Signature and Activity

Precipitation is crucial for infiltrating groundwater from the surface to the subsurface and is collected in the aquifer [29]. Phosphate fertilizers, heavy metals and natural organic matter typically contain excess heavy metals, inorganic contaminants and radionuclides that can penetrate groundwater from the agricultural sector and contaminate the groundwater region [30]. Nevertheless, groundwater is contaminated by uranium through geogenic activity. The rocks and soils present in uranium ore are the major sources of groundwater contamination [25, 31]. Uranium contamination is due to the lithology and geology of an area. The primary minerals that hold uranium are converted to secondary minerals, which leads to the leaching of uranium into the groundwater [32]. Uranium is present as vandates, silicates, phosphates and oxides. Additionally, during rock–water interactions, phosphate and carbonate rocks tend to leach uranium into soil, groundwater and surface water. In certain uranium mine tailings areas, high concentrations of uranium in groundwater have been observed. In some areas, the fertilizers used in the area contain radionuclides, which can contaminate the groundwater. The US EPA and WHO have developed regulations and stated that the maximum contamination level of uranium in drinking water is set to 30 ppb for balanced health [33].

Uranium is found in surface waters mainly because of the weathering of soils and rocks, and its amount generally reflects the lithology of weathered bedrock [34]. In riverine water worldwide, the average uranium concentration was found to be 0.19 μg/L. Most rivers draining small catchments are likely to record relatively high uranium concentrations [35]. The uranium concentration is naturally less than 4 μg/L in river water. In certain basins, streams display increased uranium concentrations greater than 10 μg/L [36]. Only very little information is available on the geogenic fluctuations of uranium in small streams, and increased attention has been given to uranium in agricultural drainage areas [37, 38]. The uranium concentration in groundwater is reported to vary much more, with values ranging from 0.03 to 130 μg/L. The uranium concentration in groundwater is constrained by the oxidation reduction potential [39]. The uranium is more soluble under oxidizing conditions than under reducing conditions. The presence of limestone or alkaline conditions favors uranium mobility in groundwater through carbonate formations. The two major principal mechanisms behind uranium mobility in groundwater are

(i) weathering of geogenic source material and (ii) desorption from mineral surfaces [40]. Anthropogenic and geogenic factors increase uranium mobility in groundwater, leading to the degradation of drinking water quality. Researchers have reported that nitrate in groundwater promotes uranium mobility. In certain shallow aquifers, uranium mobility is favored during the recharge of oxidizing water into the aquifer. Redox conditions drive selenium and uranium in groundwater. The uranium and selenium are found in the U roll front deposits. In sedimentary basins, deposits enriched with organic matter play an important role in U accumulation in the form of detrital material, peat and lignite formations [39–41].

At the head of the Essoner River valley (seine Basin), a maximum uranium concentration of 22 μg/L (100-fold the average river concentration) was occasionally detected in the framework of a radicological national survey. In parallel, isotope analyses conducted in the stream river revealed AR values below 0.5, indicating a large deficit in 234 U compared with 238 U. To date, such a low AR (half that of 1.17) has not been described in other surface waters around the world. The ground catchment of the Euf River is part of the Beauce Limestone Aquifier System (BLAS), one of the major French aquifiers in the Paris Basin [42]. In this reservoir, high Se concentrations (above 10 μg/L) are occasionally found in groundwater for drinking water. The role of river and groundwater interactions was examined in this chapter to provide new insight into the characteristics of geogenic uranium in small streams and to improve the understanding of typical observations of freshwater [43, 44].

3 Geogenic Sources and Variability of Uranium in Surface and Groundwater

Groundwater serves as a primary resource for drinking and domestic needs across the globe. However, in many regions, this vital resource is compromised by elevated levels of salinity, heavy metals, radionuclides, and major and trace ions, posing significant health risks to the population [33, 45, 46]. With increasing reliance on groundwater for everyday consumption, especially in regions with limited surface water access, ensuring water quality has become a public health priority [47]. Currently, groundwater contamination is a growing concern worldwide, stemming from both natural geogenic processes and intensified anthropogenic activities [48]. Among the most concerning inorganic pollutants are uranium (U) and fluoride (F^-), which contaminate groundwater through both natural mineral leaching and human-induced sources [49]. Uranium- and fluoride-bearing minerals, often embedded in host rocks or ore zones, release these elements into groundwater under favorable chemical and hydrological conditions. Additionally, excessive application of fertilizers and over-extraction of groundwater further exacerbate the spread of these contaminants. Rani et al. [50] emphasized that both rural and urban communities in India heavily depend on groundwater for drinking purposes, which increases vulnerability to exposure. The pressures of population growth and rapid industrial and agricultural expansion have compounded the issue, transforming groundwater contamination into a critical public health crisis [51]. To mitigate these threats, it is essential to evaluate groundwater quality in detail, with a focus on the hydrogeochemical behavior of aquifers and the spatial distribution of harmful constituents [52]. High fluoride concentrations in groundwater are often linked to the dissolution of minerals such as fluorite, apatite, biotite, and hornblende from geological formations. Fluoride levels may also rise due to human activities, particularly the widespread use of phosphate-based fertilizers in agricultural practices [50, 53].

4 Biogenic and Toxicity Profiles of Uranium

Research has shown that 90 μg of uranium is usually present in the human body during the consumption of food, and most uranium (95%) is expelled through the metabolism of the kidney. Excessive intake of uranium above the threshold level has a significant impact on human life. The health hazard or lifetime average daily dose (LADD) ratio was calculated by the daily intake of water as well as the weight and age of the person [54, 55]. Uranium exposure for a longer period of time through oral intake changes human metabolism and functions, leading to severe health hazards. The regular intake of uranium through drinking water into bone affects bone function and metabolism and increases the elimination of calcium and phosphate from the body. It has been reported that reproductive organs are affected by the intake of uranium, which leads to reproductive toxicity, causing oxidative stress and inflammation. In certain cases, uranium can cause neurotoxicity, and uranium exposure to the brain affects neurons and can cause anxiety, neurological disorders, oxidative stress, and neuronal inflammation and disrupt sleep. Finally, it can lead to paralysis, lung cancer, and kidney and thyroid cancer.

Biological reduction of U(VI) typically occurs via dissimilatory pathways, where the reduction process supports cellular respiration and growth, although uranium itself is not a necessary nutrient. To date, no specific U(VI)-reducing enzyme has been identified. Instead, the reduction is often facilitated by components of the electron transport chain that serve broader metabolic roles. Notably, c-type cytochromes have been implicated in U(VI) reduction across Desulfovibrio, Shewanella, and Geobacter species [56].

The cellular localization of U-reducing cytochromes appears to be associated with the periplasmic space, the outer membrane, and possibly the extracellular matrix [55]. Biochemical studies suggest that these enzymes reduce U(VI) to U(V) through a single-electron transfer mechanism. The transient U(V) species then undergoes disproportionation, yielding both U(VI) and U(IV) [57]. Furthermore, Geobacter species possess an additional mechanism involving conductive extracellular pili, which facilitates increased U(VI) reduction rates and prevents the intracellular precipitation of reduced uranium, thus preserving cell function [58]. Disruption of c3 cytochrome genes in Shewanella mutants led to decreased but not abolished U(VI) reduction, suggesting redundancy in the electron transfer pathways and overlap with other processes, such as Fe(III) reduction [59].

In subsurface environments, redox processes follow a sequence governed by redox potential—from oxygen to nitrate, manganese(IV), iron(III), and sulfate. The U(VI)/U(IV) couple has a redox potential close to that of Fe(III)/Fe(II), making it sensitive to shifts in redox chemistry and aqueous uranium speciation. The speciation of uranium in solution critically influences its bioavailability. While most metal-reducing microbes can reduce U(VI) in carbonate-complexed forms, certain uranyl-organic or ternary complexes, especially those involving calcium, can inhibit U(VI) reduction in some sulfate-reducing strains, although the exact inhibition mechanisms remain unclear [54, 55, 60]. Experimental data from Shewanella oneidensis show that increased bicarbonate concentrations reduce U(VI) reduction rates, highlighting the sensitivity of the process to geochemical parameters such as carbonate and calcium ion concentrations and pH [61]. Although the presence of Ca-uranyl-carbonate complexes may suppress the kinetics of U(VI), microbial activity persists under such conditions [62]. Consequently, the efficacy of microbial uranium reduction under field conditions is inherently influenced by environmental factors, including pH, mineral interactions, and the composition of microbial communities. Under laboratory conditions, U(VI) reduction typically results in the formation of a fine, black precipitate—biogenic uraninite [UO_2(s)]. First identified in Geobacter metallireducens cultures, uraninite has also been observed in association with numerous other microbial species [63]. This precipitate often forms extracellular polymeric substances [64]. Biogenic uraninite is nanocrystalline (2–10 nm) and structurally distinct from bulk UO_2, showing a contracted lattice structure at the surface due to a high surface-to-volume ratio, whereas the interior remains crystalline and stoichiometrically consistent with UO_2 [65].

5 Remediation Strategies for Uranium and Fluoride Contamination

At circumneutral pH, U(IV) oxidation is governed primarily by abiotic processes, with dissolved oxygen (O_2), nitrate (NO_3^-), and solid-phase oxidants such as manganese (Mn) and iron (Fe) oxides acting as the major oxidizing agents. Reactive oxygen species, including peroxides formed from radiolytic decay, also contribute to U oxidation, although their presence is generally restricted to intermediate- and high-level radioactive waste systems and falls beyond the focus of this discussion. In environmental and engineered remediation contexts, interest in U redox behavior has largely centered on the stability of biogenic uraninite—nanocrystalline UO_2 precipitated through microbial reduction of U(VI). While some studies have explored biotic oxidation pathways, abiotic oxidation processes are typically more kinetically favorable under pH conditions. A key topic of debate has involved whether the nanoscale nature of biogenic UO_2 significantly enhances its solubility owing to increased surface area and lattice defects [66] or if it retains solubility characteristics akin to those of bulk, stoichiometric UO_2 due to structural stability [67].

Controlled experiments comparing synthetic and biogenic UO_2 under flow-through conditions revealed that, when normalized to surface area, their oxidation rates were comparable [68, 69]. Although smaller particle sizes increase oxidation rates, particle aggregation can mitigate this effect, limiting the influence of particle size on uranium mobility [67]. Additionally, biogenic UO_2 may incorporate environmental dopants such as Mn, Ca, or rare earth materials into its fluorite crystal structure, thereby reducing its solubility and oxidative susceptibility similar to what is observed in natural uraninite, which is often nonstoichiometric [70]. Mn(II)-doped biogenic UO_2 at rates 38 times lower than those of undoped UO_2 [71], indicating the stabilizing influence of trace metal incorporation. Numerous studies have demonstrated the oxidation of U(IV) by dissolved oxygen, nitrate, and intermediates of denitrification (e.g., nitrate, nitric oxide, and nitrous oxide), as well as by Fe(III) [72, 73].

References

1. Chabaux F, Riotte J, Dequincey O (2003) U-Th-Ra fractionation during weathering and river transport. Rev Mineral Geochem 52:533–576
2. Ollivier P, Radakovitch O, Hamelin B (2011) Major and trace element partition and fluxes in the Rhône River. Chem Geol 285:15–31
3. Zebracki M et al (2017) U isotopes distribution in the lower Rhône River and its implication on radionuclides disequilibrium within the decay series. J Environ Radioact 178–179:1–12
4. Palmer MR, Edmond JM (1993) Uranium in river water. Geochim Cosmochim Acta 57:4947–4955
5. Windom H, Smith R, Niencheski F, Alexander C (2000) Uranium in rivers and estuaries of globally diverse, smaller watersheds. Mar Chem 68:307–321
6. Camacho A et al (2010) Distribution of uranium isotopes in surface water of the Llobregat River basin (Northeast Spain). J Environ Radioact 101:1048–1054
7. Salminen R, et al. (2005) Geochemical Atlas of Europe. Part 1—Background information, methodology and maps. Retrieved from http://weppi.gtk.fi/publ/foregsatlas/index.php
8. Snow DD, Spalding RF (1994) Uranium isotopes in the Platte River drainage basin of the north American high plains region. Appl Geochem 9:271–278
9. Gardner CB et al (2022) Carbonate weathering, phosphate fertilizer, and hydrologic controls on dissolved uranium in rivers in the US corn belt: disentangling seasonal geogenic- and fertilizer-derived sources. Sci Total Environ 817:160455
10. Mangini A, Sonntag C, Bertsch G, Müller E (1979) Evidence for a higher natural uranium content in world rivers. Nature 278:337–339
11. Banning A, Demmel T, Rüde TR, Wrobel M (2013) Groundwater uranium origin and fate control in a river valley aquifer. Environ Sci Technol 47:13941–13948
12. Riedel T, Kübeck C (2018) Uranium in groundwater: a synopsis based on a large hydrogeochemical data set. Water Res 129:29–38
13. Nolan J, Weber KA (2015) Natural uranium contamination in major U.S. aquifers linked to nitrate. Environ Sci Technol Lett 2:215–220

14. Rosen MR, Burow KR, Fram MS (2019) Anthropogenic and geologic causes of anomalously high uranium concentrations in groundwater used for drinking water supply in the southeastern San Joaquin Valley, CA. J Hydrol 577:124009
15. Wu Y, Wang Y, Guo W (2019) Behavior and fate of geogenic uranium in a shallow groundwater system. J Contam Hydrol 222:41–55
16. Cumberland SA, Douglas G, Grice K, Moreau JW (2016) Uranium mobility in organic matter-rich sediments: a review of geological and geochemical processes. Earth Sci Rev 159:160–185
17. Pregler A et al (2019) Increased uranium concentrations in ground and surface waters of the Swiss plateau: a result of uranium accumulation and leaching in the Molasse Basin and ancient wetlands? J Environ Radioact 208–209:106026
18. Read D et al (1993) The migration of uranium into peat-rich soils at Broubster, Caithness, Scotland, UK. J Contam Hydrol 13:291–308
19. Bassil J et al (2016) Investigation of the nature and origin of the geological matrices rich in selenium within the hydrogeological experimental site of Poitiers, France. CR Geosci 348:598–608
20. Israelson C, Björck S, Hawkesworth CJ, Possnert G (1990) Direct U-Th dating of organic and carbonate-rich lake sediments from southern Scandinavia. Earth Planet Sci Lett 97:1–12
21. Riotte J, Chabaux F (1999) ^234U/^238U activity ratios in freshwaters as tracers of hydrological processes: the Strengbach watershed (Vosges, France). Geochim Cosmochim Acta 63:1263–1275
22. Chabaux F, Riotte J, Clauer N, France-Lanord C (2001) Isotopic tracing of the dissolved U fluxes of Himalayan rivers: implications for present and past U budgets of the Ganges-Brahmaputra system. Geochim Cosmochim Acta 65:3201–3217
23. Chabaux F, Bourdon B, Riotte J (2008) U-series geochemistry in weathering profiles, river waters and lakes. Radioact Environ 13:49–104
24. Dahlkamp FJ (1993) Uranium Ore Deposits. Springer, Berlin
25. Durand S, Chabaux F, Rihs S, Duringer P, Elsass P (2005) U isotope ratios as tracers of groundwater inputs into surface waters: example of the Upper Rhine hydrosystem. Chem Geol 220:1–19
26. Fleischer RL (1980) Isotopic disequilibrium of uranium: alpha-recoil damage and preferential solution effects. Science 207:979–981
27. Pierret MC, Stille P, Prunier J, Viville D, Chabaux F (2014) Chemical and U-Sr isotopic variations in stream and source waters of the Strengbach watershed (Vosges mountains, France). Hydrol Earth Syst Sci 18:3969–3985
28. Kaufman MI, Rydell HS, Osmond JK (1969) $^{234}U/^{238}U$ disequilibrium as an aid to hydrologic study of the Floridan aquifer. J Hydrol 9:374–386
29. Michel RL, Kraemer TF, Cecil LD (2009) Surface water, unsaturated zone, and glacial systems. In: Radioactivity in the environment, vol 16, pp 239–264
30. Prasad M et al (2021) Uranium contamination in drinking water as a health concern in Uttarakhand, India. J Radioanal Nucl Chem 331:1933–1940
31. Paces JB, Wurster FC (2014) Natural uranium and strontium isotope tracers of water sources and surface water–groundwater interactions in arid wetlands–Pahranagat Valley, Nevada, USA. J Hydrol 517:213–225
32. Prapanchan VN, Subramani T, Karunanidhi D, Gopinathan P (2024) Groundwater quality assessment for drinking and irrigation purposes and its human health risks in the Sevathur mine region, South India. Desalin Water Treat 320:100883
33. Paikaray S, Chander S (2021) Geochemical variations in uranium and fluoride enriched saline groundwater around a semi-arid region of SW Punjab, India. Appl Geochem 136:105167
34. Ma M, Wang R, Xu L, Xu M, Liu S (2020) Emerging health risks and underlying toxicological mechanisms of uranium contamination: lessons from the past two decades. Environ Int 145:106017
35. Sankhla MS, Kumar R (2018) Fluoride contamination of water in India and its impact on public health. ARC J Forensic Sci 3:10–15

36. Karunanidhi D et al (2020) The effects of geochemical processes on groundwater chemistry and the health risks associated with fluoride intake in a semi-arid region of South India. RSC Adv 10:4840–4855
37. Aravinthasamy P et al (2019) Fluoride contamination in groundwater of the Shanmuganadhi River basin (South India) and its association with other chemical constituents using GIS and multivariate statistics. Geochemistry 80:125555
38. Aravinthasamy P et al (2019) Geochemical evaluation of fluoride contamination in groundwater from Shanmuganadhi River basin, South India: implication on human health. Environ Geochem Health 42:1937–1963
39. Balarama V, Rani A, Rathore DPS (2022) Uranium in groundwater in parts of India and world: a comprehensive review of sources, impacts to the environment and human health, analytical techniques, and mitigation technologies. Geosyst Geoenviron 1:100043
40. Kaishwa SJ, Marwa EM, Msaky JJ (2018) Uranium natural levels in soil, rock and water: assessment of the quality of drinking water in Singida Urban District, Tanzania. J Water Health 16:542–548
41. Bjorklund G et al (2020) Uranium in drinking water: a public health threat. Arch Toxicol 94:1551–1560
42. Cary L, Joulian C, Battaglia-Brunet F, Decouchon E (2018) Arsenic and selenium in groundwater of Beauce and Sologne: characterization across 26 communes. BRGM Report BRGM/RP-67590-FR
43. Dar FA, Kurella S (2023) Fluoride in drinking water: an in-depth analysis of its prevalence, health effects, advances in detection and treatment. Mater Today Proc. https://doi.org/10.1016/j.matpr.2023.05.645
44. Selvaganapathi R et al (2023) Evaluating hydrogeochemical controls and noncarcinogenic health risk assessment of fluoride concentration in groundwater of Palacode and Pennagaram taluk, Dharmapuri district, Tamil Nadu, India. Environ Monit Assess 195:1472
45. Burgos WD et al (2002) Theoretical and experimental considerations related to reaction-based modeling: a case study using iron(III) oxide bioreduction. Geomicrobiol J 19:253–287
46. Karthikeyan P et al (2018) Assessment of heavy metals in the surface sediments of the Emerald Lake using spatial distribution and multivariate techniques. Environ Monit Assess 190:668
47. Babuji P, Thirumalaisamy S, Duraisamy K, Gopinathan P (2023) Human health risks due to exposure to water pollution: a review. Water 15:2532
48. Chander S et al (2023) δ^18O and δ^2H isotopes, trace metals and major ions in groundwater around uranium and fluoride contaminated Indus valley quaternary alluvial plain, SW Punjab, India: implications for hydrogeochemical processes, irrigation use and source. Appl Geochem 152:105652
49. Bacquart T et al (2015) Multiple inorganic toxic substances contaminating the groundwater of Myingyan Township, Myanmar: arsenic, manganese, fluoride, iron, and uranium. Sci Total Environ 517:232–245
50. Rani A et al (2023) Hydrochemical characteristics and potential health risks of nitrate, fluoride, and uranium in Kota district, Rajasthan, India. Environ Sci Pollut Res Int 30:82485–82505
51. Shukla S, Saxena A (2021) Appraisal of groundwater quality with human health risk assessment in parts of Indo-Gangetic alluvial plain, North India. Arch Environ Contam Toxicol 80:55–73
52. Gopinathan P et al (2020) A geo-spatial approach to perceive the groundwater regime of hard rock terrain: a case study from Morappur area, Dharmapuri district, South India. Groundwater for Sustainable Development 10:100316
53. Gopinathan P et al (2023) Environmental impact and health risk assessment due to coal mining and utilization. Environ Geochem Health 45:6915–6922
54. Suzuki Y, Suko T (2006) Geomicrobiological factors that control uranium mobility in the environment: update on recent advances in the bioremediation of uranium-contaminated sites. J Mineral Petrol Sci 101:299–307
55. Wall JD, Krumholz LR (2006) Uranium reduction. Ann Rev Microbiol 60:149–166

56. Lovley DR, Widman PK, Woodward JC, Phillips EJP (1993) Reduction of uranium by cytochrome c3 of Desulfovibrio vulgaris. Appl Environ Microbiol 59:3572–3576
57. Renshaw JC et al (2005) Bioreduction of uranium: environmental implications of a pentavalent intermediate. Environ Sci Technol 39:5657–5660
58. Cologgi DL, Lampa-Pastirk S, Speers AM, Kelly SD, Reguera G (2011) Extracellular reduction of uranium via Geobacter conductive pili as a protective cellular mechanism. Proc Natl Acad Sci USA 108:15248–15252
59. Payne RB, Gentry DM, Rapp-Giles BJ, Casalot L, Wall JD (2002) Uranium reduction by Desulfovibrio desulfuricans strain G20 and a cytochrome c3 mutant. Appl Environ Microbiol 68:3129–3132
60. Ganesh R, Robinson K, Reed G, Sayler G (1997) Reduction of hexavalent uranium from organic complexes by sulfate- and iron-reducing bacteria. Appl Environ Microbiol 63:4385–4391
61. Sheng L, Fein JB (2014) Uranium reduction by Shewanella oneidensis MR-1 as a function of $NaHCO_3$ concentration: surface complexation control of reduction kinetics. Environ Sci Technol 48:3768–3775
62. Brooks SC et al (2003) Inhibition of bacterial U(VI) reduction by calcium. Environ Sci Technol 37:1850–1858
63. Lovley DR, Phillips EJP, Gorby YA, Landa ER (1991) Microbial reduction of uranium. Nature 350:413–416
64. Marshall MJ et al (2006) C-type cytochrome-dependent formation of U(IV) nanoparticles by Shewanella oneidensis. PLoS Biol 4:e268
65. Burgos WD et al (2008) Characterization of uraninite nanoparticles produced by Shewanella oneidensis MR-1. Geochim Cosmochim Acta 72:4901–4915
66. Suzuki Y, Kelly SD, Kemner KM, Banfield JF (2002) Nanometre-size products of uranium bioreduction. Nature 419:134
67. Bargar JR, Bernier-Latmani R, Giammar DE, Tebo BM (2008) Biogenic uraninite nanoparticles and their importance for uranium remediation. Elements 4:407–412
68. Ulrich KU et al (2008) Dissolution of biogenic and synthetic UO_2 under varied reducing conditions. Environ Sci Technol 42:5600–5606
69. Ulrich KU et al (2009) Comparative dissolution kinetics of biogenic and chemogenic uraninite under oxidizing conditions in the presence of carbonate. Geochim Cosmochim Acta 73:6065–6083
70. Finch RJ, Ewing RC (1992) The corrosion of uraninite under oxidizing conditions. J Nucl Mater 190:133–156
71. Veeramani H et al (2009) Effect of Mn(II) on the structure and reactivity of biogenic uraninite. Environ Sci Technol 43:6541–6547
72. Ginder-Vogel M, Criddle CS, Fendorf S (2006) Thermodynamic constraints on the oxidation of biogenic UO_2 by Fe(III) (hydr)oxides. Environ Sci Technol 40:3544–3550
73. Lovley DR, Phillips EJP (1992) Reduction of uranium by Desulfovibrio desulfuricans. Appl Environ Microbiol 58:850–856

Unveiling Uranium Behavior in Polluted Water: Seawater, Wastewater, and Industrial Water

Bhupender Singh, Krishan Kant, Nandita Kanwar, Naresh Tanwer, Rajesh Kumar, Sakshi Yadav, and Kamlesh Kumar Sharma

Abstract Unlocking the behavior of uranium in seawater requires an understanding of its chemical speciation, primarily as a highly soluble uranyl ion (UO_2^{2+}), which stabilizes complexes with phosphate, sulfate, carbonate, etc., to increase mobility. Despite its high solubility, uranium can precipitate as uranite under reducing conditions. The redox mechanism affects its speciation; reducing and oxidizing conditions produce less soluble U(IV) and prefer U(VI). Uranium may adsorb onto particles and sediments, especially in lower settings, and then be absorbed back into the water when environmental conditions change. Because of its bioavailability, it may be hazardous to people, aquatic organisms, and other creatures. Geographical and temporal fluctuations in seawater river supplies, hydrothermal vents, and changes in seawater chemistry influence uranium concentrations. The industries that release uranium-contaminated water are metal mining, nuclear power plants, phosphate fertilizer manufacturing, and uranium mining. Thus, the increased need for military, industrial and agricultural activities will result in higher uranium concentrations in nearby water bodies. Furthermore, using too much groundwater might lead to a higher uranium content. Human activities, including agricultural practices involving the use of chemical fertilizers (containing phosphate, potassium,

B. Singh · K. K. Sharma (✉)
Department of Applied Sciences and Humanities, IIMT College of Engineering, Greater Noida, Uttar Pradesh, India

K. Kant
Maharshi Dayanand University, Rohtak, Haryana, India

N. Kanwar
SSB Hospital, Kailash Colony, Delhi, India

N. Tanwer
Industrial Pollution Control-1 Division, Central Pollution Control Board, New Delhi, India

Department of Environmental Sciences, Maharshi Dayanand University, Rohtak, Haryana, India

R. Kumar · S. Yadav
Formerly Department of Physics, Gurugram University, Gurugram, Haryana, India

M. K. Jindal, K. Patni (eds.), *Water Quality-Uranium Solubility Nexus: Unlocking the Secrets of Uranium Solubility*, Earth and Environmental Sciences Library, https://doi.org/10.1007/978-3-031-98909-4_14

etc.), residues from uranium mining, nuclear testing, etc., increase uranium levels in water bodies. High levels of uranium in the ocean cause detrimental bioaccumulation in marine life, disrupting ecosystems and posing risks to reproductive and renal systems; control and monitoring are key because consuming contaminated seafood and long-term exposure to uranium radiation increase the risk of kidney damage, cancer, and radiation sickness in humans. Comprehending these actions is essential for controlling uranium pollution in wastewater and safeguarding the public's health. This chapter focuses on revealing uranium behavior in polluted water.

Keywords Uranium solubility · Ocean · Wastewater · Chemical speciation · Industrial waste

1 Introduction

Natural radioactivity has always been present in the environment since the beginning of the world's genesis due to primordial radionuclides such as uranium, thorium, and potassium [1]. Although uranium has the potential to be a valuable energy source, it poses a significant risk (chemical and radiological effects) when it enters water bodies [2–7]. Environmental variables affect soluble uranium in ground and surface water, including temperature, redox (oxidation–reduction) conditions, pH, and complexing agents such as carbonates [8]. The two oxidation states of uranium most often found in the environment are U(IV) and U(VI). The U(VI) form, sometimes called the uranyl ion (UO_2^{2+}), is highly soluble, particularly under oxidizing circumstances. In contrast, the U(IV) form is primarily insoluble in water and tends to remain stationary [9]. Uranium is usually found as U(VI) in oxygen-rich settings such as shallow groundwater or surface water, forming highly soluble and mobile uranyl-carbonate complexes. On the other hand, uranium usually resides as U(IV) in reducing environments, such as deep underground or in places with little oxygen. In these settings, uranium forms solid minerals such as uraninite (UO_2), which are much less soluble and hence less mobile. The pH of the water also affects uranium solubility [10]. Because stable uranyl–carbonate complexes develop in alkaline (pH over 7) environments, particularly in waters with high concentrations of dissolved carbonates or bicarbonates, uranium solubility increases noticeably. Although uranium can also be more soluble in acidic environments (pH less than 6), its mobility frequently decreases because of its propensity to adsorb onto mineral surfaces. The behavior of uranium can also be influenced by other ions found in water, including calcium, phosphate, and sulfate. While phosphate can cause uranium to precipitate as minerals such as autunite, decreasing its solubility, calcium can form stable uranyl-calcium-carbonate complexes [11]. Uranium can also be adsorbed by iron and manganese oxides, which are frequently present in soils, reducing its mobility even further. Given the toxicity of uranium, it is crucial to regularly check the amount of the naturally occurring radionuclide uranium (U) in surface and groundwater resources to minimize exposure to uranium and prevent excessive doses [12]. This method summarizes the U (total U) concentrations in surface and groundwater resources worldwide. In oxic surface waters, U(+6) is the predominant form of U, but in anoxic waters, U(+4) is the predominant form. A practical method of measuring uranium in all water sources must be used to obtain accurate findings [13]. Even though there are more than 4 billion metric tons of uranium dissolved in the ocean, the comparatively low uranium concentration in seawater (3.3 ppb) necessitates the development of highly effective extraction techniques for this strategy to be financially competitive with conventional terrestrial mining [14]. Numerous adsorbent materials have been investigated, including high surface area polymeric fibers with uranium-semiselective amidoxime-based functional ligands [15]. Although uranium extraction from seawater may soon be technically and financially possible owing to advancements in adsorbent design, any significant adverse environmental impacts might prohibit the implementation of such technology. Marine life may be poisoned by adsorbents, their leachates, or adsorbents with bound metals. As a result, when choosing materials for additional development, it is crucial to consider the early detection of hazardous or inert behavior.

Several analytical techniques are available to ascertain the amount, presence, or concentration of uranium and its isotopes in various media. In humans, uranium is found in soil, water, and ambient air. Australia, Canada, China, Kazakhstan, Namibia, Niger, Russia, and Uzbekistan were the top producers by weight in 2019. At least seven of its more than 100 mineral types are mineable worldwide. The U-235 isotope is the main reason that uranium is mined [16]. In addition, uranium is utilized in various applications, such as gyroscope wheels, glass tinting agents, ceramic glazes, chemical catalysts, radioactive source shields, X-ray tube targets, military armor, and kinetic penetrators. Now, look at some historical background of uranium. Since the Roman era, uranium compounds have been utilized as colorants. In a pitchblende sample, Martin Heinrich Klaproth identified uranium as a chemical element. He reported the findings of his investigation in 1789 [17]. Pitchblende is an impure uranium oxide composed of U_3O_8 and uraninite (UO_2), the most reduced oxide. By partially dissolving it in strong acid, Klaproth demonstrated that the solutions produced precipitates that differed from those of known elements. However, previous mineralogists considered this mineral to be complicated iron and tungsten oxide or iron and zinc oxide.

One essential raw material for the nuclear industry is uranium (U). Numerous sectors have extensively used it, including radiotherapy, diagnostic imaging, industrial flaw detection, agricultural irradiation and breeding, cultural relics and archaeology, and more. The UCW (uranium-contaminated water) produced during the various phases of uranium mining, processing, smelting, and use must be managed and treated promptly to prevent it from entering surface water and contaminating soil, sediment, and groundwater through underground seepage and surface runoff. UCW pollution has a detrimental effect on agricultural output and quality in the biosphere, jeopardizes aquatic organism life, and poses a significant risk to the ecological environment and local population health [18, 19]. In addition to other metal complexes or oxides, uranium is found mainly in wastewater in two valence forms: U(IV) and U(VI). U(VI) has a high migration capacity and is more soluble than the others are. U(VI) is a free uranyl ion that may react with phosphate to form uranyl phosphate in acidic settings. Additionally, it can create complexes that are difficult to adsorb and breakdown with hydroxyl ions, carbonate, humus, and colloids [20]. Extended-term exposure to uranium may readily result in malignant alterations in human cells and tissues because of its high radioactivity, extended half-life, and significant biological and chemical toxicity. Additionally, uranium decays to generate the very radioactive elements radium and radon. After entering the body through the skin, food chain, or breath, uranium can irreparably harm the human nervous, circulatory, reproductive, digestive, and endocrine systems [21]. Therefore, treating UCW effectively and safely via proper techniques and enriching and separating uranium in UCW into "second uranium minerals" is crucial from a scientific and practical standpoint for human health, environmental preservation, and the sustainable growth of nuclear energy.

The world's surface water contains 160 million tons of uranium, which can be extracted endlessly at a pace of a few thousand tons yearly. A literature review indicated that uranium concentrations in the open ocean range from 0.3 to 7 $\mu g\,l^{-1}$ [22]. It is present in marine environments such as the Arctic, Antarctic, Pacific, Atlantic, and other oceans. Before the Fukushima accident, approximately 8000 tons of uranium-contaminated water flowed annually through the Kuroshio Current on the west side of the North Pacific Ocean Basin. Today, more than 5.2 million tons of uranium-contaminated water pass through it [23]. It is crucial to understand the behavior of uranium in contaminated waterways, including seawater, wastewater, and industrial effluents, to evaluate the hazards to the environment and human health and to create remediation plans. Complex chemical interactions, such as organic matter, competing metal ions, fluctuating pH levels, salinity, and redox conditions, significantly affect uranium mobility and solubility in these fluids. Because saltwater contains a large amount of carbonate and chloride, uranium generally remains in the soluble hexavalent state, making it stable yet accessible. The behavior of uranium in wastewater and industrial water varies significantly depending on the presence of industrial chemicals, chelating agents, and reducing agents, which can either increase or decrease its mobility. The dynamic character of uranium chemistry in contaminated aquatic habitats emphasizes the necessity of site-specific research and customized treatment strategies. A thorough understanding of local water chemistry and pollution sources is necessary for effective monitoring and management. We can better safeguard ecosystems, guarantee water safety, and promote sustainable environmental management techniques by identifying the variables influencing uranium behavior in these water systems.

2 Water Pollution by Uranium Contamination

2.1 Pollutant Water in Natural Ways

There are different amounts of uranium, a naturally occurring radioactive metal, in the crust of the Earth. It is crucial to understand that uranium is released into the environment (Fig. 1) by natural geological and biological processes, even while

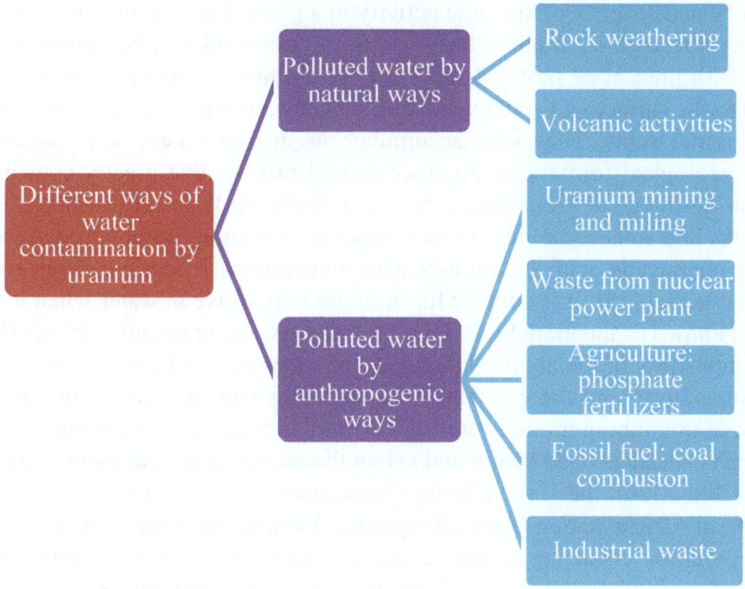

Fig. 1 Representation of polluted water and the corresponding sources

manmade sources, including mining, nuclear power generation, and industrial waste, account for a large portion of the worldwide concern over uranium pollution. The natural geochemical cycling of uranium depends heavily on these processes, which have been ongoing for millions of years. The weathering of uranium-rich rocks is one of the most common natural sources of uranium release [24]. At tiny levels, uranium is frequently found in rocks, including granite, shale, and phosphate-bearing strata. These rocks gradually deteriorate due to natural weathering caused by rainfall, temperature changes, wind erosion, biological activity, and mechanical disintegration. When they breakdown, uranium trapped in the mineral matrix is liberated and made accessible to the environment. Uranium may either attach to soil particles, stay solid, or dissolve into surface water or groundwater, depending on particular environmental factors, including pH, redox potential, and organic matter.

This process is particularly active in areas with significant bedrock uranium concentrations, such as mountain ranges, volcanic regions, and sedimentary basins, where ongoing exposure to natural processes can result in a slow but persistent release of uranium into neighboring ecosystems. In addition to rock weathering, volcanic activity is another important natural process that contributes to the release of uranium. Magma from deep beneath the Earth's mantle is rapidly ejected to the surface during volcanic eruptions, where it frequently carries trace metals such as uranium. The magma's uranium is converted into gases, lava, and volcanic ash before being scattered over large landmasses or released into the atmosphere. After settling, uranium-containing particles might be absorbed into the soil or washed into bodies of water, increasing the amount of uranium in the area. Depending on the frequency and severity of volcanic activity in a given location, this process may be episodic; nonetheless, uranium levels in the environment may be notably higher in places with present or past volcanic activity. Sometimes, uranium-rich fluids are carried to the surface of Earth by volcanic systems, thermal springs, fumaroles, and hydrothermal vents. They may accumulate in surface waters or be absorbed by nearby geological formations. Another natural process that greatly contributes to uranium discharge is groundwater leaching, particularly in areas where uranium-rich rocks and soils come into direct contact with running water. Groundwater can interact chemically with uranium-bearing minerals as it seeps through rock and sediment layers [25]. It is simpler for uranium to dissolve in water when it is more soluble, known as the uranyl ion (UO_2^{2+}), under oxidizing circumstances. The mineral composition of the aquifer, the length of water and rock contact, the water pH and temperature, and the existence of competing ions are factors that affect the amount of uranium in groundwater. The natural uranium concentrations in groundwater can exceed advised health and safety thresholds in several parts of the world. Parts of India [26], the United States (particularly the Colorado Plateau), Canada, and several African nations have all seen this. Even in the absence of industrial pollution, naturally occurring increased uranium levels are a public health risk since groundwater is frequently utilized for irrigation and drinking. Furthermore, dissolved uranium may be transported to rivers, lakes, and wetlands by groundwater that rises to the surface through seeps or springs, impacting the quality of surface water and the health of ecosystems. Forest fires are another significant but less well-known natural source of uranium emissions. Over time, small amounts of uranium can be absorbed and retained by vegetation, soil, and organic matter, mostly through the deposition of dust particles in the air and root absorption from the soil. This stored uranium can be released back into the environment in various ways during a wildfire when organic materials burn. For example, ash particles deposited on land or transported by wind over great distances may include uranium. Furthermore, certain chemicals may volatilize due to the high temperatures observed during fires, which might allow trace amounts of uranium to reach the atmosphere. In addition to contributing to the mobility and redistribution of uranium in the environment, post-fire landscapes are frequently susceptible to erosion and surface runoff, which can carry uranium-rich ash and soil into neighboring streams and lakes. This process may become more significant to the natural uranium cycle if wildfires grow more frequently and intensely globally due to climate change. These natural processes—volcanic activity, groundwater leaching, forest fires, and rock weathering—collectively constitute continuous environmental mechanisms incorporating uranium into the biosphere. The consequences of these natural processes vary; uranium mobilization, transportation, and deposition are influenced by the local geology, hydrology, climate, and ecosystem type. Studying and comprehending these sources is vital because they establish the background or baseline amounts of uranium in the environment, which are necessary to differentiate between anthropogenic contamination and natural fluctuations. The development of suitable land and water use plans in impacted areas, risk evaluations, and public health policies is aided by monitoring natural uranium flow. In summary, although the discussion of uranium contamination is frequently dominated by human activity, nature plays an important and enduring part in the global uranium cycle. Long before industrial activities started, these naturally occurring sources helped shape the chemical landscape of our world by releasing uranium steadily into soils, water bodies, and the atmosphere. In addition to being crucial for environmental research, understanding and recognizing these natural sources are essential for public health and the efficient management and preservation of natural resources.

2.2 *Pollutant Water Via Anthropogenic Ways*

Fuel fabrication, processing, and waste disposal in the nuclear industry can release uranium and other radioactive elements, leading to environmental or occupational exposure risks. Countries such as the USA, Great Britain, and France have deployed depleted uranium (DU) munitions, which can result in contamination from uranium-rich particles in the environment. Uranium particles have been detected near uranium processing facilities, highlighting the risks associated with uranium use in the military and nuclear industries [27]. Industrial processes, mining operations, nuclear power generation, agriculture, military actions, and waste management are the main human (Artificial)-caused sources of uranium discharge. Among them, the most significant contributors include the mining and processing of uranium. Massive amounts of rock are treated throughout the uranium extraction process to separate uranium, creating enormous amounts of trash known as tailings. In addition to retaining leftover uranium, these tailings also include heavy metals and radioactive decay products that, if improperly handled, might seep into nearby soil and water systems. Airborne contamination can also result from radioactive particles being carried into the atmosphere by dust produced during mining and milling [28]. Nuclear power facilities are another significant source of liquid and gaseous waste containing uranium. These effluents may release traces of uranium and other radioactive isotopes into the atmosphere or adjacent aquatic bodies. Past nuclear accidents such as those reported by Fukushima and Chornobyl indicate that inadvertent leaks, however uncommon, can cause severe environmental pollution even when regular discharges are often controlled and handled. Aside from the nuclear industry, uranium leakage is also a result of agriculture. Phosphate fertilizers are widely used in agriculture and are made from phosphate rocks that naturally contain uranium. The application of uranium to soil can cause it to wash off into adjacent water bodies or leach into groundwater, increasing uranium concentrations in agricultural areas over time [28]. Additionally, military operations have contributed to uranium pollution, especially when depleted uranium (DU) is used in military vehicles and armour-piercing ammunition. Fine uranium particles may be released into the environment when DU bombs break and burn upon impact. These particles pose long-term hazards to ecosystems and human health because they can linger in soil and be carried by wind or water. The environmental uranium load was further increased by atmospheric nuclear weapon testing in the middle of the twentieth century, which discharged large amounts of uranium and its radioactive isotopes into the high atmosphere. These particles then disperse across the world. The burning of coal and other fossil fuels is another man-made source. Traces of uranium are found naturally in coal [29], and when coal is burned in power plants, fly ash or stack emissions may release uranium into the atmosphere. If the resulting ash is not disposed of appropriately, it can pollute water and land. Uranium contamination can also be caused by industrial waste from industries such as metallurgy, glass manufacturing, ceramics, and pigments. Facilities that use uranium-based compounds may produce solid waste or wastewater that releases uranium into the environment or municipal sewage systems if it is not adequately handled. One significant but frequently disregarded source of uranium release is sewage. Particularly in areas where groundwater naturally contains high quantities of uranium, people may consume uranium through food and drinking water [30] and then expel it in their urine. Small amounts of uranium isotopes or tracers may also be released into sewage systems by research facilities, medical facilities, and industrial effluents. Uranium may occasionally enter treatment facilities from polluted groundwater that seeps into sewage infrastructure. Different ways of uranium contamination in wastewater are shown in Table 1. These activities are mainly industrial, mining, and nuclear-related activities.

2.2.1 Nuclear Power Plants

Nuclear power plants, often referred to as nuclear power stations, nuclear-generating stations, or atomic power stations, are thermal power plants that use nuclear reactors as their heat source. Heat is utilized to create steam, which powers a steam turbine coupled to a generator that generates electricity, as is common in thermal power plants. According to the International Atomic Energy Agency, as of September 2023, 57 nuclear power reactors were being built, and 410 nuclear power reactors were operating in 32 nations worldwide [31]. Most nuclear power facilities use thermal reactors with enriched uranium in a once-through fuel cycle. When the proportion of atoms that absorb neutrons increases such that a chain reaction can no longer be maintained, usually after 3 years, the fuel is removed. After cooling in onsite spent fuel pools for several years, it is moved to long-term storage. The spent

Table 1 Different ways of uranium contamination in wastewater

Source Type	Specific Source	Discharge Medium
Natural	Rock weathering, volcanic activity	Soil, water
Mining	Uranium mining & milling	Soil, water, air
Energy	Nuclear power plants	Air, water
Agriculture	Phosphate fertilizers	Soil, water
Fossil Fuels	Coal combustion	Air (fly ash), water
Industrial	Uranium-containing product industries	Water, waste streams

fuel is classified as high-level radioactive waste despite its small volume. Radioactive wastewater produced by nuclear power plant operations can pollute rivers, groundwater, and other water sources if not adequately handled, increasing the possibility of radionuclide exposure for humans [32]. Radio wastewater must be treated effectively to safeguard the environment and public health from the damaging effects of uranium and other radioactive elements. Approximately 20% of the power utilized in the US is produced by nuclear reactors. The most common fuel in nuclear reactors at power plants is uranium. Fission is the process by which uranium atoms split, generating nuclear energy. A massive quantity of energy is released as heat during fission. The steam produced by this heat powers a steam turbine. An electric generator that produces power is attached to the turbine. Radioactive waste is generally divided into low-, moderate-, and high-level waste. Since other nations employ more thorough classification, the radioactive concentration and half-life—the amount of time it takes for the waste to lose half of its radioactivity—are the common principles used to identify the type of waste.

2.2.2 Nuclear Weapons Testing

Uranium enrichment is one of the two techniques used to create nuclear explosive material. The creation of plutonium is the other. For many years, gaseous diffusion was the only method used for uranium enrichment, and it was used mainly in the United States. The USSR, China, France, and the UK had the greatest capacity and dominated the global enrichment market for a long period of time [33]. However, when new methods became more competitive with gaseous diffusion and other nations entered the market, this strong position gradually started to wane in the middle of the 1970s. The research and development (R&D) programs that produced the new processes were sparked by the 1960s expectation of a sizable and expanding global market for enrichment services. At the time, most people thought that nuclear energy would become widely used and that the light-water moderator and coolant design—using enriched uranium as fuel—would become the predominant power reactor architecture. Many nations have invested primarily in developing indigenous enrichment capacity due to the potential for a burgeoning enrichment industry and the growing demand for energy independence.

2.2.3 Agricultural Waste and Sewage

Agricultural activities are a significant source of uranium contamination in groundwater, primarily because of the use of phosphate fertilizers. These fertilizers contain uranium impurities. The long-term use of these fertilizers can increase uranium levels in agricultural soils, which may leach into groundwater under certain conditions. Phosphate fertilization is a primary route for uranium contamination of soils, and the extensive use of uranium-containing fertilizers can significantly increase uranium concentrations in the soil, affecting the water quality in surrounding areas

[34]. This is especially true in regions with heavy agricultural activity, such as Germany, where uranium levels in rivers and streams in agrarian catchments are ten times higher than those in forests [35]. U(VI) is found in oxidizing environments and exists as uranyl ions (UO_2^{2+}), which can form complexes with various elements, such as hydroxides and carbonates, making uranium migration in the environment more pronounced. Multiple studies have highlighted the influence of carbonate complexes on uranium adsorption and mobility [36].

3 Uranium Behavior in Seawater

To understand the adsorption mechanism completely and create appropriate adsorbents, it is necessary to know the chemical state of uranium in seawater. It is vital to investigate uranium speciation in the ocean in the following ways because most earlier research has focused only on freshwater settings. Uranyl compounds, or compounds containing UO_2^{+2}, have long been recognized to have significant solubility in sodium carbonate solutions. Development of stable complexes, as indicated by the discontinuities at the locations that corresponded to the carbonate ratio, uranium mole ratio, two ratios, one ratio, and three ratios, one. Starik and Kolyadin predicted that uranium would exist in saltwater in the following ways on the basis of the components present [37]: certain of its constituents are soluble, as mentioned in Fig. 2.

On the basis of uranium tests conducted with CO_3 in seawater, the uranium level in seawater as a complex with colloidal particles decreases with decreasing pH (to less than 7.5). In the pH range from 3–6, uranium adsorption on negative- and positive-ion exchangers further demonstrated a balance between the UO_2^{2+} hydrolysis products and the complex of positive ions, which was significantly altered when an ion exchanger was added to the solution. When the pH is greater than 7.5, the uranyl ion becomes anionic.

For a positive ion exchanger, U adsorption may increase the partial pressure of CO_2 over a solution with a particular pH, as mentioned in the literature. However, uranium adsorption on glass followed the reverse pattern. Cations, which are byproducts of hydrolysis, have a particular affinity for glass. A literature review revealed that uranium in saltwater does not form organic complexes and that uranyl ions exist as carbonate complexes at pH values higher than 7.5. Because seawater has a far higher concentration of carbonate ions than uranyl ions, the complex will form a uranyl tricarbonate complex. Kennedy also reported that both artificial and natural saltwater media do not dissolve uranyl ions [38]. The pH of the artificial seawater was brought back to its starting point of 8.0 via sodium hydroxide after removing the bicarbonate. The uranyl ions are insoluble in seawater and are free from organic complexing agents. Only uranyl tricarbonate complexes are available in saltwater because of the elevated level of CO_3^{2+} in seawater compared with uranyl ions. Uranyl ion and uranyl complexes in seawater are determined by considering the stability constants of the potential reactive ions in an appropriate proportion [39].

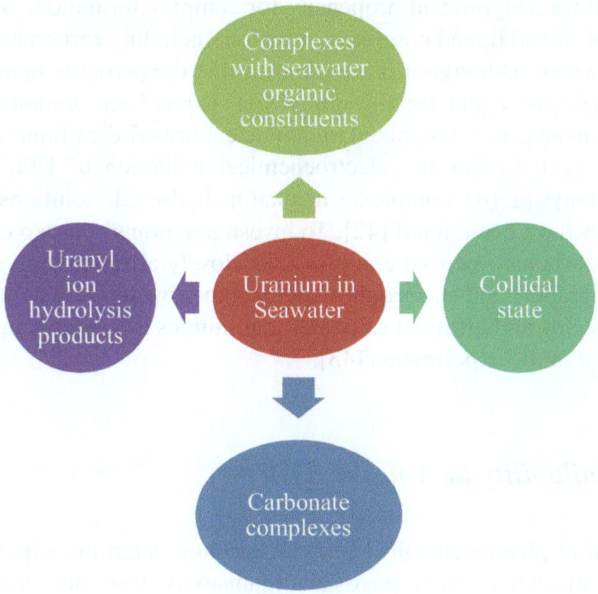

Fig. 2 Different forms of uranium exist in seawater

3.1 Stability of Uranyl Complexes in Seawater

Seawater has a constant content of 3.3 g/L dissolved uranium. It manifests mainly as dissolved uranium(VI) compounds with various ligands. The anionic uranyl tricarbonate complex $UO_2(CO_3)_3^{4-}$ is now considered the most common chemical form of dissolved uranium in saltwater. The complex might be more accurately defined as $Na_2UO_2(CO_3)_3^{2-}$ because of the ion-pairing processes. According to recently released data, uranyl-hydroxo complexes are the most common uranium species in seawater [39]. This claim is most likely the result of an uncritical or even incorrect use of an inflated uranyl–hydroxide association constant value. However, because of its inherent geochemical cycle, the distribution of dissolved uranium species in seawater is highly significant. Nearly 20 years ago, Van Baleen and Marler determined the hydrogen peroxide concentration in seawater, which was 10^{-7} mol/L [40]. Over the last 10 years, many systematic studies have been conducted, and this concentration of surface seawater has been verified. The photic layer of the ocean is where hydrogen peroxide is most frequently formed photochemically. Several forms of hydrogen peroxide photochemical production in surface waters have been reported in the literature [41]. In addition to some algae, specific chemical molecules stimulated by light also create hydrogen peroxide. Peroxide concentrations can reach with exceptionally high organic production. Light alone is not what causes peroxide to develop. In the dark, a particular phytoplankton, Hymenomonas carterae, produces H_2O_2. H_2O_2 is produced by a large variety of enzymes and is released into the water by some bacteria. Because uranium and peroxide have a significant propensity for complex formation, we investigated the potential of mixed ligand complex formation, including carbonate and peroxide ligands in seawater. Although it is well known that the peroxide ligand is the most effective complexing agent for the uranyl ion, it has been demonstrated that an excess of carbonate causes the mixed mononuclear uranyl-dicarbonato peroxo complex $UO_2(CO_3)_2(O_2)^{4-}$. For the electrochemical reduction of UO_2^{2+} -H_2O_2 1 M Na_2CO_3 and uranyl-peroxo complexes in alkaline hydroxide solutions, several electrode behaviors have been noted [42]. To investigate uranyl-peroxo complexes and mixed uranyl-carbonato-peroxo complexes as closely as possible to saltwater conditions, polarographic and spectrophotometric experiments were conducted. Model electrolyte systems and artificial saltwater that mimics the natural aquatic environment were used in the experiments [43].

3.2 Bioavailability in Aquatic Systems

The speciation or physicochemical form of uranium determines its bioavailability or capacity to attach to or move across an organism's cell surface. Numerous physicochemical forms of uranium can be found in aquatic systems, such as free metal ions (U^{4+} or UO_2^{2+}), complexes with inorganic ligands (such as uranyl carbonate or uranyl phosphate), and humic substances (HS) in dissolved, colloidal, and/or particulate forms (such as uranyl fulvate or humate) [44]. Thus, understanding the U distribution among its many physicochemical forms (i.e., speciation) is essential for comprehending how U interacts with the cell surfaces of aquatic species (i.e., bioavailability). Potential ecological hazards arise from surface waters and sediments contaminated with anthropogenic U, primarily from the nuclear fuel cycle (such as the mining, milling, and reprocessing of waste) but also from the burning of petrol fuels (such as coal) and the production and use of phosphatic fertilizers (such as superphosphate). To assess such hazards, at least initially, the total U content in water and/or sediments is measured and compared to existing norms or guidelines (where available) for aquatic ecosystem protection. Even though U is radioactive, its chemical toxicity to aquatic life matters most for the ecosystem [45]. Three oxidation states of U can be found in surface waters: U^{4+}, UO_2^{2+} (U[VI] or uranyl ions), and U. U occurs as U^{4+} or UO^{2+} in anoxic waters (low redox potential). In contrast to UO^{2+}, which produces soluble complexes, U(IV) has a strong propensity to precipitate (such as uraninite and UO_2[s]) and remain immobile [46]. The most crucial part of U transport during weathering involves stable, easily soluble ionic and/or neutral complexes that are highly mobile and form in oxic environments when U is present as UO_2^{2+}. The pH-Eh parameters and the availability and concentration of complex ions determine the dominant U species [47]. Since hydrolytic processes can restrict solubility or affect particle sorption, they significantly impact U redox and complexation reactions. The average U concentration in aquatic sediments is 3 mg kg^{-1}, with a typical range of 0.5 to 5 mg kg^{-1} [48]. Particle size, mineral makeup (which reflects local geology), and water physicochemistry—especially in freshwater settings—can affect U concentrations. Sediments taken from anoxic habitats, such as swamps or deep ocean basins, have been shown to have significant U enrichment (up to 100 times) [49].

4 Uranium Behavior in Wastewater and Industrial Water

In addition to being an essential resource for the nuclear industry, uranium (U), a naturally occurring radioactive heavy metal, is also used in diagnostic imaging, industrial fault detection, agricultural irradiation, radiotherapy, and the study of cultural artifacts and archaeology. To prevent UCW, which is produced throughout different phases of uranium mining, processing, and use, from spreading through surface runoff or subsurface seepage to contaminating soil, sediment, and groundwater, it is essential to manage and treat UCW. The natural environment, aquatic life, human health, and crop productivity and quality are all at risk due to uranium poisoning in water. The two oxidation forms of uranium that are commonly found in wastewater are U(IV) and U(VI), with U(VI) being more soluble and mobile. In acidic environments, U(VI) manifests as a free uranyl ion that can react with hydroxyl ions, carbonate, humus, and colloids to generate uranyl phosphate or combine with phosphate to form uranyl phosphate, which makes it more difficult to absorb and breakdown [50]. Uranium is extremely radioactive, has a long half-life, and is poisonous. Long-term exposure through eating, skin contact, or inhalation can result in cancer and irreversible harm to the human nervous, circulatory, reproductive, digestive, and endocrine systems. Thus, human health, environmental preservation, and the long-term growth of nuclear energy all depend on the efficient treatment of UCW. The term emerging contaminants (ECs) refers to unregulated substances that pose potential health risks and are found in water systems. These contaminants, such as endocrine-disrupting chemicals (EDCs), pharmaceutical and personal care products (PPCPs), and persistent organic pollutants (POPs), have recently become the focus of research because of their low concentrations and potential hazards [51]. ECs are commonly found in different water sources and are often not fully degraded in sewage treatment systems. As a result, residues enter the environment through effluent discharge, contaminating surface and groundwater and sometimes even drinking water. ECs in sewage sludge, which are sometimes used for agricultural fertilization, further exacerbate water pollution [52].

4.1 Role of Organic Matter and Reducing Agents

Organic materials and reducing agents significantly influence the geochemical behavior and environmental mobility of uranium. Through complexation, organic materials such as plant waste and microbial biomass may bind uranium and change its solubility [53]. Hydrogen sulfide, ferrous iron, and organic carbon are examples of reducing agents that aid in the transformation of soluble hexavalent uranium (U^{6+}) into its less soluble tetravalent form (U^{4+}), which precipitates as uraninite (UO_2) [54]. Uranium must be immobile in sediments by this reduction process to prevent it from migrating into groundwater. Therefore, the ambient chemistry of uranium and its safe confinement depend heavily on organic matter and low circumstances. Various components, such as strata, sandstone, breccia, and granitic plutons, contain relatively high levels of organic matter (OM) [55]. Owing to the presence of the above components, there are significant amounts of uranium ore on every continent. The presence of uranium in organic matter-rich sedimentary deposits. Given their potential for U and other elements, OM-rich deposits will be increasingly considered as studies and explorations focus on a broader range of geological settings. As demonstrated by the extraction of U and vanadium on the U.S. Colorado Plateau, coproduct mining may increase the economic viability of mining operations [56]. Although U in OM-rich sedimentary deposits is important, little is known about the relative stability and mobility of resident uranium.

There are two fundamental theories concerning why uranium mobility in OM-rich deposits is unknown. First, the macrogeological context of OM-rich deposits differs; their composition is variable, and the interactions between these heterogeneous components are complicated. For example, distinct deposits may include various ions, elements, and microorganisms that can interact in habitats with varying Eh-pH levels. In addition to geological formations (roll front deposits, for example). Second, because of technical constraints, OM-rich deposits have historically been challenging to analyze [47]. With fresh analytical methods and the development of characterization approaches, a richer understanding of deposit heterogeneity has been gained, improving the clarity of constituent interactions. A synthesis of pertinent field and laboratory-based research is still lacking to determine the scope and importance of multicomponent relationships involving uranium. These connections entail interactions between uranium, OM, and other components or complexes that may affect the mobility and cycling of uranium in the crust.

The physical and chemical factors affecting the mobility of uranium, as well as its stability, change over geologic time. In both dissolved and particulate/mineral forms, uranium can cycle from source to sink throughout geological time periods due to environmental processes such as chemical and mechanical rock weathering. The primary determinants of uranium incorporation or release from mineral formations are its oxidation state and mineral solubility. The U(IV) state of uranium is less soluble. Additionally, it forms more stable compounds than does the more readily mobilized +6 oxidation state (U(VI)). The oxidation or reduction of U is controlled by an extensive variety of parameters, including organic and microbial enzymatic routes and inorganic physicochemical variables such as pH, pO_2, and pCO_2 [57]. Therefore, multicomponent connection modeling and consideration are necessary to understand uranium fate.

4.2 Influence of Chelating Agents and Surfactants

Chelating agents and surfactants strongly impact the mobility and bioavailability of environmental uranium. Chelating compounds, such as citrate and EDTA, combine with uranium ions to create stable complexes that increase their solubility and make moving through water systems easier [58]. This may increase the mobility of uranium, increasing its availability for microbial and plant absorption. Natural and artificial surfactants change the surface tension and how uranium interacts with the surfaces of minerals or soil. They can facilitate uranium desorption from particles, further encouraging their migration. These substances are essential for uranium extraction, cleanup, and environmental risk assessment [59]. A special family of chemicals known as chelating surfactants combines the characteristics of surfactants and chelating agents. These substances are important in sectors where controlling metal ions is essential since they may reduce surface tension and improve cleaning effectiveness while forming stable complexes with metal ions. The surfactant headgroup usually chelates, providing a multipurpose solution for various industrial, medicinal, and personal care applications. When scientists started looking at the possibility of fusing chelating and surfactant functions into single molecules in the 1950s and 1960s, chelating surfactants were born. As a result, different chelating surfactants with better performance and adaptability than conventional surfactants or chelating agents alone were synthesized and characterized.

4.3 Precipitation and Adsorption Behavior

Published reports indicate that uranium has essential properties and is used as fuel in nuclear reactors. However, it also poses a risk of radiotoxicity due to its toxic radioactive emissions. Uranium is often found in nature as a highly soluble species in hexavalent states and as an insoluble species in trivalent states [60]. In saltwater and radioactive wastewater, UO_2^{2+} is significantly more stable and soluble. Uranium is present in the Earth's crust in rock and soil [41]. The granitic and syenite rocks in the igneous category have relatively high levels of uranium. The mean uranium level in the Earth's crust is reported to be nearly 2.7 mg/L. Geological conditions, lithology, geomorphology, and other locations affect uranium content. In terms of chemistry, metallic uranium is very reactive and can be alkaline or acidic. Additionally, uranium concentrations in groundwater are projected to be relatively in the range of 0.1–1 ppb. With an atomic number of 92, uranium has two isotopes: ^{235}U (0.72%) and ^{238}U (99.27%). Only ^{235}U is typically fissile, whereas all uranium isotopes are radioactive and fissionable. Australia, Kazakhstan, and Canada have large deposits of uranium. The Australian Ranger mines and Olympic Dam are the largest and most significant mines [61]. The Athabasca Basin region of Canada also has high-grade uranium resources. Most other uranium mines are in Canada's Cigar Lake and McArthur River basins. The Chu-Sarysu Basin in Kazakhstan has more than half of the nation's known uranium deposits [62]. However, India has no uranium deposits; imports from other countries meet the uranium requirements of nuclear power plants. Some Indian regions, such as Andhra Pradesh and Telangana between Sresailam and Seshachalam, are thought to have uranium mines [63]. However, tons of radioactive waste are produced by more than 10 nuclear power plants across India and dumped into waterways without being adequately treated.

5 Conclusion

This chapter covers the behavior of uranium in polluted water. Uranium exists in seawater in the following ways on the basis of the components present: certain of its constituents are soluble; therefore, it exists in the colloidal state, compounds with seawater colloidal particles, carbonate complexes and soluble uranyl ion hydrolysis products, and complexes with seawater organic constituents. The stability of uranyl complexes and their bioavailability in aquatic systems revealed the behavior of uranium in seawater. Organic materials and reducing agents significantly influence the geochemical behavior and environmental mobility of uranium. Through complexation, organic materials such as plant waste and microbial biomass may bind uranium and change its solubility. Hydrogen sulfide, ferrous iron, and organic carbon are examples of reducing agents that aid in the transformation of soluble hexavalent uranium (U^{6+}) into its less soluble tetravalent form (U^{4+}), which precipitates as uraninite (UO_2). Chelating compounds, such as citrate and EDTA, combine with uranium ions to create stable complexes that increase their solubility and make moving through water systems easier. This may increase the mobility of uranium, increasing its availability for microbial and plant absorption. Precipitation and adsorption behavior indicate the uranium distribution in wastewater and industrial waste.

References

1. UNSCEAR (2000) United Nations Scientific Committee on the effect of atomic radiation, Annex B: exposures from natural radiation sources. United Nations, New York
2. IAEA (International Atomic Energy Agency) (2005) Worldwide marine radioactivity studies (WOMARS): radionuclide levels in oceans and sea, IAEA-TECDOC-1429. IAEA, Vienna, p 187
3. Singh B, Kant K, Sharma D, Kanwar N, Tanwer N, Kumar R (2025) Elevated radiation levels and their potential causes in regions of the marine environment. In: Jindal MK, David A, Khandaker MU (eds) Radiation status in the marine world, Springer oceanography. Springer, Cham. https://doi.org/10.1007/978-3-031-82403-6_8
4. Tanwer N, Kant K, Singh B, Laura J, Khosla B (2024) Uranium in groundwater is an emerging concern in Rajasthan, India: a study of distribution, potential sources, health impacts, and management strategies. In: Environmental metagenomics, water quality and suggested remediation

measures of polluted waters: a combined approach. Elsevier. https://doi.org/10.1016/B978-0-443-13659-7.00006-0
5. Tanwer N, Arora V, Kant K, Singh B, Laura JS, Khosla B (2024) Prevalence of Uranium in groundwater of rural and urban regions of India. In: Madhav S, Srivastav AL, Izah SC, van Hullebusch E (eds) Water resources management for rural development. Elsevier, pp 213–234. https://doi.org/10.1016/B978-0-443-18778-0.00002-7
6. Singh B, Kant K, Garg M (2021) Distribution of radon and uranium levels in groundwater of District Palwal, Southern Haryana, India. Radiat Protect Dosimet 196(3–4):241–247. https://doi.org/10.1093/rpd/ncab158
7. Singh B, Kant K, Tanwer N (2024) A study of seasonal variation and impact of geology on gamma radiation levels of district Faridabad of State Haryana, India. J Radioanal Nucl Chem 333:5379–5390. https://doi.org/10.1007/s10967-024-09574-8
8. Agbasi JC, Chukwu CN, Nweke ND, Uwajingba HC, Khan MYA, Egbueri JC (2023) Water pollution indexing and health risk assessment due to PTE ingestion and dermal absorption for nine human populations in Southeast Nigeria. Groundw Sustain Dev 21:100921. https://doi.org/10.1016/j.gsd.2023.100921
9. Ao B, Qiu R, Tang J, Chen J (2021) New theoretical insights into the actual oxidation states of uranium in the solid-state compounds. J Nucl Mater 543:152563. https://doi.org/10.1016/j.jnucmat.2020.152563
10. Schnug E, Lottermoser BG (2013) Fertilizer-derived uranium and its threat to human health. Environ Sci Technol 47(6):2433–24342. https://pubs.acs.org/doi/10.1021/es4002357
11. Latta DE, Kemner KM, Mishra B, Boyanov MI (2016) Effects of calcium and phosphate on uranium(IV) oxidation: comparison between nanoparticulate uraninite and amorphous UIV–phosphate. Geochim Cosmochim Acta 174:122–142. https://doi.org/10.1016/j.gca.2015.11.010
12. Hussain R, Hussain H (2011) Natural occurring radionuclide materials. https://doi.org/10.5772/20562
13. Smedley PL, Kinniburgh DG (2023) Uranium in natural waters and the environment: distribution, speciation and impact. Appl Geochem 148:105534. https://doi.org/10.1016/j.apgeochem.2022.105534
14. Liu X, Xie Y, Hao M, Chen Z, Yang H, Waterhouse GIN, Ma S, Wang X (2022) Highly efficient electrocatalytic uranium extraction from seawater over an amidoxime-functionalized in–N–C catalyst. Adv Sci. https://doi.org/10.1002/advs.202201735
15. Song Y, Deng B, Wang K, Zhang Y, Gao J, Cheng X (2024) Highly-efficient adsorbent materials for uranium extraction from seawater. J Environ Chem Eng 12(5):113967. https://doi.org/10.1016/j.jece.2024.113967
16. World Nuclear Association WNA (2024) World uranium mining production. HOME/Information Library/nuclear fuel cycle/mining-of-uranium/World Uranium Mining Production. https://world-nuclear.org/information-library/nuclear-fuel-cycle/mining-of-uranium/world-uranium-mining-production
17. US Department of Energy. Nuclear fuel facts: Uranium. https://www.energy.gov/ne/nuclear-fuel-factsuranium#:~:text=Uranium%20was%20discovered%20in%201789,uran%E2%80%9D%20after%20the%20planet%20Uranus
18. USEPA. US Environmental Protection Agency. Climate change impacts on agriculture and food supply. https://www.epa.gov/climateimpacts/climate-change-impacts-agriculture-and-food-supply
19. Xiong X, Liu J, Xiao T et al (2025) Remediation of uranium-contaminated water and soil by biochar-based materials: a review. Biochar 7:41. https://doi.org/10.1007/s42773-025-00438-2
20. Acharya C, Chandwadkar P, Joseph D et al (2013) Uranium (VI) recovery from saline environment by a marine unicellular cyanobacterium, *Synechococcus elongatus*. J Radioanal Nucl Chem 295:845–850. https://doi.org/10.1007/s10967-012-1912-x

21. Nelson F, Michelson DC, Holloway JH (1974) Separation of U, Np, & Pu via ion exchange. In: Radiochemistry of neptunium, National Research Council, Nuclear science series. National Academy of Sciences, Washington, DC
22. Teh-Lung K, Mathieu GG, Knauss KG (1977) Uranium in open ocean: concentration and isotopic composition. Deep-Sea Res 24(11):1005–1017. https://doi.org/10.1016/0146-6291(77)90571-910.1016/0146-6291(77)90571-9
23. NOAA (2024). Marine Debris Program. Office of Response and Restoration National Oceanic and Atmospheric Administration. https://marinedebris.noaa.gov/our-work/emergency-response/japan-tsunami-marine-debris
24. Li L, Chen J, Chen T, Chen Y, Hedding DW, Li G, Li L, Li T, Robinson LF, West AJ, Wu W, You C-F, Zhao L, Li G (2018) Weathering dynamics reflected by the response of riverine uranium isotope disequilibrium to changes in denudation rate. Earth Planet Sci Lett 500:136–144. https://doi.org/10.1016/j.epsl.2018.08.008
25. Babu MNS, Somashekar RK, Kumar SA et al (2008) Concentration of uranium levels in groundwater. Int J Environ Sci Technol 5:263–266. https://doi.org/10.1007/BF03326020
26. Balaram V, Rani A, Rathore DPS (2022) Uranium in groundwater in parts of India and world: a comprehensive review of sources, impact to the environment and human health, analytical techniques, and mitigation technologies. Geosyst Geoenviron 1(2). https://doi.org/10.1016/j.geogeo.2022.100043
27. Okafor-Yarwood I (2014) Use of depleted uranium weapons in contemporary military interventions. Asian J Peacebuild 2:111–125. https://doi.org/10.18588/201405.000022
28. Chiluba B (2018) Critical review of dust in the mining environment: a focus on workers and community health. https://doi.org/10.20944/preprints201811.0602.v1
29. Pak I, Pak D, Zhmagul N, Tebaeva A (2021) Natural radioactivity of coal in the context of radioecological safety and rational use. Izv Vyssh Uchebn Zaved Gornyi Zh:97–106. https://doi.org/10.21440/0536-1028-2021-1-97-106
30. Ashish A, Bangotra P, Dillu V et al (2024) Human exposure to uranium through drinking water and its detrimental impact on the human body organs. Environ Geochem Health 46:397. https://doi.org/10.1007/s10653-024-02150-9
31. IAEA (2024) Operating experience with nuclear power stations in member states 2024 edition, 55th edn. International Atomic Energy Agency, Vienna
32. Ma H, Shen M, Tong Y, Wang X (2023) Radioactive wastewater treatment technologies: a review. Molecules 28(4):1935. https://doi.org/10.3390/molecules28041935
33. Krass AS, Boskma P, Elzen B, Smit WA (1983) Uranium enrichment and nuclear weapon proliferation. Taylor & Francis Ltd, London/New York. ISBN 0-85066-219-2
34. Yamazaki IM, Geraldo LP (2003) Uranium content in phosphate fertilizers commercially produced in Brazil. Appl Radiat Isot 59(2–3):133–136. https://doi.org/10.1016/S0969-8043(03)00159-3
35. Birke M, Rauch U, Lorenz H (2009) Uranium in stream and mineral water of the Federal Republic of Germany. Environ Geochem Health 31:693–706. https://doi.org/10.1007/s10653-009-9247-4
36. Krause S, Singh J, Dehbandi R, Kumar PV, Kumar P, Toor AP, Srivastava A (2025) Assessment and mitigation of heavy toxic elements with emphasis on uranium in the Malwa Region of Punjab, India. ACS ES&T Water 5
37. Starik IE, Kolyadin LB, Nikolaev DS (1959) Physico-chemical conditions of occurrence of microquantities of uranium in ground waters. U.S. Department of Energy Office of Scientific and Technical Information
38. Chopra M, Choudhury N (2014) Effect of uranyl ion concentration on structure and dynamics of aqueous uranyl solution: a molecular dynamics simulation study. J Phys Chem B. https://doi.org/10.1021/jp506477s
39. Djogić R, Branica M (1991) Dissolved uranyl complexed species in artificial seawater. Mar Chem 36(1–4):121–135. https://doi.org/10.1016/S0304-4203(09)90058-5

40. O'Sullivan DW, Neale PJ, Coffin RB, Boyd TJ, Osburn CL (2005) Photochemical production of hydrogen peroxide and methyl hydroperoxide in coastal waters. Mar Chem 97(1–2):14–33. https://doi.org/10.1016/j.marchem.2005.04.003
41. Bakather OY, Zouli N, Abutaleb A, Mahmoud MA, Daher A, Hassan M, Eldoma MA, Alasweda SO, Fowad AA (2020) Uranium (VI) ions uptake from liquid wastes by Solanum incanum leaves: biosorption, desorption and recovery. Alex Eng J 59(3):1495–1504. https://doi.org/10.1016/j.aej.2020.03.013
42. Keech P, Noel J, Shoesmith D (2008) The electrochemical reduction of hydrogen peroxide on uranium dioxide under intermediate pH to acidic conditions. Electrochim Acta 53:5675–5683. https://doi.org/10.1016/j.electacta.2008.03.008
43. Whitfield M (1973) A chemical model for the major electrolyte component of seawater based on the Bronsted-Guggenheim hypothesis. Mar Chem 1(4):251–266. https://doi.org/10.1016/0304-4203(73)90016-9
44. Zhanming L, Peipei S, Chenxi Z, Nali Z, Nan X, Dongrui L, Yuxi G, Jiating Z (2024) Translocation and transformation of uranium along the aquatic food chain: new insights into uranium risks to the environment. J Hazard Mater 478:135499. https://doi.org/10.1016/j.jhazmat.2024.135499
45. Kraemer LD, Evans D (2012) Uranium bioaccumulation in a freshwater ecosystem: impact of feeding ecology. Aqua Toxicol 124–125:163–170. https://doi.org/10.1016/j.aquatox.2012.08.012
46. Shilov VP, Yusov AB, Peretrukhin VF, Delegard CH, Gogolev AV, Fedosseev AM, Kazansky LP (2007) Oxidation of U(IV) by atmospheric oxygen in pH 1.5–7.4 aqueous solutions. J Alloys Compd 444–445:333–338. https://doi.org/10.1016/j.jallcom.2007.02.082
47. Cumberland SA, Douglas G, Grice K, Moreau JW (2016) Uranium mobility in organic matter-rich sediments: a review of geological and geochemical processes. Earth-Sci Rev 159:160–185. https://doi.org/10.1016/j.earscirev.2016.05.010
48. Crawford SE, Lofts S, Liber K (2017) The role of sediment properties and solution pH in the adsorption of uranium(VI) to freshwater sediments. Environ Pollut 220(Part B):873–881. https://doi.org/10.1016/j.envpol.2016.10.071
49. Markich SJ (2002) Uranium speciation and bioavailability in aquatic systems: an overview. Sci World J 2:707–729. https://doi.org/10.1100/tsw.2002.130
50. Kalintsev A, Guan Q, Brugger J, Migdisov A, Etschmann B, Ram R, Liu W, Mei Y, Testemale D, Xu H (2023) Nature and coordination geometry of geologically relevant aqueous uranium(VI) complexes up to 400 °C: a review and new data. J Hazard Mater 452:131309. https://doi.org/10.1016/j.jhazmat.2023.131309
51. Bi R, Zhou C, Jia Y, Wang S, Li P, Reichwaldt ES, Liu W (2019) Giving waterbodies the treatment they need: a critical review of the application of constructed floating wetlands. J Environ Manag 238:484–498. https://doi.org/10.1016/j.jenvman.2019.02.064
52. Wang F, Xiang L, Leung KS-Y (2024) Emerging contaminants: a one health perspective. Innovation (Camb) 5(4):100612. https://doi.org/10.1016/j.xinn.2024.100612
53. Liu Y, Wu S, Nguyen TAH, Chan T-S, Lu Y-R, Huang L (2022) Biochar mediated uranium immobilization in magnetite rich Cu tailings subject to organic matter amendment and native plant colonization. J Hazard Mater 427:127860. https://doi.org/10.1016/j.jhazmat.2021.127860
54. Ganesh R, Robinson K, Reed G, Sayler G (1997) Reduction of hexavalent uranium from organic complexes by sulfate- and iron-reducing bacteria. Appl Environ Microbiol 63:4385–4391. https://doi.org/10.1128/AEM.63.11.4385-4391.1997
55. Mabagala FS, Mng'ong'o ME (2022) On the tropical soils; the influence of organic matter (OM) on phosphate bioavailability. Saudi J Biol Sci 29(5):3635–3641. https://doi.org/10.1016/j.sjbs.2022.02.056
56. Radwany MR, Barton IF (2022) The process mineralogy of leaching sandstone-hosted uranium-vanadium ores. Miner Eng 187:107811. https://doi.org/10.1016/j.mineng.2022.107811
57. Eduah JO, Arthur A, Dogbatse JA, Amoako-Attah I, Afful EA (2024) Ecological effects of soil physicochemical properties and copper speciation on the microbial properties associated

with land use management in cacao production. Environ Technol Innov 33:103538. https://doi.org/10.1016/j.eti.2024.103538
58. Li L, Li R, Guo R (2024) Preparation and evaluation of a combination of chelating agents for the removal of inhaled uranium. Molecules:5759. https://doi.org/10.3390/molecules29235759
59. Wei Y, Chen J, He G, Lu A, Ji S, Zhou J (2025) Fabrication of recoverable plant polyphenol-based surfactants for efficient removal of uranyl ions from skin and cotton fabric. Sep Purif Technol 356, Part A:129790. https://doi.org/10.1016/j.seppur.2024.129790
60. Holmberg K, Bilén F, Bordes R (2025) Development of amino acid-based surfactants: from synthesis to applications. Curr Opin Colloid Interface Sci 75:101884. https://doi.org/10.1016/j.cocis.2024.101884
61. Pizarro J, Sainsbury B, Hodgkinson J, Loechel B (2017) Australian uranium industry climate change vulnerability assessment. Environm Dev 24:109–123. https://doi.org/10.1016/j.envdev.2017.06.002
62. Dahlkamp F (2009) Kazakhstan. In: Uranium deposits of the world. Springer, Berlin, Heidelberg. https://doi.org/10.1007/978-3-540-78558-3_6
63. Reddy Y, Sadasivaiah B, Rajakullaiswamy K, Indira P, Pullaiah T (2013) Avifauna of Thummalapalle uranium mining area, Andhra Pradesh, India. World J Zool 8:62–66. https://doi.org/10.11609/JoTT.o3318.6556-65

Water Chemistry Influence on Uranium Solubility Across Diverse Water Systems: A Comparative Study

Benjamin Uchechukwu Nwaka, Emmanuel Obinna Ogueri, I. Fabian Ezema, and Mamta Bisht

Abstract Mankind is exposed to uranium concentrations through drinkable water, inhalation and terrestrial sources, which deliver radiation to cells and tissues. Acute and chronic exposure to uranium radionuclides can lead to severe biological and radiation health risks. The influence of water chemistry on uranium solubility has received considerable attention; however, comparative studies in different water systems, such as groundwater and surface water, including aquatic and marine environments, are still needed. One of the radionuclide contaminants of water bodies is uranium, which is found in diverse water systems (groundwater and surface water bodies), and its significant concentrations depend on natural and man-made activities. Rock–water interactions controlled by hydrochemical and hydrogeological conditions, discharge of effluents in water bodies from industries and factories, the use of agrochemicals for crop production, and the use of chemical and nuclear weapons for conflict and warning are the major sources of uranium and its decay series in water bodies. The solubility and dissolving capacity of uranium in water depend on the physical and chemical characteristics and modified conditions of water chemistry. Water chemistry parameters that influence or control uranium solubility in different water systems include temperature, water pH, oxidation state,

B. U. Nwaka (✉)
Environmental Radiation and Health Research Group, Department of Physics, Alvan Ikoku Federal University of Education, Owerri, Nigeria
e-mail: benjamin.nwaka@alvanikoku.edu.ng

E. O. Ogueri
Environmental Radiation and Health Research Group, Department of Health Education, Alvan Ikoku Federal University of Education, Owerri, Nigeria
e-mail: emmmauael.ogueri@alvanikoku.edu.ng

I. Fabian Ezema
Nano Research Group, Department of Physics and Astronomy, University of Nigeria, Enugu, Nigeria
e-mail: fabian.ezema@unn.edu.ng

M. Bisht
ICAR-Indian Institute of Water Management, Bhubaneswar, Odisha, India

M. K. Jindal, K. Patni (eds.), *Water Quality-Uranium Solubility Nexus: Unlocking the Secrets of Uranium Solubility*, Earth and Environmental Sciences Library, https://doi.org/10.1007/978-3-031-98909-4_15

redox potential, and dissolved carbonate and phosphate contents. Increasing the temperature of the water system influences the reaction rate. The pH of water is one of the factors controlling the speciation of uranium in water. Among the oxidation states of uranium(II), uranium(III), uranium(VI), uranium(V) and uranium(IV) ions, the oxidation state of uranium(VI) ions is the most stable in aqueous solutions and is highly mobile and present as uranyl ions (UO_2^{+2}). Water chemistry influences uranium adsorption/desorption by changing the surface charge and solubility of minerals, the uranium oxidation state and speciation, and the speciation of aqueous and surface complexes.

Keywords Complexes · Oxidation states · Uranium solubility · Water bodies · Water chemistry

1 Water as Inorganic and Universal Solvent

Water is one of the most common substances, as it is a good solvent for many substances. While water is an inorganic and polar solvent containing inorganic elements (hydrogen and oxygen), benzene is an organic and nonpolar solvent consisting of organic compounds. Benzene is attracted by a nonuniform electrostatic field, unlike water and ethanol, which are deflected or not attracted by a nonuniform electrostatic field. Water and ethanol, which are deflected by nonuniform electrostatic fields, are called polar substances. Substances such as benzene, which are not deflected by nonuniform electrostatic fields, are called nonpolar substances. Polar substances consist of polar molecules, and a polar molecule is defined as a molecule that contains positive and negative poles, as there is a partial positive charge at one point in the molecule and a partial negative charge at another point in the molecule. Natural water includes spring water, rainwater, well water, stream water, river water, saltwater lake water, sea water and ocean water with distinct characteristics [1–4].

The amount of uranium in water bodies is variable and depends on environmental activities and the surrounding bedrock. Uranium enters drinkable groundwater when it dissolves minerals that contain uranium. Uranium enters surface water by the deposition of particles from air or by the erosion of rock and soil in surface water. Uranium in surface water can disperse over large distances to lakes, ponds, streams, rivers, seas and oceans. Uranium emits high linear energy transfer (LET) alpha particles that are densely ionized along their track length. These alpha particles in soft tissues can travel through 40–70 micrometers (μm) in soft tissue, incrementally transferring their kinetic energy to the series of atoms and molecules with which they interact along their short straight paths [5]. Water chemistry influences uranium solubility across different water systems [6–11].

The transformation of uranium in water involves the formation of complexes with anions, cations, and ligands and oxidation–reduction reactions. The oxidized forms of uranium are relatively more soluble than the reduced forms are and can be leached from rocks and soils to migrate to the environment. The precipitation of soluble uranium occurs in strongly reducing environments. In humans, uranium is absorbed in all tissues regardless of the exposure route. In general, the absorption of uranium increases with increasing solubility [5].

2 Physical, Chemical and Radiological Properties of Water

Water, as a unique inorganic solvent, has several physical and chemical properties that make it very important for living on Earth, as it supports various physical, chemical, biological, ecological, and industrial processes as well as aquatic and marine life; however, it may be contaminated with the anthropogenic radionuclides cesium–137 (^{137}Cs) and the natural radionuclides polonium–210 (^{210}Po) [2, 12, 13]. Radiation exposure through drinking water results from naturally occurring uranium and thorium series radionuclides in addition to radioisotopes of potassium–40 in drinkable water, as the activity concentrations of radionuclides in groundwater and surface water are primarily due to radioisotopes of potassium–40, uranium and thorium-bearing rocks and soil mineral deposits [4]. Some of the physical properties of water, including density, viscosity, surface tension, boiling and freezing points, melting points, heat capacity and latent heat, transparency and refractive indices, compressibility and bulk modulus, electrical conductivity and acoustic properties, are studied. Some of the chemical properties of water, such as chemical formula and structure, polarity and dipole moment, hydrogen bonding, solubility and dissolving capacity, pH and acid–base chemistry, redox reactions, complexation and chelation, electrochemical properties and isotopic composition, are studied.

3 Chemical Processes in Water

Dissolution and precipitation, acid–base reactions, redox reactions, and complexations are some of the chemical processes in water. Dissolution and precipitation are critical processes in water chemistry that address the interaction between water and solid phases. Acid–base reactions involve the transfer of protons between substances and are crucial in understanding many water chemistry processes, including the oxidation of organic matter. Complexation explains the formation of complexes between metal ions and ligands, and it affects the solubility and reactivity of many substances.

4 Water Chemistry and Its Application

Water chemistry is the study of the chemical composition and properties of water. It is a complex interdisciplinary concept that involves understanding the interactions between water and its surroundings, including the atmosphere, geology and living organisms. Water chemistry is present in different aquatic and marine environments, such as freshwater environments, marine water environments, groundwater environments and wastewater environments, all of which have unique, dissimilar or distinct water chemistry characteristics. Freshwater environments include streams, rivers, lakes and wetlands; marine water environments include estuaries, seas and oceans. Groundwater environments include borehole water and spring water, which have unique water chemistry characteristics, including the presence of dissolved gases and minerals, whereas wastewater environments include sewage and industrial effluent, which have distinct water chemistry characteristics.

Water chemistry has a good number of applications, such as in monitoring water quality and detecting changes in water chemistry, in designing and operating water treatment systems, and in environmental remediation, as it is essential in designing and implementing environmental remediation strategies. Water chemistry is also applied in aquatic ecology, where water chemistry affects the survival and growth of aquatic organisms.

The physical and chemical factors influencing uranium solubility include temperature; oxidation state; pH; redox potential or electrochemical potential (Eh); and carbonate, phosphate and sulfate concentrations. Water chemistry significantly controls uranium solubility by affecting uranium oxidation states, complexation with dissolved ions (particularly carbonates), pH levels and redox potential (Eh) with oxidizing, alkaline, carbonate-rich environments, largely leading to high uranium solubility due to the formation of soluble uranyl complexes. In contrast, reducing conditions tend to precipitate uranium as less soluble minerals such as uraninite [1, 9].

5 Water Chemistry Parameters

Water chemistry parameters are radiological, physical, chemical, and biological characteristics that describe the composition and properties of water. They are grouped into radiological parameters, physical parameters, chemical parameters, biological parameters, radiological parameters (Fig. 1) and other parameters, such as total organic carbon (TOC), which measures the concentration of organic carbon in water; ultraviolet (UV) absorbance, which measures the absorbance of ultraviolet light or rays in water, taste and odor; and the aesthetic quality of water. These parameters are essential for understanding the quality and behavior of water in various environments, including natural ecosystems, drinking water supplies and industrial processes.

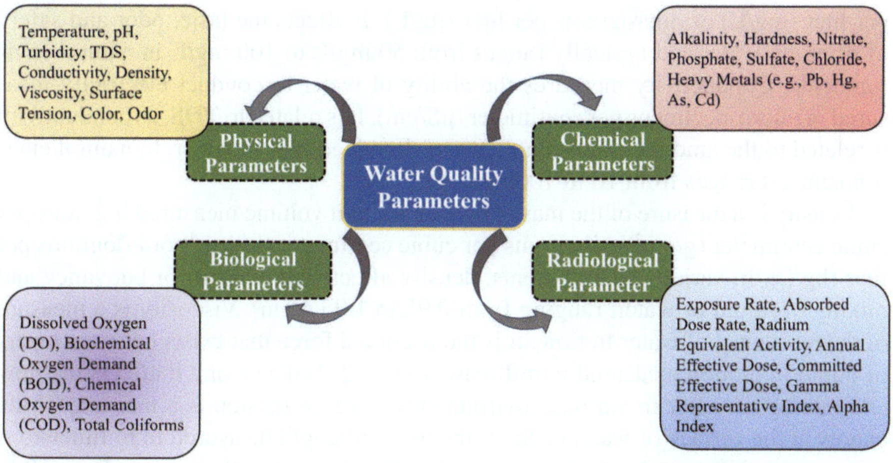

Fig. 1 Water chemistry parameters

Moreover, water chemistry influences uranium adsorption/desorption by changing the surface charge and solubility of minerals, uranium oxidation states and speciation, as well as the speciation of aqueous and surface complexes. Water chemistry parameters that influence or control uranium adsorption include natural organic matter, redox potential, pH, dissolved carbonate and phosphate.

5.1 Physical Parameters of Water

The physical parameters of water are crucial for determining the behavior and properties of water in various environments. The physical parameters that are commonly used in water chemistry are temperature, pH, turbidity, total dissolved solids (TDS), conductivity, density, viscosity, surface tension, color and odor [8]. Temperature is a measure of the thermal energy of water with units of Celsius (°C), Fahrenheit (F) and kelvin (K). The system international (S.I.) unit of temperature is kelvin (K). In the natural environment, temperatures ranging from 0 °C to 31 °C and 27 °C are regarded as room temperature. Temperature affects the physical and chemical properties of water, such as viscosity, density, and solubility. The pH is a measure of the acidity and alkalinity of water and ranges from 0 to 14. A range of 1–6 is acidic, 7 is neutral, and a range of 8–14 is alkaline. In natural environments, the pH ranges from 6 to 8.4, and the pH affects the chemical reactivity and solubility of substances in water. Turbidity measures the clarity and cloudiness of water in nephelometric turbidity units (NTUs). It ranges from 0 to 100 NTU in natural environments and affects drinking water quality.

The total dissolved solids (TDS) measure the concentration of dissolved particles or solids in water in parts per billion (ppb), parts per million (ppm), milligrams per liter (mg/L) or micrograms per liter (μg/L). It affects the taste, odor and safety of drinking water and typically ranges from 50 mg/L to 100 mg/L in natural environments. Conductivity measures the ability of water to conduct electricity measured in microspecimens per centimeter (μS/cm). It is related to TDS, as conductivity is related to the amount or concentration of dissolved ions in water. In natural environments, it ranges from 10 to 1000 μS/cm.

Density is a measure of the mass of water per unit volume measured in grams per cubic centimeter (g/cm^3), kilograms per cubic centimeter (kg/cm^3) or kilograms per liter (kg/L). In various environments, density affects the upthrust or buoyancy and mixing of liquid or water, ranging from 0.95 to 1.05 g/cm^3. Viscosity is a measure of the resistance of water to flow; it is the frictional force that exists between layers of liquid in motion measured in millipascal seconds (mPa.s), and it affects the flow and mixing of water in various environments. Surface tension is a measure of the energy at the surface of water or the force pet unit length measured in millinewtons per meter (mN/m) or dynes per centimeter (dyn/cm), typically ranging from 60 to 80 mN/m. Surface tension affects the behavior of water at interfaces and its ability to interact with other substances. Color is a measure of the aesthetic quality of water in color units (CU) or platinum color units. Color affects the aesthetic quality and safety of drinking water. The Oduor is also a measure of the aesthetic quality of water in an odor unit. Oduors affect the aesthetic quality and safety of drinking water.

5.2 *Chemical, Biological and Radiological Parameters of Water*

Alkalinity, hardness, nitrate, phosphate, sulfate, chloride and heavy metal concentrations are some of the chemical parameters that are measured when assessing drinking water quality for safe use. Some of the biological parameters include dissolved oxygen (DO), biochemical oxygen demand (BOD), chemical oxygen demand (COD), total coliform bacteria, and *Escherichia coli* (*E. coli*). The radiological parameters included the exposure rates, absorbed dose rates, radium equivalent activities, annual effective doses, annual committed effective doses, gamma representative indices, alpha indices, and excess lifetime cancer risks.

6 Uranium and Uranium Isotopes

Uranium is a radioactive element belonging to the actinide series of the periodic table. It is a radioactive pollutant that is more common in groundwater than in surface water and could pose very serious radiological human health challenges if not controlled. Uranium, under natural conditions, exists mainly as three isotopes, ^{238}U,

^{235}U, and ^{234}U (99.3%, 0.72% and 0.005% by mass, respectively), with ^{238}U and ^{235}U having longer half–lives with relatively high abundances than ^{234}U. The half-lives of ^{238}U, ^{235}U, and ^{234}U are approximately 4.5×10^5, 7.1×10^8, and 2.5×10^5 years, respectively.

Among the three isotopes of ^{238}U, ^{235}U, and ^{234}U, 234 U has the shortest half-life and is the most radioactive. Approximately 48.9%, 2.2% and 48.8% of the radioactivity of natural uranium is attributed to ^{234}U, ^{235}U and ^{238}U, respectively. Three uranium isotopes, ^{238}U, ^{235}U, and ^{234}U, which exist naturally, behave alike chemically but have dissimilar radioactive properties. They are primarily alpha emitters that decay through two different series headed by ^{235}U and ^{238}U series as the parent radionuclides. Isolates are used as radioactive tracers in the dating of various oceanographic processes as well as in the study of sources of dissolved solutes in groundwater, surface water and marine environments [12–14]. Some inorganic soluble uranium salts are uranyl nitrate hexahydrate, uranyl fluoride, uranyl tetrachloride, and uranyl acetate dihydrate with molecular formulas of $UO_2(NO_3)$. $6H_2O$, F_2O_2U, UCl_4 and $UO_2\ (CH_3COO)_2{,}H_2O$ respectively, and molecular weights of 502.13, 308.03, 370.84, and 424.15, respectively [5].

Uranium exists naturally in different valences or oxidation states, such as U^{+2}. U^{+3}, U^{+4}, U^{+5}, U^{+6} (+2, +3, +4, +5 and + 6, respectively). The discharge or release of uranium in water is strongly influenced by the kinetics of mineral dissolution and uranium desorption. Water passing through geologic formations can dissolve uranium; therefore, uranium exists in aqueous solutions in the oxidation states of uranium (IV; U^{+4}), uranium (V; U^{+5}) and uranium (VI; U^{+6}). Among all these oxidation states, the uranium(VI; +6; U^{+6}) oxidation state is the dominant valence state of uranium, which is the most stable in aqueous solutions and is present as uranyl cations (UO_2^{+2}) [11, 12].

Uranium is associated with a number of different mineral phases, including silicate minerals, Fe–Mn (oxy) hydroxides, oxidizable minerals and exchangeable phases of weakly absorbed uranium. The release of uranium from these minerals to water is a complicated process that is controlled by interactive geochemical reactions such as the dissolution of uranium-bearing minerals, especially silicate minerals and Fe–Mn (oxy) hydroxides, and uranium desorption from mineral surfaces. Primary uranium minerals weather and breakdown very easily as soon as they are exposed to water and oxygen to produce secondary oxidized minerals. Uranium enters water by leaching from soils and rocks.

7 Influence of Water Chemistry on Uranium Solubility across Diverse Water Systems

The natural radionuclides ^{238}U, ^{235}U, and ^{234}U are present in ground and surface water, and the activity concentrations of these natural radionuclides in water are due to the specific activities of their decay progenies [15]. Most natural water at different locations contains dissolved uranium at very low concentrations, such as in Japan, where a few micrograms per liter (2 μg/L) is the safety reference limit [16]. According to the World Health Organization (WHO), the normal tolerable limit of uranium concentrations in groundwater is 30 μg/L for drinkable water quality [11, 16–18], whereas relatively high concentrations of 60 μg/L in drinking water and 1000 μg/L have been reported in India and Russia, respectively [16, 19, 20].

Major factors influencing uranium solubility include the oxidation state; pH; redox potential or electrochemical potential (Eh); and calcium, magnesium, nitrate carbonate, phosphate and sulfate concentrations [11]. Water chemistry significantly impacts uranium solubility by affecting uranium oxidation states, complexation with dissolved ions (particularly carbonates), pH levels and redox potential (Eh) with oxidizing, alkaline, carbonate-rich environments, largely leading to high uranium solubility due to the formation of soluble uranyl complexes. In contrast, reducing conditions tend to precipitate uranium as less soluble minerals such as uraninite [1, 9].

In groundwater and soil water, uranium is controlled by geochemical factors, the discharge of uranium from minerals (through mineral dissolution and uranium desorption from mineral surfaces) to water, and the extent of the release of uranium from minerals to water is controlled by uranium-hosting minerals, the oxidation states of uranium and water chemistry [10, 15, 21, 22]. An attempt was made by Srivastava et al. [11] to correlate uranium concentration with water quality parameters for two different seasons, and the correlation data revealed that oxidation–reduction potential (ORP), nitrates, phosphate, calcium and magnesium were positively correlated with uranium concentration for the two different seasons. On the other hand, total dissolved solids (TDS), electrical conductivity, temperature, dissolved oxygen (DO), fluoride and chloride are negatively correlated. The positive correlation results revealed that uranium may be present in groundwater as dissolved salts of the studied water parameters [11].

In surface water and groundwater, the pH of water is one of the controlling factors of the speciation of uranium in water. The oxidation state UO_2^{+2} forms rapidly soluble complexes that are transported in groundwater connected to surface water through the hydrological cycle. Below a pH of 5, UO_2^{+2} is the principal species in water. In the pH range of 5–7, the main species in water are hydroxyl complexes (UO_2OH^+) and multinuclear hydroxide complexes $\left(UO_2\right)_3\left(OH\right)_5^+$, whereas above a pH of 7, the dominant species is $UO_2\left(OH\right)_3^-$ [23, 24]. Chronic exposure to uranium radionuclides can lead to severe health risks [25, 26].

7.1 *Influence of Water Chemistry on Uranium Solubility in Groundwater*

Groundwater is a global source of freshwater that is utilized not only for drinking but also for other activities, such as irrigation, domestic, and industrial uses; moreover, the quality of the water used in our environment needs to be scrutinized. Uranium is introduced into the groundwater system by natural geochemical processes (geogenic) or anthropogenic activities [9]. Geogenic sources include mineral dissolution of host rocks and sediments along the flow path of groundwater or desorption from the surface of various minerals. Cl clays, carbonates, Fe, Mn, and Al oxides and hydroxides act as strong adsorbents of uranium. Anthropogenic sources of uranium include mining and ore processing of uranium, radioactive material production and disposal, combustion of fuels and the use of phosphate fertilizers and organic manure in agricultural fields [9, 27, 28]. However, the uranium concentration is highly variable and may depend on groundwater chemistry, aquifer characteristics and bedrock geology. Redox processes control the enrichment of uranium in groundwater, and its solubility is highly sensitive to aquifer geochemistry.

Uranium ions generally exist in tetravalent (+4) and hexavalent (+6) oxidation states or form in groundwater, and their contamination results from natural geochemical processes such as mineral dissolution and desorption of adsorbed uranium from mineral surfaces [10]. The kinetics of mineral dissolution and uranium desorption principally influence the release of uranium into water. The transport of uranium in groundwater mainly depends on the oxidation state, and relatively high concentrations are found in toxic waters [7]. The hexavalent (+6) oxidation state is more inclined or susceptible to groundwater contamination by diffusion and advection processes. Under reducing conditions, the oxidation state of uranium is +4, and the stable uranium tetravalent phases are uraninite and coffinite. The solubility of the uranium tetravalent state (U^{+4}) is very low, and reducing conditions effectively diminish uranium transport in groundwater. While the oxidative dissolution of uranium tetravalent state (U^{+4}) minerals is a principal mechanism of the distribution or mobilization of uranium, the reductive precipitation of uranium hexavalent states (U^{+6}) is an effective method to stop the distribution or immobilize uranium [6, 7]. Metallic uranium is chemically reactive and can be leached into groundwater in acidic or alkaline environments. Metallic uranium is strongly absorbed into groundwater in alkaline environments. The presence of uranium in groundwater is governed mainly by ORP conditions and by pH conditions [7]. Additionally, the presence of dissolved minerals such as carbonates and prevailing redox conditions play essential roles in determining uranium solubility in groundwater systems.

7.2 *Influence of Water Chemistry on Uranium Solubility in Surface Water*

Uranium occurs in natural water in three oxidation states: uranium(IV) or (U^{+4}), uranium(V) or UO_2^+ and uranium(VI) or UO_2^{+2} or uranyl ions. In reducing surface water, uranium occurs as uranium(IV) or (U^{+4}) and uranium(VI) or UO_2^{+2} or uranyl ions. The uranium ion (cation) UO_2^{+2} is the form of uranium (VI) that is principally responsible for eliciting a toxic response in aquatic organisms. Typically, inorganic and organic complexes of uranium(VI) increase the uptake and toxicity of uranium(VI) by reducing the concentration of UO_2^{+2}. The redox and complexation reactions of uranium in surface water are strongly influenced by hydrolysis, since hydrolytic reactions may limit its solubility or influence its sorption behavior. The speciation of uranium is relatively complex in oxidized fresh surface water at pH values ranging from 5 to 9.

In seawater, dissolved uranium predominantly exists as a uranyl–tricarbonate complex [$UO_2\left(CO_3\right)_3^{-4}$]. Analytical methods and theoretical methods are available techniques for determining uranium speciation in water. One of the most important complexing agents for uranium in oxidized freshwater is carbonate, whereas phosphate is an important complexing agent when its concentration approaches 75 μg/L [29, 30]. Changes in uranium speciation are responsible for changes in the toxicity of uranium, as sorption plays a dominant role in determining the fate of uranium in freshwater systems. Generally, the sorption of uranium to particles increases with increasing pH until a threshold point is reached in the range of pH 6–8. The sorption of uranium to particles and organic matter decreases with increasing salinity [31, 32].

Surface water with high carbonate concentrations and fluctuating pH levels can experience variations in uranium solubility depending on seasonal changes and water mixing. Groundwater rich in uranium is concentrated by evaporation, and the uranium hexavalent state is chelated by $C0_2^{-2}$ to form highly soluble $UO_2\left(CO_3\right)_3^{-4}$ in water, which is a major cause of uranium enrichment in lakes [14]. Like groundwater, biogeochemical processes, redox potentials, TDS, salinity, pH concentrations and ion concentrations (anions and cations) are surface water quality factors that influence the aqueous uranium content distribution [14]. For surface water, the concentration of aqueous uranium (^{238}U) in freshwater lakes was lower than that in saline lakes. In addition, high TDS concentrations lead to high concentrations of uranium in surface water [14].

7.3 *Effects of Water Chemistry on Uranium Solubility in Marine Environments*

The marine environment includes various aquatic ecosystems and habitats in seas and the ocean. Marine ecosystems contain high concentrations of dissolved salt, which includes the open ocean, deep–Sea Ocean, and coastal marine ecosystems. In marine water systems, the solubility of uranium is influenced mostly by water quality parameters or water chemistry factors such as redox potential, pH concentrations, carbonate concentrations and the presence of dissolved ions such as calcium and sulfate, which determine the speciation of uranium, primarily the uranium(VI) oxidation state, under oxidized conditions and its ability to form soluble complexes, heavily impacting how much uranium can remain dissolved in the water column, with high carbonate concentrations and near-neutral pH leading to increased uranium solubility due to the formation of stable uranyl carbonate complexes. The major mechanism is the oxidation state of uranium(VI), which is more soluble than uranium(IV) in marine water.

At relatively high pH values, uranium is more likely to form soluble carbonate complexes in marine water, thus increasing its solubility in marine environments. Redox conditions are essential water chemistry factors influencing uranium solubility in marine environments. In marine environments, the uranium(VI) oxidation state is more soluble than the uranium(IV) oxidation state is, so a reducing environment tends to precipitate uranium out of solution as insoluble uranium(IV) compounds. High carbonate concentrations support or encourage the formation of stable uranyl carbonate complexes, which enhances uranium solubility. The presence of calcium ions can further increase uranium solubility by forming calcium–urinary carbonate complexes. OM can also influence uranium solubility by forming complexes with uranium, potentially increasing uranium transport in the water column [14].

Studies carried out in oceans and river estuaries have revealed that aqueous uranium (^{238}U) concentrations increase linearly with increasing TDS [33, 34]. In aquatic and marine environments, areas with high biological, physical, and chemical activities and nutrient inputs can influence uranium concentrations through interactions with organic matter and phytoplankton. Additionally, in deep sea regions, with low oxygen concentrations, uranium tends to precipitate more rapidly because of reduced conditions. In marine sediments where redox conditions can vary, uranium can be readily exchanged between the dissolved phase and solid phase depending on the local chemistry. The concentration of ^{238}U is related to biochemical processes in which organic matter interacts with iron and magnesium oxides and hydroxides. Studies have shown that the concentration of uranium (VI) in the aqueous phase drastically decreases during bacterial processes because of the formation of insoluble uranium (IV) compounds [14]. One of the processes of removing uranium from the ocean is diffusion across sediment water—the interface of organic-rich sediments [35].

8 Uranium Mobility/Transport in Water

Uranium transport in water bodies is largely controlled by their ability to form complexes with both inorganic ($Cl^-, F^-, PO_4^{-3}, CO_3^{-2}$) and organic ligands. Under normal conditions, uranium is present in the form of $UO_2\left(CO_3\right)_2^{-4}$. The pH of surface water containing CO_3^{-2} and HCO_3^- increases with increasing concentrations of CO_3^{-2} and HCO_3^-. As the pH decreases, $UO_2\left(CO_3\right)_2^{-4}$ decomposes slowly in water to form $UO_2\left(CO_3\right)_2^{-2}$. Under normal conditions, aqueous ^{238}U concentrations are affected by TDS and pH, which could cause different water bodies in the same region to have different aqueous ^{238}U concentrations. The major controlling factor of aqueous ^{238}U concentrations in saline lakes is pH, which is far greater than TDS [14]. The type of minerals present in surrounding rock formations can significantly influence or impact uranium solubility by providing adsorption sites or acting as sources of uranium. Organic compounds in water can complex with uranium, influencing uranium mobility and solubility. Mining, industrial waste disposal and agricultural practices can introduce additional uranium to water systems, altering the existing chemical equilibrium and impacting solubility.

The kinetics of mineral dissolution and uranium desorption strongly influence the release of uranium in water. The solubility of uranium(IV) minerals is extremely low, and reducing conditions effectively diminish the movement of uranium in groundwater and soil. The reductive precipitation of uranium(IV) is an effective method for immobilizing uranium. The oxidative dissolution of uranium(VI) minerals is a major mechanism of uranium mobilization. Under reducing conditions, the oxidation state of uranium is +4, and the stable uranium(IV) phases are mainly uraninite and coffinite. Organic complexes of uranium(IV) associated with humic materials may also retain uranium(IV) in the solid phase. Under oxidizing conditions, the predominant oxidation state of uranium is +6, also represented as uranium(VI), which exists mainly in the form of uranyl ions (UO_2^{+2}) [10].

The concentrations of uranium in natural water are usually low, typically less than 4 μg/L in river water, approximately 3.3 μg/L in open seas water, and usually less than 5 μg/L in groundwater; however, higher concentrations can occur in groundwater and surface water, with extremes in the mg/L range [36].

The ability of uranium to be absorbed, mobilized, immobilized and precipitated is determined mainly by its oxidation state and mineral solubility. The environmental physics and chemistry of uranium are influenced by its oxidation–reduction potential (ORP) because of its redox dependence. There are three main oxidation states of uranium in nature, namely, uranium(IV), uranium(V) and uranium(VI); however, uranium(IV), which is the reduced form, and uranium(VI), which is the oxidizing form, control most uranium geochemistry [9]. The solubility of uranium(VI) is extremely high, and oxidizing conditions effectively enhance uranium movement. Uranium mobilization is characterized by the various complexes formed in the water system depending on the pH and PCO_2. At low pH values (less than 5), the dominant species of uranium is uranyl ions (UO_2^{+2}); however, uranyl ions from complexes are dominant. With carbonate, phosphate and calcium at high pH values. Uranium concentrations are also affected by recharge and discharge from aquifers, and stable isotopes (^{18}O and ^{2}H) are widely used to gain insight into aquifer recharge systems. These environmental tracers are recognized as recorders of important processes such as evaporation, transpiration, recycling, and mixing of water [9].

In some cases, uranium pollution of groundwater is caused by anthropogenic sources such as uranium mining, production and disposal of radioactive sources, processing of uranium ores and natural geochemical processes. More often, in groundwater, uranium is introduced more often by natural geochemical processes. Natural geochemical processes such as mineral dissolution and desorption of absorbed uranium from mineral surfaces result in uranium contamination in groundwater [10]. Comparative studies of the current influence of water chemistry on uranium solubility across different water systems are relatively rare. Mineral identification and knowledge of the effects of water chemistry on uranium solubility in diverse water systems (streams, rivers, seas and oceans) are essential for understanding, predicting and managing geochemical/geogenic uranium pollution in diverse water systems and environments. Studies have shown that Fe–Mn (Oxy) hydroxides and silicate minerals are the principal uranium-hosting minerals and that a significant fraction of uranium exists as absorbed ions on minerals [10].

The release of uranium in the environment and ground water is controlled by a number of interactive processes, which include the dissolution of uranium-bearing minerals, uranium desorption from mineral surfaces, the formation of aqueous uranium complexes, and the reductive precipitation of uranium [10].

9 Summary

Water quality parameters broadly include radiological, physical, chemical, biological, and radiological parameters. Physical and chemical parameters are called physicochemical parameters, such as viscosity, surface tension, temperature, pH, TDS, EC, ORP, and salinity, among others, and include several major and minor ions, trace elements and heavy elements [17]. The total dissolved solid (TDS) comprises inorganic salts and small amounts of organic matter that are dissolved in water. The range of TDS above 900 mg/L is considered poor for drinking, as it can contain pollutant solids from sewage, urban runoff and industrial wastewater.

Electrical conductivity (EC) is the ability of a solution to conduct electric current. pH is the scale for monitoring the acidity and basicity of aqueous solutions. The oxidation–reduction potential (ORP) is the ability of a lake or river to clean itself or measure water's ability to oxidize contaminants. A higher ORP simply indicates a greater number of oxidizing agents; hence, a lower ORP is preferred for consumption. One possible reason for the low concentration of uranium in water could be the low availability of rocks containing uranium, which directly indicates that there may be a low probability of finding uranium in water even when complex forming factors are available. An evaluation of the solubility of uranium revealed that pH, complexation, and redox potential, among other physical and chemical properties of water, are the main parameters affecting the solubility of uranium in diverse water systems. Under oxidizing conditions, uranium exists as uranium (VI) as a uranyl cation (UO_2^{+2}), whereas under reducing conditions, uranium transforms to an insoluble uranium (IV) state. The oxidizing state of uranium (uranium hexavalent; U^{+6}) and the reducing state of uranium (uranium tetravalent; U^{+4}) control most of the uranium geochemistry.

10 Conclusion

Uranium is an actinide element with three naturally occurring radioisotopes, ^{234}U, ^{235}U, and ^{238}U, and is mobile at acidic pH values (2–4), principally as uranyl cations (UO_2^{+2}). Uranium occurs in the oxidation states of uranium(II), uranium(III), uranium(IV), uranium(V) and uranium(VI); however, the dominant states or forms in nature are uranium(IV) and uranium(VI). Uranium can be found in both the solid and liquid phases. The water chemistry of uranium in aqueous solutions is complicated, and of the five oxidation states of uranium, only three oxidation states, namely, uranium(IV) or (U^{+4}), uranium(V) or UO_{2^+} and uranium(VI) or UO_2^{+2}, exist in natural water depending on the surrounding conditions. In reducing surface waters, uranium exists as U^{+4} and UO_{2^+} (uranium IV and uranium V). Uranium (IV) has a strong tendency to precipitate and remain immobile, whereas uranium (V) forms soluble but relatively unstable complexes. In oxidized surface waters, uranium occurs as (UO_2^{+2}) and forms stable, readily soluble, cations, anions and neutral complexes that are highly mobile. Uranium mobility in water depends on redox conditions, oxidation states, pH, complexation, and solute chemistry, which are shaped by geological conditions. The redox reactions and complexation reactions of uranium in surface water are strongly influenced by hydrolysis since hydrolytic reactions may limit its solubility or influence its sorption behavior. Measureable uranium concentrations are always assumed to represent dissolved uranium(VI) species. While uranium(IV) or minerals of uranium(IV) are generally insoluble, uranium(VI) is rather soluble in aqueous solutions. Uranium cations form strong complexes with carbonate, sulfate phosphate and hydroxide. Complexation of aqueous uranium has been found to increase the solubility of uranium minerals. Another mechanism of increasing aqueous uranium is the release of uranium from the mineral structure followed by complexation by aqueous ligands. Furthermore, chelating effects are another mechanism of increasing aqueous uranium. Under these conditions, oxygen-containing ligands such as hydroxide, phosphate and carbonate bind to uranium in the mineral structure and subsequently release uranium complexes into the solution. To a great extent, water pH can influence the solubility, sorption, hydrolysis, complexation and colloid formation of uranium. Reports have revealed that the concentration of uranium in natural water is usually low, typically less than 4 μg/L in river water, approximately 3.3 μg/L in open seas water, and usually less than 5 μg/L in groundwater; however, higher concentrations can occur in groundwater and surface water, with extremes in the mg/L range.

References

1. Balaram V, Rani A, Rathore DPS (2022) Uranium in groundwater in parts of India and world: A comprehensive review of sources, impact to the environment and human health, analytical technique and mitigation technologies. Geosyst Geoenviron 1(2):100043. https://doi.org/10.1016/j.geogeo.2022.100043
2. International Atomic Energy Agency (IAEA). (1995). Sources of radioactivity in the environment and their relative contributions to overall dose assessment from marine radioactivity (MARDOS). IAEA – TECDOC – 838; IAEA Vienna ISSN: 1011–4289
3. Nwaka BU, Avwiri GO, Ononugbo CP (2018) Radiological risks associated with gross alpha and beta activity concentrations of water resources within salt water lakes, Ebonyi State, Nigeria. Int J Trop Dis Health 30(1):1–10. https://doi.org/10.9734/IJTDH/2018/39066

4. Ononugbo CP, Nwaka BU (2017) Natural radioactivity and radiological risk estimation of drinking water from Okposi and Uburu salt lake areas, Ebonyi State, Nigeria. Phys Sci Int J 15(3):1–15. https://doi.org/10.9734/PSIJ/2017/31625
5. Elsa N, Krestine G, Ole L (2013) Uranium, inorganic and soluble salts: Evaluation of health hazards and proposal of a health based quality criteria for drinking water. Environmental Policy Project No. 1527. The Danish Environmental Protection Agency Denmark; pp 1–44. ISBN: 978-87-93026-82-7, www.mst.dk/english
6. Bajwa BS, Kumar S, Sahoo SK, Singh S, Tripathi RM (2015) Uranium and other heavy toxic elements distribution in the drinking water samples of SW Punjab, India. J Radiat Res Appl Sci 10:13–19. https://doi.org/10.1016/j.jrras.2015.01.002
7. Chandrasekar T, Sabarathinam C, Viswanathan PM, Rajendiran T, Mathivanan M, Natesan D, Samayamanthula DR (2021) Potential interplay of uranium with geochemical variables and mineral saturation states in ground water. Appl Water Sci 11:75. https://doi.org/10.1007/s13201-021-01396-3
8. Nwaka BU, Avwiri GO, Eze JO (2022) Analyses of physicochemical characteristics and heavy metals risk assessments in locally processed salts and refined iodized salt brand consumed in Ebonyi State, Nigeria. FUNAI J Sci Technol 5(1):48–63. www.fjst.org
9. Ritu B, Karanveer DD (2022) Occurrence and behavior of uranium in the groundwater and potential health risk associated in semi—arid region of Punjab, India. Groundw Sustain Dev 12:100731. https://doi.org/10.1016/j.gsd.2022.100731
10. Samrat MA, Cheng T (2014) Uranium release from sediment to groundwater: influence of water chemistry and insights into release mechanism. J Contam Hydrol 164:71–87. https://doi.org/10.1016/j.jconhyd.2014.06.001
11. Srivastava M, Srivastava PK, Kumar D, Kumar A (2022) A systemic study of uranium groundwater and the correlation with other water quality parameter. Water Supply 22(3):2478–2492. https://doi.org/10.2166/ws.2021.458
12. Nwaka BU, Ogueri EO, Jindal MK, Amala D, Avwiri GO, Jibiri NN, Amakom CM, Ezema FI (2025) Chapter 6: Impact of radioactive elements and oceanographic factors on marine ecosystems. In: Radiation status in the marine world, Springer oceanography. Springer, Cham. https://doi.org/10.1007/978-3-031-82403-6
13. Nwaka BU (2019) Introduction to ionizing radiation measurement and quantities. LAMBERT Academic Publishing. ISBN-13:978-613-9-87463-7, www.amazon.com
14. Zhao C, Zhange P, Xiangzhong L. NY, Tan l, Edwards RL, Yao X, Cheng H (2020) Distribution characteristics and influencing factors of uranium isotopes in saline lake water in the northeast of Qaidam Basin. Minerals 10(1):74. https://doi.org/10.3390/min10010074
15. Aliyu A, Shallhokma AE, Garba I (2022) Determination of gross alpha and gross beta activities and corresponding radiological risk parameters in water samples from Dzamah Community, Hong Nigeria. J Radiat Nucl Appl 7(2):7–13. https://doi.org/10.18576/jrna/070202
16. Jindal MK, Pandit SA, Karunakara N, Changrashekara SN, Salim SKD, Srinivasan R (2023) High uranium dose from groundwater in a granitic terrain in the eastern part of Karnataka, Indian. J Radioanal Nucl Chem. https://doi.org/10.1007/s10967-023-09053-6
17. World Health Organization (WHO) (2008) Guidelines for drinking water quality 3rd edition (incorporating 1st and 2nd addenda) Volume 1 Recommendation, World Health Organization Library. World Health Organization, Geneva
18. World Health Organization (WHO) (2012) Uranium in drinking water. Background document for development of World Health Organization guidelines for drinking—water quality. WHO, Geneva, pp 1–21
19. Atomic Energy Regulatory Board (AERB) of India (2004) Limits on uranium in drinking water Atomic Energy Regulatory Board of India, Mumbai; AERB/VC/16/2004/80/373
20. Sahoo SK, Jha VN, Patra AC, Jha SK, Kulkarni MS (2020) Scientific background and methodology adopted on derivation of regulatory limits for uranium in drinking water—a global perspective. Environ Adv 2:100020

21. John SOG, Olukokun SF, Mathuthu M (2025) Assessment of radioactivity concentrations and associated radiological health risk in natural spring mineral bottled drinking water from South Africa. Water 17(2):156. https://doi.org/10.3390/w17020156
22. Singh KP, Chandra S, Prasad M, Panwar P, Joshi A, Sharma S et al (2025) Measurement of natural radionuclides and health risk assessment of soil samples of the main central thrust region in Garhwal Himalaya, India. Environ Monit Assess 197(2):128
23. Deshmukh P, Sar SK, Jindal MK (2023) Magnetic based green biocomposite for uranium exclusion from aqueous solution. J Radioanal Nucl Chem 332:297–310
24. Kimura T, Serrano GJ, Nakayama S, Takahashi K, Takeishi H (1992) Speciation of uranium in aqueous solutions and in precipitates by photoacoustic spectroscopy. Radiation Acta 58(1):173–178
25. International Commission on Radiological Protection (ICRP) (2007) Recommendations of International Commission on Radiological Protection, ICRP publication 103. Ann ICRP 37(2–4):1–332. https://doi.org/10.1016/j.icrp.2007.10.003
26. United Nations Scientific Committee on Effects of Atomic Radiation (UNSCEAR) (2000) UNSCEAR, sources and effects of ionizing radiation. 1 Annex B, 84–140, Report to General Assembly with Scientific Annexes, New York
27. Nwaka BU, Jibiri NN (2018) Activity concentrations of gamma dose rates levels in poultry feeds and manure used in Ibadan, Nigeria. Glob J Sci Front Res 18(4):53–66
28. Nwaka BU, Jibiri NN (2018) External and internal radiation doses from chemical fertilizers used in Ibadan, Oyo State, Nigeria. Glob J Sci Front Res 18(4):73–81
29. Langmuir D (1978) Uranium-solution-mineral equilibria at low temperatures with applications to sedimentary ore deposits. Geochim Cosmochim Acta 42:547–569
30. Langmuir D, Herman JS (1980) The mobility of thorium in natural waters at low temperatures. Geochim Cosmochim Acta 44:1753–1766
31. Van Den Berg CMG (1993) Complex formation and the chemistry of selected trace elements in estuaries. Estuaries 16:512–520
32. Willet IR, Bond WJ (1995) Sorption of magnesium, uranium and radium by highly weathered soils. J Environ Qual 24:834–845
33. Lü E, Lei Z, Min C, Yusheng Q, Na X, Weifeng Y, Yanping L, Yipu H (2007) Size-fractionated uranium isotopes in surface waters in the Jiu-long River Estuary in China. Acta Oceanol Sin 29:59–68
34. Owens SA, Buesseler KO, Sims KWW (2011) Re-evaluating the ^{238}U salinity relationship in seawater: implications for the^{238}U-^{234}Th disequilibrium method. Mar Chem 127:31–39
35. Klinkhammer GP, Palmer MR (1991) Uranium in the oceans: where it goes and why. Geochim Cosmochim Acta 55(7):1799–1806. https://doi.org/10.1016/0016-7037(91)90024-Y
36. Smeldly PL, Kinniburge DG (2023) Uranium in natural waters and the environment: distribution, speciation and impact. Appl Geochem 148:105534. https://doi.org/10.1016/j.apgeochem.2022.105534

Current Scenario of Soluble Uranium in Water

Understanding the Combined Effects of Multiple Water Quality Parameters

Taufiq Ahamad, Rohit Singh Sajwan, Manoj Kumar Jindal, and Shubham Sharma

Abstract Monitoring water quality is crucial, as it directly impacts the health and well-being of ecosystems and the human population. The water quality parameters pH, redox potential, total dissolved solids, and competing ions are critically examined for their roles in modifying uranium chemistry within the water. As water quality parameters predominantly influence uranium mobilization in water, monitoring and understanding the influence of various water quality parameters is essential for measuring the hydrochemistry of water. The issue of uranium contamination in water (groundwater, surface water, etc.) is a concern at the global scale because of its deteriorating health effects. Notably, uranium is soluble in water, which results in both radiological and chemical toxicity in the human environment. Uranium is mobile in its oxidized state (U^{+6}), forms at neutral to alkaline pH levels, and is facilitated by the formation of stable complex salts. Under such conditions, uranium often coexists with other similarly regulated anions and oxyanions, such as fluoride, arsenic, vanadium, and molybdenum. This chapter aims to explain the combined effect of soluble uranium in water with the interpretation of multiple water quality parameters that collectively influence its behavior and impact. Through a comprehensive literature survey of recent research and case studies, we aim to understand the complexities behind uranium mobility in water, the influence of water quality

T. Ahamad (✉)
Department of Physics, School of Life and Applied Sciences, Maya Devi University, Dehradun, India

R. S. Sajwan
Department of Chemistry, School of Applied and Life Science, Uttaranchal University, Dehradun, India

M. K. Jindal
Center for Sustainable Drinking Water, Indian Institute of Technology Jodhpur, Jodhpur, Rajasthan, India

S. Sharma
Department of Physics, H.N.B. Garhwal University, S.R.T. Campus Badshahi Thaul, Tehri Garhwal, India

M. K. Jindal, K. Patni (eds.), *Water Quality-Uranium Solubility Nexus: Unlocking the Secrets of Uranium Solubility*, Earth and Environmental Sciences Library, https://doi.org/10.1007/978-3-031-98909-4_16

parameters, and bioavailability in aquatic environments, which is pivotal for devising effective water management strategies and mitigating the risks associated with uranium contamination. Our discussion further explores the implications of these findings for water safety monitoring and public health, highlighting the need for integrated approaches in addressing water quality issues in uranium-affected areas around the world.

Keywords Uranium solubility · Hydrochemistry · Water quality parameters · Strategies for monitoring uranium contamination · Radiological and chemical risk

1 Background and Significance of Multiple Water Quality Parameters

Potable water quality has long been a concern in several areas around the world, and in India, and there is a definite detrimental effect on humans. It is now challenging for authorities to supply a clean drinking water source for sustainable development in the coming future. Some dissolved cations, such as Cu, Fe, Se, and Zn, in natural water samples can play a beneficial role in human biology, whereas others, such as Pb, Hg, As, U, rare earth elements, and platinum group elements, can affect human health [1]. Several previous studies have linked positive exposure to uranium in water with chronic diseases such as kidney and lung cancer and several other health problems [2]. In addition to uranium contamination, water quality also suffers from multiple water quality issues, such as pH, redox potential, total dissolved solids, arsenic and fluoride contamination, etc. [3]. Another recent concern is the issue of uranium contamination in water (groundwater, surface water, etc.) and its deteriorating health effects. Uranium is sparingly soluble in water, which supports both radiological and chemical toxicity in the human environment [4].

Water quality plays an important role in maintaining the health and well-being of the human population. It is also crucial for aquatic ecosystems and other industrial processes that are working to improve human health. Water quality is determined by physiochemical characteristics, which include physical and chemical properties. These parameters include hardness, pH, total dissolved solids (TDS), redox potential, electrical conductivity and the presence of heavy metal ions. Determining these factors is essential to avoid potential health complications [5–7]. Furthermore, water quality can also be categorized in terms of biological, radiological or aesthetic quality. Understanding the various parameters that define water quality is essential for maintaining, managing and improving water resources. Several key physicochemical properties play a significant role in determining water quality.

Groundwater and water sources have been proven to be dependable sources for meeting human water demand in many areas of the world. Groundwater is considered a 'safe source' of potable water for the population if it has a low microbial load and other contamination. However, water resources around the population are commonly exposed to pollution, which may affect water quality [8]. Moreover, water quality generally refers to the chemical, physical, biological, radiological and aesthetic characteristics of water bodies [9, 10]. Good water quality for human life is crucial for consumption in balanced ecosystems [11]. According to the WHO, approximately 80% of diseases in the human environment are caused by a contaminated water supply. Various diseases, such as typhoid, cholera, and other bacterial or viral infections, spread through contaminated or polluted water [12]. Water is vital to aquatic life, human health, and the environment. The majority of people are afflicted with numerous water-borne illnesses, and people may be affected by the presence of uranium in water (Fig. 1).

The water quality index (WQI) is widely considered and accepted as the most effective method for assessing water quality, which indicates how safe the water is for portability [14]. Coyte et al. [15] reported that human-made pollutants and naturally occurring toxic minerals adversely affect the water quality in Rajasthan, India, at a frightening rate.

Water quality parameters vary seasonally due to various factors, such as rainfall, groundwater, oxygen level, hardness and temperature. pH plays an important role in maintaining aquatic life in both plants and animals. Various studies have revealed significant variation in these parameters seasonally [16]. The pH level is affected by season, as rivers and ponds are affected by geogenic as well as human activities [17]. Aquatic biota are influenced by variations in water temperature. Mazhar and Ahmad [18] studied the variation in water quality parameters in different regions

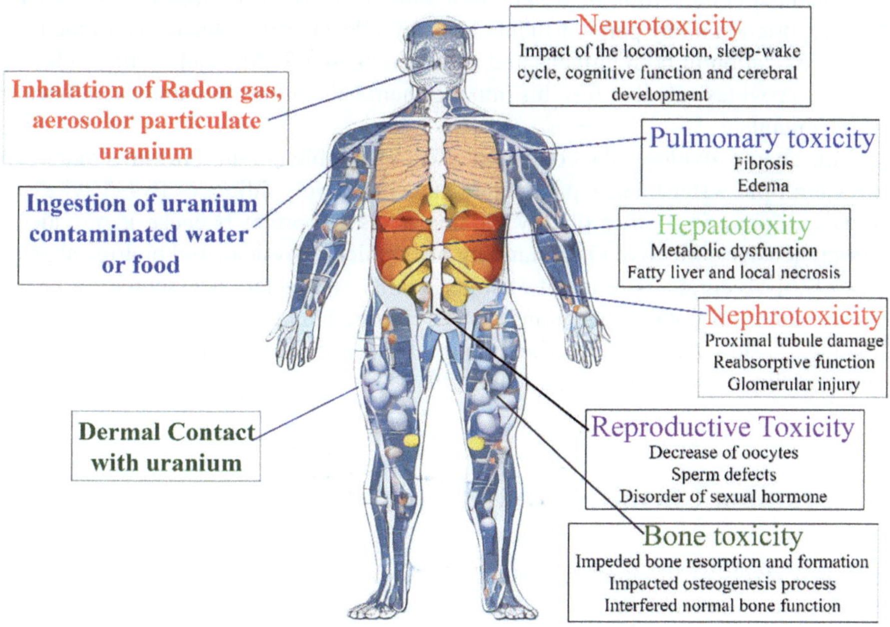

Fig. 1 Effects on the human body due to uranium occurrence in water [13]

and concluded that water quality impacts the profile of water bodies. Temperature variation and geogenic and anthropogenic activities contribute to changes in water quality parameters [13]. Water temperature displays noticeable seasonal fluctuations that impact dissolved oxygen levels and other metabolic activities in aquatic organisms. Studies carried out by Shrestha and Kazama [19] stress that variations in temperature are among the primary drivers of seasonal changes in the water quality of a region and that higher temperatures are due to heat absorption in the ecosystem. Dissolved oxygen (DO) and biological oxygen demand (BOD) are sensitive indicators of water quality and are usually influenced by seasonal variations. During the winter season, relatively high levels of DO are generally observed due to the increased solubility of oxygen. Pesce and Wunderlin [20] studied hypoxic conditions in summer and reported significant seasonal variations in DO and BOD in summer. Water turbidity and the pollutant load in water are crucially dependent on rainfall in a region. A study conducted by Delpla et al. [21] demonstrated significant correlations between turbidity and rainfall in surface waters. Floods occurring in a particular season affect the aquatic composition by introducing sediments and contaminants into water bodies. Furthermore, the concentrations of trace metals have shown seasonal variation, which is influenced by geogenic and anthropogenic activities. According to Kazi et al. [22], the solubility of trace elements increases during the wet season. This was due to industrial discharge and mining activities.

Water quality, especially groundwater water quality, is significantly affected by the heavy use of fertilizers in the agricultural sector [23]. Local geological parameters, hydrological characteristics and intense anthropogenic activities affect the geochemical composition of water. High amounts of fluoride in groundwater have been reported in many parts of India [24], and in Rajasthan, it leads to chronic fluoride intoxication in both humans and domestic animals [25]. Each of these characteristics provides valuable insights into the chemical composition and overall health of inhabitants.

This chapter explores the combined effect of soluble uranium in water and interprets multiple water quality parameters that collectively influence its behavior and impact. Monitoring water quality parameters is important because it provides a comprehensive collective understanding of chemical, physical, and biological properties [26]. Accurate assessment of water quality parameters is essential for ensuring the safety of potable water and water resource management [27, 28]. The data obtained during the assessment help to establish standard regulations and improve water quality at local and global scales [29].

2 Uranium Chemistry and Mobility in Water

Elevated levels of uranium in water are of concern in many areas because of its radioactive nature and chemical toxicity [30]. Uranium may be transported because it is oxidized and leached into surface water or groundwater. This transportation of uranium could lead to direct exposure to humans who ingest uranium-contaminated water. Furthermore, uranium can enter the food chain through plants growing in contaminated sediment or soil and animals consuming contaminated vegetation. Humans are exposed to uranium by consuming either contaminated vegetation or contaminated animals. This increases the potential dose from direct ingestion exposure [31]. The concentration of dissolved uranium entering the soil profile is governed by the weathering reactions among the uranium metal particles, their oxidation products, and water. The local soils/rocks may attenuate the uranium concentrations in the soil matrix because of the reactions between the dissolved uranium and the minerals. These reactions include mineral precipitation and adsorption reactions, which support the reduction in uranium concentration in soil pores filled with water [32].

Uranium is found in certain types of soil/rock in three different forms: background, resistant and interstitial [33, 34]. It occurs in +4 and +6 oxidation states in soil/rock as well as in aqueous forms. The reduction process of uranium from U6+ to U4+ is the desired mechanism known as the deletion mechanism, which results in the solubility of U4+ species to a lesser extent in water and thus does not account for much of the transport of uranium in groundwater [35]. Thus, the solubility of uranium in water plays a significant role when actual approaches for its remediation are allowed. The following reactions represent the dissolution of amorphous silica (1), the dissociation of silicic acid (2), and complex formation with uranyl ions (3) [36].

$$SiO_2\left(amorphous\right) \rightleftharpoons Si\left(OH\right)_4 \tag{1}$$

$$Si\left(OH\right)_4 \rightleftharpoons Si\left(OH\right)_3 O^- + H^+ \tag{2}$$

$$UO_2^{2+} + Si\left(OH\right)_3 O^- \rightleftharpoons UO_2Si\left(OH\right)_3 O^+ \tag{3}$$

Uranium generally occurs in tetravalent and hexavalent ionic forms, and the mobility of uranium in water bodies is dependent mainly on its oxidation state. Studies revealed that higher uranium concentrations were shown in toxic waters [37, 38]. The geochemistry of water bodies is more significant because the chemical composition varies, and its influence on groundwater can increase due to increases in population, agricultural processes and industrial development [39, 40]. Additionally, owing to hydrogeochemical processes, the uranium concentration in the water body system of an aquifer varies in its characteristics while moving along its path from recharge to discharge areas [41].

2.1 Path of Uranium Solubility in Water

Water distribution systems provide a clean and reliable source of potable water to communities. Different agencies across the globe typically develop several models to perform long-term planning, design new components, and resolve hydraulic or water quality problems for communities [42]. Water body classification via various techniques plays a pivotal role in the effective management of water resources available on earth. The classification of water systems is essential for understanding their distribution and characteristics and ultimately informing water usage [43]. Water sources are natural or man-made bodies of water that provide water for consumption, agriculture, industry, and enjoyment. Surface water (rivers, lakes, and reservoirs), groundwater (wells, springs, and aquifers), and precipitation-harvested rainfall are some of these sources. Each of these sources may have unique features and issues regarding water quality, which are determined by variables such as land use, pollution, and natural geological conditions. The following freshwater sources outline the ultimate transport of uranium in various ways.

2.1.1 Groundwater

Groundwater is water found under the Earth's surface and stored in aquifers, which are layers of rock, sand, or gravel capable of holding water. It provides a major supply of freshwater for drinking, agriculture, and industry, particularly in areas where surface water is limited or contaminated. Groundwater is often obtained from wells and springs, and in many regions, it accounts for a significant amount of the community and ecosystem water supply. Groundwater is normally protected by the earth and rock layers above it, whereas surface water is more vulnerable to pollution from direct human activity. It is susceptible to pollution from agricultural runoff, septic systems, industrial waste, and landfills. Heavy metals (e.g., arsenic and lead), pesticides, nitrates from fertilizers, and pathogens from human or animal waste are among the most common pollutants in groundwater [44]. These contaminants may gradually penetrate groundwater, particularly in locations with permeable or broken bedrock. Compared with surface water, groundwater is often of greater quality, but monitoring is still needed to maintain its safety for human consumption and environmental health [45]. Testing often involves evaluating factors such as pH, dissolved oxygen, salinity, and pollutants such as heavy metals, nitrates, and volatile chemical compounds. Groundwater quality is critical for sustainable water management and public health.

2.1.2 Surface Water

Surface water is water that is found on the Earth's surface in bodies such as rivers, lakes, reservoirs, and wetlands. Precipitation, rainwater runoff, and groundwater flow are the primary sources of surface water replenishment. Surface water, however, is more prone to pollution than groundwater because of its direct contact with environmental elements and human activity. Industrial discharge, agricultural runoff including fertilizers and pesticides, untreated sewage, and stormwater runoff are among the most common causes of pollution. These pollutants may impair water quality, making it dangerous for consumption, leisure, and aquatic life. Furthermore, surface water bodies are susceptible to eutrophication, which occurs when excess nutrients (mostly nitrogen and phosphorus) cause hazardous algal blooms that decrease oxygen levels, degrade water quality, and pose health hazards [46]. Monitoring and regulating surface water quality are critical to ensuring its safety for diverse applications. Regular water quality studies assist in detecting pollution sources and advise efficient water treatment, conservation techniques, and sustainable management of surface water resources [47].

Lakes Lakes are natural water reservoirs that constitute a vitally important resource for humans. Lakes are generally defined as large natural bodies consisting of inland waters. Additionally, man-made lakes called reservoirs are constructed to store water for utility purposes. Most lakes are of catastrophic origin; that is, they are formed by tectonic, volcanic, glacial, and fluvial processes occurring on Earth. Physical processes play an important role in understanding the complex interrelationships among ecosystem components, water quality, and contaminant exchanges in lakes [48, 49].

Rivers and Stream Surface water flows from very small intermittent streams to very large water bodies, such as the Ganges or Mississippi River. Owing to their physical, chemical, and biological characteristics, rivers are unique [50]. River hydrology is influenced mainly by the climate, i.e., the level of precipitation and the contributing drainage area [51]. Streams often originate in headwater tributaries, swamps, and lakes and eventually form rivers, which then drain many kilometers into estuaries, where freshwater meets the salty ocean [52]. Rivers are also influenced by many components of the hydrological cycle (precipitation, runoff, groundwater flow, and evaporation), the catchment area, geographical setting (e.g., topography, forest cover, geology, etc.), and human land use practices [50].

2.1.3 Other Sources

In addition to surface water and groundwater, other important water sources include rainfall, desalinated water, and recycled water. Rainwater harvesting includes collecting rainwater from surfaces such as roofs and highways, making it an important resource in locations with limited access to freshwater. While rainfall is generally pure, it may be polluted by contaminants in the air or on collecting surfaces, necessitating filtering and treatment for safe use. Desalinated water is created by removing salts and pollutants from saltwater or brackish water, and it is an important source of freshwater in dry places, although the process is energy intensive and expensive. Recycled or reclaimed water is defined as wastewater that has been cleansed and purified for reuse, generally for irrigation, industrial usage, or even potable reasons. In light of global issues such as climate change and population increase, these alternative water sources are becoming more important in solving water shortages and supporting sustainable water management [53, 54].

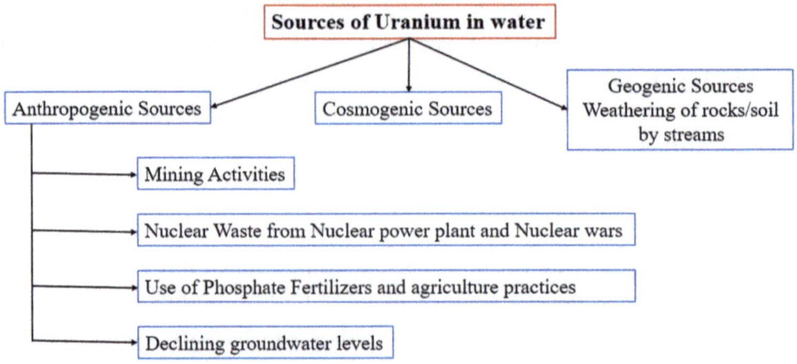

Fig. 2 Different sources of uranium in water [13]

2.2 *Uranium Sources*

The presence of uranium in water bodies may be due to various sources, and uranium in water bodies originates from both natural and anthropogenic sources (Fig. 2). These are described as *natural sources:* Uranium is found in small amounts in nature in rocks, soil, water, air, plants, and animals. Uranium concentrations in natural waters are usually low, with less than 4 micrograms per liter in river water and less than 5 μg per liter in groundwater [13]. *Anthropogenic sources:* Another important source of uranium is anthropogenic activities, i.e., uranium can also originate from human activities, such as the use of phosphate fertilizers, mine waste, fly ash from power plants and military use. Therefore, uranium can be present in water for these reasons [55]. *Geogenic sources:* The presence of uranium in the human environment can also be due to uranium being extracted from rocks through water–rock interactions. Oxidizing conditions can increase the solubility of the extracted uranium in water [13]. *Over-exploitation of groundwater:* Over-exploiting groundwater can lead to higher uranium concentrations. Uranium is a complex chemical that can form solids and complexes with common macroions such as calcium, magnesium, sodium, and phosphate.

2.3 *Basic Chemistry of Uranium in Aquatic Systems*

Uranium, which has the greatest half-life, is a naturally occurring element and is present within many minerals. Figure 3 shows the most common model describing the chemistry of uranium solubility in water bodies: the free ion activity model and the conceptual model [37]. It is founded on the core idea of the interaction of soluble metal compounds with cells [56]. This model was established in the late twentieth century as a result of advanced aqueous equilibrium models that calculate species in aqueous solutions across various chemistries in water aquifers. The transport and transformation of uranium is a complex process in real aquatic

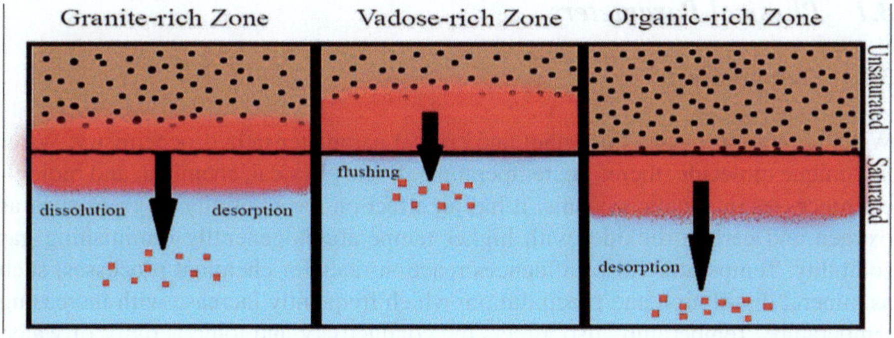

Fig. 3 Uranium mobility and solubility in water in different geological settings (reproduced from [37] with permission from Elsevier Publishing)

environments ranging from low to high. It is imperative to highlight the properties of uranium within the food chain of aquatic systems [57]. Uranium is found in rocks and is present in both ground and surface water bodies. The solubility of uranium depends on the chemistry of the water. The average concentration of uranium in the Earth's crust is 2.8 μg/g. It is mainly concentrated in acidic rocks. It is further incorporated into other important minerals, viz. monazite, uraninite and zircon [58]. The trash amount of uranium is also present in terrestrial and ocean water aquifers.

3 Hydrochemistry: Factors Influencing Uranium Solubility

Mineralized deposits are widespread in the environment. Such deposits lead to the distribution of uranium in soil and water [31]. Hydrochemical methods focus mainly on the chemical characteristics of water in natural systems, including surface water and groundwater. It plays a crucial role in understanding water quality, its suitability for different uses, and the processes influencing its composition. It includes part of atmospheric chemistry since the deposition of substances from the atmosphere constitutes an important source for some substances dissolved in continental waters. The interaction between water and minerals in the ground produces additional changes in the chemical composition of the water. These interactions are caused by geochemical processes such as dissolution [59]. When groundwater emerges and accumulates on the surface of the earth in lakes and water courses, the chemical composition of the water may change again due to new equilibrium conditions of the dissolved gases. However, a frequently much more pronounced change is caused by biological activity in the water, an activity covered by the subject area of limnology, which is also relevant to hydrochemistry [60, 61].

All these properties differentiate it from more than 15 million other chemical species. We can understand its uniqueness by relating water molecules with the microscopic world to the macroscopic world in which we live.

3.1 Physical Parameters

3.1.1 Temperature

Water temperature is an important and critical parameter in hydrochemistry. Water is a unique molecule; therefore, temperature affects physical, chemical, and biological processes in aquatic systems. It has an effect on the solubility of gases such as oxygen and carbon dioxide, with higher temperatures generally diminishing gas solubility. Temperature also influences reaction rates for chemical processes, such as mineral dissolution and precipitation, which frequently increase with increasing temperature. Temperature also affects the conductivity and ionic activity of water, which influences its overall chemistry. In the carbonate system, changes in the dissociation constants of carbonic acid play a significant role in influencing both the pH and buffering capacity [62]. Understanding these alterations can provide valuable insights into the dynamics of the system. Microbial activity and nutrient cycling are strongly influenced by temperature changes, making temperature a crucial factor in ecosystem dynamics [63]. Additionally, temperature gradients can cause thermal stratification in large water bodies, affecting the distribution of dissolved substances and biological activity. Accurately measuring and understanding water temperature is crucial for evaluating water quality and predicting the behavior of chemical and biological systems in aquatic environments.

3.1.2 Electrical Conductivity

Electrical conductivity (EC) is an important metric in hydrochemistry that provides insight into water quality. It represents the ability of water to conduct electrical current, which is affected by the number of dissolved ions. Understanding EC can help us evaluate the overall health of aquatic habitats and advise appropriate water management techniques. It directly measures the total dissolved solids (TDS) in water, providing information about its overall ionic strength and salinity. Higher EC values clearly indicate greater quantities of dissolved ions such as calcium, magnesium, sodium, chloride, sulfate, and bicarbonate. It is influenced by elements such as the geological makeup of the surrounding region, human activity, and the amount to which water and rock interact. Temperature has a significant effect on electrical conductivity (EC) because warmer water allows ions to flow more freely [64]. As a result, it is critical to modify measurements to account for temperature fluctuations and ensure that everything is standardized to 25 °C. Electrical conductivity (EC) is widely used in hydrochemical research to classify water types, monitor pollution levels, and evaluate groundwater recharge and mixing processes. This application is critical for determining water quality and sustainability. Electrical conductivity (EC) is widely used in hydrochemical research for classifying water types, monitoring pollution levels, and assessing groundwater recharge and mixing processes. This application is critical for determining water quality and sustainability.

3.1.3 TDS

The total dissolved solids (TDS) content is a key parameter in hydrochemistry and represents the total concentration of dissolved inorganic and organic substances in water. It is composed of cations such as calcium, magnesium, sodium, and potassium, as well as anions such as chloride, sulfate, carbonate, and bicarbonate, along with minor organic matter. The TDS is an essential indicator of water quality, as it directly reflects the salinity, mineralization, and suitability of water for various uses, such as drinking, agriculture, and industrial applications.

The TDS is influenced by natural factors such as the geological composition of the aquifer, water–rock interactions, and climate, as well as anthropogenic activities such as agricultural runoff, wastewater discharge, and industrial effluents [65]. It is typically measured either gravimetrically by evaporating water and weighing the residue or indirectly via electrical conductivity, as there is a strong correlation between EC and TDS. High TDS levels can affect the taste of water, its corrosiveness, and its usability, making it a critical parameter in hydrochemical assessments and water resource management.

3.1.4 Turbidity

Turbidity is an essential metric in hydrochemistry. It quantifies the cloudiness or haziness of water produced by the presence of suspended particles such as clay, silt, organic matter, bacteria, and other fine elements. This optical feature indicates how much light is scattered or absorbed by these particles instead of being transmitted through the water. Turbidity is often employed as an indirect measure of water quality, particularly in surface water and drinking water sources. High turbidity levels may restrict sunlight penetration, impacting photosynthetic activity and changing aquatic habitats [66]. It may also interfere with disinfection operations, since suspended particles hide germs from exposure to disinfectants such as chlorine. Turbidity is influenced by natural causes such as soil erosion, runoff, and algae blooms, as well as artificial sources such as construction operations, agricultural runoff, and wastewater discharge. Turbidity monitoring is critical in hydrochemical studies to analyze sediment movement, pollution levels, and the efficacy of water treatment systems, ensuring water safety and sustainability for their intended usage.

3.1.5 Color

In hydrochemistry, color is an important aesthetic and qualitative criterion for assessing water quality. It is induced by the presence of dissolved organic materials, such as humic and fulvic acids, and inorganic compounds, such as iron, manganese, or other metal ions. The natural water color varies depending on the source, with surface waters frequently having a yellowish-brown tint owing to organic debris from decaying flora, whereas groundwater may have coloration due to mineral dissolution [67]. Water color is normally evaluated visually via conventional color scales or spectrophotometric procedures, which are represented in platinum–cobalt units (PCUs) [68]. Excessive color, although not physically damaging, may have an impact on water aesthetics, customer acceptability, and usefulness. Color removal is critical in water treatment operations because it frequently corresponds with the presence of organic debris, which may produce disinfection byproducts during chlorination. Monitoring color is critical in hydrochemical studies, particularly for drinking water and industrial uses.

3.2 Chemical Parameters

The chemical characteristics of hydrochemistry are vital for understanding the composition and quality of water, as well as its interactions with its surroundings. These factors assist in determining the appropriateness of water for drinking, agricultural, industrial, and ecological applications. The key chemical parameters include the following:

3.2.1 pH

In hydrochemistry, pH is a crucial chemical measure that determines the acidity or alkalinity of water, with logarithmic values ranging from 0 (very acidic) to 14 (highly alkaline). It reflects the concentration of hydrogen ions (H^+) and plays a pivotal role in controlling chemical reactions, solubility, and speciation of dissolved substances. Natural water normally has a pH of 6.5–8.5, which is altered by geological composition, biological processes, and human activities. Low pH values may promote water corrosivity, whereas high pH levels can cause scaling owing to the precipitation of calcium and magnesium ions. Furthermore, pH is important for sustaining aquatic ecosystem health since most species thrive within certain pH ranges. It also influences carbonate equilibrium, which influences alkalinity and buffering capacity. Accurate pH monitoring is important in water treatment, where processes such as coagulation and disinfection are pH dependent [69].

3.2.2 Dissolved Organic Matter

Dissolved organic matter (DOM) is organic material that dissolves in water and may come from natural sources such as plant degradation and microbial processes, as well as human activities such as wastewater discharge and agricultural run-off [70]. DOM is a complex combination of chemicals that includes humic substances, carbohydrates, and proteins. It is frequently referred to as dissolved organic carbon. DOM is important for monitoring water quality since it indicates organic pollution levels and the influence of numerous environmental processes. It is an important component of the global carbon cycle, influencing the mobility and toxicity of trace metals via complexation. Furthermore, DOM may react with disinfectants in water treatment, resulting in the creation of potentially hazardous byproducts. It also absorbs light, decreasing water clarity, which impacts aquatic ecosystems, notably photosynthesis. DOM is often assessed via procedures such as DOC analysis, absorbance, fluorescence, and biodegradability assays [71].

3.2.3 Dissolved Ions

Dissolved ions are important chemical parameters in hydrochemistry because they influence the chemical composition, reactivity, and overall quality of water. These ions, which include major cations such as calcium (Ca^{2+}), magnesium (Mg^{2+}), sodium (Na^{+}), and potassium (K^{+}), as well as major anions such as chloride (Cl^{-}), sulfate (SO_4^{2-}), bicarbonate (HCO_3^{-}), and nitrate (NO_3^{-}), are sourced primarily from natural weathering processes, soil leaching, and atmospheric deposition, along with anthropogenic contributions such as agricultural runoff and industrial discharges. The concentration and balance of these ions are important in understanding water chemistry because they affect water hardness, buffering capacity, pH stability, and the solubility of other compounds. Furthermore, dissolved ions interact with bacteria and aquatic species, influencing biological processes and ecosystem health. Monitoring dissolved ions aids in determining water quality, identifying pollution sources, and guiding water treatment procedures [72].

3.2.4 Heavy Elements

Heavy elements in hydrochemistry often refer to metal ions with greater atomic masses, which are frequently found at tiny levels in natural waters. These elements include important nutrients, such as iron (Fe), manganese (Mn), and copper (Cu), and harmful compounds, such as lead (Pb), arsenic (As), cadmium (Cd), mercury (Hg), and chromium. While certain heavy metals are required in tiny quantities for biological functions, their concentrations are tightly controlled. However, excessive levels, which are often caused by industrial activity, mining, and agricultural runoff, may pose major environmental and health hazards. Heavy elements tend to accumulate in water systems owing to their comparatively limited mobility in comparison with lighter components. They may interact with dissolved organic materials to generate complexes that alter their bioavailability and toxicity [73]. Heavy metals that exceed specified concentrations may be detrimental to aquatic life, disrupt food webs, and cause long-term ecological damage. Human exposure to these elements via polluted drinking water or the food chain may result in a variety of health problems, including brain impairment, renal failure, and cancer [74]. There are regulatory measures in place to manage their levels in water, guaranteeing the health of both the environment and humans.

3.2.5 Pesticides

Pesticides are chemical compounds used to manage, prevent, or destroy pests, including insects, weeds, fungi, and rodents. Herbicides, insecticides, and fungicides are often employed in agricultural and pest control, but they may infiltrate water systems via runoff, leaching, and atmospheric deposition. Pesticides are a major issue in water quality research since many are hazardous to aquatic life and may represent a danger to human health via polluted drinking water or the food chain. Pesticides in water may be divided into two categories: persistent and nonpersistent. Persistent pesticides, such as DDT, may remain in the environment for extended periods of time, accumulating in sediments and biota, whereas nonpersistent pesticides breakdown relatively quickly. Pesticides in aquatic systems may disturb ecosystems by reducing the reproduction and survival of aquatic creatures, disrupting food webs, and possibly inflicting long-term environmental impacts.

3.2.6 Dissolve Oxygen

Dissolved oxygen (DO) is an important chemical characteristic in hydrochemistry that indicates the amount of oxygen gas (O_2) dissolved in water. It is necessary for the respiration of aquatic species, including fish, invertebrates, and microbes. Temperature, salinity, air pressure, water turbulence, and biological activity are some of the elements that determine DO levels. Cold water may store more dissolved oxygen than warm water does; however, oxygen solubility decreases in saltwater or contaminated water. DO concentrations are used to assess water quality and ecosystem health. Higher DO levels often indicate a well-oxygenated environment, which is advantageous for the survival of aerobic species. Low DO levels, on the other hand, may cause hypoxic or anoxic conditions, which are harmful to aquatic life, resulting in fish death and disruption of biochemical cycles. Low DO concentrations are often caused by organic pollution, since microbial breakdown of organic materials depletes oxygen [75]. DO is also important in water treatment operations since it assists in the elimination of impurities during the biological treatment phases. Monitoring DO levels is critical for maintaining freshwater and marine habitats and ensuring the efficacy of water treatment facilities. It is normally measured by electrochemical sensors or optical techniques, and the findings are usually given in milligrams per liter (mg/L) or as a percentage of saturation [76].

3.3 *Biological Parameters*

The biochemical oxygen demand (BOD) is an important biological metric in hydrochemistry that quantifies the quantity of oxygen needed by microbes to break down organic materials in water. BOD is often used to determine the degree of organic pollution in water, with greater BOD readings indicating higher quantities of organic matter, which may deplete dissolved oxygen and damage aquatic life. BOD is measured over 5 days (BOD_5) under conventional circumstances (temperature of 20 °C, dark, and neutral pH), indicating the oxygen utilized by microbial organisms during the breakdown of organic material.

BOD is a crucial water quality measure for determining the effects of organic pollution from wastewater, agricultural runoff, or industrial discharge on aquatic habitats. When organic matter is put into a water body, microbes consume oxygen to digest it, and a high BOD leads to a decrease in available dissolved oxygen, resulting in hypoxia or even anoxia, which can damage aquatic species [77]. BOD is also utilized in the design and monitoring of wastewater treatment systems since it assesses the effectiveness of treatment operations in eliminating organic matter before release into receiving waters. The BOD test may be supplemented with additional biological parameters, such as chemical oxygen demand (COD) and total organic carbon (TOC), to evaluate the water quality and pollutant levels [78].

3.3.1 Bacteria

Bacteria are important biological criteria for monitoring water quality because their presence and quantity might indicate contamination, especially from fecal sources. Total coliforms, fecal coliforms, and *Escherichia coli* (*E. coli*) are among the most commonly used markers; the latter is a direct marker of fecal contamination. The presence of *E. coli* in water often indicates the possible presence of dangerous organisms such as Salmonella and Shigella, which may pose major health hazards such as gastrointestinal illnesses and cholera. Monitoring these bacteria is vital for ensuring the safety of drinking and recreational water. Common techniques for detection include the membrane filter technique, the most likely number (MPN) test, and polymerase chain reaction (PCR). Detecting fecal coliforms or *E. coli* is crucial for assessing the efficacy of water treatment and detecting contaminated sources [79].

3.3.2 Viruses

In the context of hydrochemistry, viruses are minute infectious pathogens that may be found in water bodies, posing major health hazards, particularly when they are polluted by human or animal waste. Waterborne viruses, such as enteric viruses, are pathogens that may cause a variety of illnesses in humans, including gastroenteritis, hepatitis, and pneumonia. Noroviruses, rotaviruses, adenoviruses, and enteric hepatitis viruses are some of the most common waterborne viruses. Unlike bacteria, viruses cannot reproduce in water but may survive for long periods of time, especially in untreated or insufficiently treated waters. The presence of viruses in water is a major problem in areas with poor sanitation because they may be introduced by polluted wastewater, rainfall runoff, or agricultural runoff. Monitoring viruses in water is critical for maintaining public health, especially in drinking water sources, recreational waterways, and water used in food production. While typical bacterial markers such as *Escherichia coli* may indicate contamination, they may not completely represent viral contamination since viruses and bacteria frequently behave differently in the environment [80].

3.3.3 Protozoa

Protozoa are single-celled eukaryotic microorganisms that can be found in water systems, and they play a significant role in hydrochemistry owing to their influence on water quality and their propensity to spread illnesses. Protozoa may exist in both free-living and parasitic forms, and certain species are known to cause watery infections in people. Protozoan infections, including *Giardia lamblia*, Cryptosporidium parvum, and Entamoeba histolytica, are especially problematic in water quality evaluations because they are resistant to chlorine disinfection, making them difficult to eliminate from drinking water and wastewater treatment procedures. Giardia and Cryptosporidium are among the most prevalent protozoan infections found in water, and they may cause gastrointestinal disorders called giardiasis and cryptosporidiosis, respectively [81]. These protozoa contain cysts that are very resistant to environmental stresses, including UV radiation and chlorination, enabling them to stay in water for long durations. Protozoan infections may cause symptoms such as diarrhea, dehydration, and abdominal cramps, making their presence in water a major public health problem, particularly in places with insufficient sanitation or water treatment facilities. Protozoa also play a role in nutrient cycling in aquatic settings by devouring bacteria and organic particles, contributing to microbial population control and organic matter breakdown [82].

3.3.4 Algae

Photosynthetic organisms known as algae may be found in a wide range of aquatic habitats, including freshwater and marine systems. They play an important role in hydrochemistry because they contribute to primary production, oxygen generation, and nutrient cycling in water bodies. Algae are widely categorized into tiny phytoplankton and macroscopic species such as seaweed and freshwater plants. Microscopic algae, such as diatoms, green algae, and cyanobacteria, are especially significant in terms of water quality because they serve as the foundation for aquatic food webs and are sensitive to changes in water chemistry, including nutrient levels.

Algal blooms, which develop when excess nutrients, notably nitrogen and phosphorus, enter water bodies via agricultural runoff, wastewater discharge, or rainfall, are serious hazards in water systems. These nutrient overloads may cause algae to develop rapidly, resulting in thick blooms that deplete oxygen levels in the water and produce toxins that are detrimental to aquatic life and human health. Harmful algal blooms (HABs), which are often dominated by cyanobacteria (blue–green algae), may generate toxins such as microcystins, which are poisonous to both aquatic creatures and people, presenting major threats to drinking water supplies and recreational waterways [83].

Algae also influence the physical features of water, such as its color and turbidity. The presence of algae may cause water discoloration, which can affect the aesthetic quality of water for drinking, swimming, and other recreational activities. Furthermore, as algae die and disintegrate, the process consumes dissolved oxygen, possibly resulting in hypoxic conditions that endanger aquatic life [84].

4 Contaminated Water and Combined Effects of Water Quality Parameters

Contaminated waterways are bodies of water that have been contaminated by a variety of dangerous chemicals, making them unfit for their intended use, such as drinking, irrigation, or enjoyment. Both natural processes and human activities may contaminate water, introducing contaminants that have an impact on water quality.

Contaminated waterways are major health concerns and may contribute to the destruction of aquatic ecosystems, making water quality monitoring and pollution management crucial. Efforts to reduce pollution include minimizing pollutant sources, establishing efficient wastewater treatment, and monitoring water quality via different metrics, such as turbidity, pH, dissolved oxygen, and concentrations of particular pollutants [85].

4.1 Combined Effects of pH and Redox Potential

The chemical behavior and quality of water are significantly influenced by the combined effects of pH and redox potential (also known as oxidation–reduction potential (ORP)). These two characteristics interact and, combined, govern the speciation, mobility, and reactivity of numerous chemicals in aquatic systems, including metals, nutrients, and pollutants. The ORP is a measurement of a water sample's capacity to donate or absorb electrons during chemical processes. It represents the equilibrium of the oxidizing and reducing agents in water. A high positive ORP (oxidizing conditions) implies that water contains more oxidizing agents (such as oxygen), which may increase the decomposition of organic compounds, kill microorganisms, and facilitate metal oxidation. A low ORP (reducing conditions) suggests that water is more likely to enable the reduction of chemicals such as nitrate, sulfate, and certain metals, possibly resulting in the mobilization of pollutants such as arsenic or manganese [86].

The relationship between pH and ORP is critical because it affects many environmental processes, such as the transformation of pollutants, the behavior of heavy metals, and the functioning of aquatic ecosystems. For example, in an acidic environment (low pH), the oxidation of sulfides may result in the generation of sulfuric acid, which further lowers the pH and leads to deterioration of the water quality. In contrast, under alkaline conditions (high pH), the presence of reducing agents might impact the release of hazardous chemicals such as ammonia or dissolved methane. Heavy metal toxicity and mobility, such as those of mercury and cadmium, are strongly pH- and ORP dependent, with their concentrations and speciation shifting substantially with various pH and redox conditions. Understanding how these characteristics interact is critical for forecasting and regulating contaminants, biological activity, and the general health of aquatic ecosystems [72].

4.2 Role of Dissolved Organic Matter in Uranium Complexation

The complexation of uranium in aquatic systems is significantly influenced by dissolved organic matter (DOM), which affects its mobility, bioavailability, and toxicity. Humic substances, proteins, carbohydrates, and other organic molecules are all examples of DOM, which is an abbreviation for dissolved organic matter. These organic molecules may bind to uranium and create stable complexes that affect the chemical behavior of the metal, especially in groundwater and surface water. Uranium, a naturally occurring radioactive element, occurs in two oxidation states in aqueous systems: uranium (VI) in the form of the uranyl ion (UO_2^{2+}) and uranium (IV), which are less soluble and are typically found in solid phases such as uraninite [87]. Uranium speciation is substantially influenced by pH, redox conditions, and the presence of ligands, notably DOM. The binding of uranium to DOM may have a substantial influence on its solubility and mobility in the environment.

Complexation with DOM: DOM functions as a ligand, binding to uranium and generating dissolved uranium–DOM complexes. The organic ligands of DOM, especially those with functional groups such as carboxyl, phenolic, and hydroxyl groups, have a strong affinity for metal cations such as uranium. The binding of these organic molecules to uranium may improve its solubility, preventing it from precipitating out of solution and hence increasing its mobility in aquatic settings [88]. As a consequence, uranium that would normally be trapped in sediment may remain in the dissolved phase, possibly polluting water supplies.

Environmental parameters such as pH, redox potential, and DOM content play critical roles in determining uranium complexation. In acidic environments, uranium in its soluble UO_2^{2+} form is more prone to interact with DOM. In contrast, under neutral to alkaline circumstances, uranium may precipitate out of solution as uraninite or other solid phases, decreasing its availability for complexation with DOM. Understanding the interactions between DOM and uranium is therefore essential for determining the environmental and health concerns associated with uranium pollution. The distribution, mobility, and toxicity of uranium are affected by complexation processes, which have substantial consequences for water quality management, environmental remediation, and public health [89].

4.3 Influence of Competing Ions on Uranium Speciation

The effect of competing ions on uranium speciation is an important component in controlling the chemical behavior and mobility of uranium in aquatic systems. Uranium (VI) in the form of a uranyl ion (UO_2^{2+}) may form complexes with different cations and anions in water. Uranium solubility and mobility may be affected by competing cations, including calcium (Ca^{2+}), magnesium (Mg^{2+}), and iron (Fe^{3+}). Calcium and magnesium, for example, may create stable uranyl–carbonate complexes, which reduce the concentration of free uranyl ions and restrict uranium mobility. Anionic ligands, including carbonate (CO_3^{2-}), sulfate (SO_4^{2-}), and phosphate (PO_4^{3-}), actively interact with uranium, affecting its speciation. Carbonate ions, in particular, may form highly soluble uranyl–carbonate complexes, which improve uranium mobility in alkaline environments. However, the presence of competing ions such as phosphate or sulfate may change these interactions, resulting in the precipitation of uranium as less soluble complexes and lowering its bioavailability. These interactions are critical for managing uranium's destiny in polluted ecosystems, particularly groundwater, where its mobility might pose environmental and health hazards. Understanding the function of competing ions in uranium speciation is critical for forecasting uranium transit and creating appropriate remediation solutions for uranium pollution [90].

The effect of competing ions on uranium speciation has important consequences for the fate and transport of uranium in polluted settings. In locations impacted by uranium mining, nuclear waste disposal, or natural uranium deposits, the presence of competing ions may either accelerate or inhibit uranium mobility, influencing the spread of contamination in groundwater or surface water. Understanding these interactions is critical for efficient environmental remediation efforts because it may anticipate uranium behavior under various geochemical settings and guide choices regarding the treatment and management of polluted water sources.

4.4 Effects of Biological Parameters on Water Quality

Biological parameters are important indicators of water quality because they reflect the existence and activity of living creatures in an aquatic environment. These characteristics, which include biological oxygen demand (BOD), microbial presence (such as bacteria, viruses, and protozoa), and aquatic life variety (e.g., algae and macroinvertebrates), are used to evaluate water bodies' general health and capacity to sustain life. The abundance and diversity of macroinvertebrates (such as insects, worms, and crustaceans) are essential biological factors used to assess water quality.

These organisms are sensitive to changes in water chemistry and ambient circumstances, which makes them useful bioindicators. A diversified and prolific macroinvertebrate population usually signifies high water quality, but a reduction in variety or the dominance of pollution-tolerant species might suggest deterioration in water quality. Aquatic life, particularly fish populations, also contributes to measuring the health of water bodies, with the presence of specific species frequently indicating the level of pollution or habitat deterioration [91]. Monitoring these biological characteristics is important for assessing the overall health of aquatic habitats. High levels of viruses or pollutants may have a substantial influence on human health and biodiversity, whereas low levels of dissolved oxygen, excessive algae, or biodiversity loss can suggest catastrophic ecological instability. Managing biological factors in water bodies is critical for sustaining safe and healthy ecosystems, particularly in the context of increasing pollution and climate change [92].

4.5 Radioactive Matters as Water Contaminants

Radioactive elements may contaminate water in substantial amounts, creating both environmental and health problems. These pollutants, often originating from natural sources, industrial activity, or nuclear accidents, may infiltrate water systems via numerous channels, such as direct discharge, runoff from polluted regions, or groundwater penetration. Because of their radiation and bioaccumulation potential, the presence of radioactive chemicals in water may have long-term environmental and human health consequences.

Radioactive contamination in water may arise from natural causes, such as radon (a radioactive gas from uranium decay), or from human activities, including nuclear power plants, the mining of uranium and other radioactive elements, and the incorrect disposal of radioactive waste [93]. Accidental releases, such as the Chernobyl accident and the Fukushima Daiichi nuclear disaster, have shown the catastrophic possibility of large-scale radioactive poisoning of water supplies. Radioactive elements such as radium, thorium, and cesium may be transported into neighboring water bodies by industrial effluents, mining waste, and agricultural runoff [94].

Radionuclides These are unstable isotopes that produce radiation when they decay into more stable forms. Uranium, radium, strontium-90, cesium-137, and plutonium are some of the radionuclides most often discovered in polluted water. These pollutants may pollute surface water (rivers, lakes) and groundwater via precipitation, runoff, or leaching [31, 95].

Radon Radon, a radioactive gas that may dissolve into groundwater, is a major hazard in locations with high uranium concentrations. It may also collect in indoor air from groundwater sources, creating health concerns when inhaled [96].

The primary health problems associated with radioactive contamination of water are due to radiation exposure, which may have both short-term and long-term health impacts. The major health problems associated with radioactive contamination of water are radiation exposure, which may produce both short-term and long-term health consequences. The consumption of polluted water, for example, may increase the risk of liver, bone, lung, and kidney cancer, especially from uranium and radium. Radioactive pollution in aquatic creatures may impair reproductive systems, growth, and development, resulting in population reduction and biodiversity loss [97].

Radioactive materials in water may last for a long time, depending on the substance's half-life. Uranium and radium, for example, have lengthy half-lives, which means that they may persist in the environment and represent a danger to future generations. The presence of these compounds in water may lead to the pollution of whole ecosystems and the food chain. Fish, plants, and other aquatic organisms may absorb radioactive contaminants, which are subsequently passed to predators, including people, who ingest contaminated species [98]. The long-term stability of radioactive elements in aquatic habitats makes contamination difficult to treat.

5 Empirical Evidence

Global mining activities such as gold and uranium mining have significantly contributed to the introduction of uranium and other heavy minerals into water systems through the leaching and oxidation of minerals [99, 100]. A study conducted in Mongolia by Narangarvuu et al. [101] indicated that mining activities related to urbanization impact the physical and chemical characteristics of rivers. This evidently affects the solubility of uranium complexes in water.

5.1 Case Study: Uranium Occurrence in a Mining-Affected Water Body

Mining dumping contains an elevated number of heavy element and uranium contents resulting from gold mining activities across the world [102]. The presence of uranium in South Africa's gold mining areas is profound, stemming from over a century of activity, and gold mining in South Africa has created vast amounts of tailings, which now cover large areas in densely populated regions. Before the 1950s, uranium was considered a byproduct of the gold ore metallurgical process, and it was not extracted and remained in the deposited waste tailings. After the 1950s, uranium metal emerged as a crucial strategic resource, leading South Africa to establish itself as a key producer through the extraction of gold ore. Winde and Sandham [102] concluded that mining operations help the migration of uranium into the biosphere in various ways, including tailing seepage, erosion, and the discharge of mine water in the environment. A literature review and field studies were carried out at selected mining sites across all seven active goldfields in the

Witwatersrand Basin, South Africa. This study focuses on two main aspects: (a) assessing the scope and potential sources of uranium contamination in streams linked to mining operations and (b) the mechanisms and pathways that lead to uranium contamination in nearby environments [103]. The focus is on diffuse stream contamination originating from tailings dams and the potential hazards it poses for end users [104].

Mining residues and waste dumping are the primary sources of uranium contamination in the basin area. Uranium enters water systems through two channels: direct discharge and diffuse seepage. Studies have revealed that uranium contamination substantially occurs in surface water and groundwater in and around mining sites. Field surveys followed by chemical analysis revealed that the uranium concentration in the acidic seepage solution ranged from 0.5 mg/L to 30 mg/L, which exceeds international standards [102]. Research indicates that sediments deposited near streams or rivers have uranium concentrations significantly higher than the natural background concentrations [105]. The sediment samples collected from the Wonderfontein Spruit wetland stream and Witwatersrand basin near mining activities have shown significant uranium concentrations of up to 500 ppm, which is higher than the regional background concentration, i.e., 1–2 ppm [102].

The increased contamination of uranium in water bodies has serious implications for human health and environmental concerns. Uranium exposure through water intake may lead to both chemical and radiological risks. These factors further pose risks to kidney damage and cancer when ingested through the intake of drinking water [106]. Mining sites are particularly vulnerable because the communities residing near mining sites generally rely on untreated groundwater and use this groundwater in daily use for drinking and other purposes. This leads to increased uranium solubility in the drinking water of mining-affected areas [102]. The occurrence of uranium in groundwater due to mining activities in the Witwatersrand basin highlights the environmental and health implications of such studies, which represent the consequences of historical mining practices.

Comparative Analysis of Different Geographic Regions

For the comparative analysis, we have taken two studies for our reference: one, which was carried out in the Mansa region of Punjab, India, by Sharma and Singh [107], and other recent studies conducted in Garhwal and Kumaun Himalaya, Uttarakhand [16, 108–111]. Both regions have different geographic conditions and lie in the northern part of India. In both studies, the uranium concentration was measured along with several physicochemical parameters. These two studies revealed that the uranium concentration in water varies significantly. The highest uranium concentration reported in the present study was 211.2 ppb among the samples collected in the survey of the 6182 habitations, and the highest TDS value was 1498.6 ppm [107]. Major uranium concentrations were analyzed in drinking water samples that were above the WHO threshold limit of 30 ppb. He observed large variations in uranium concentrations in groundwater, and the concentrations often exceed the permissible limits set by the WHO for drinking water. Even some other parts of Punjab presented high uranium concentrations, which supports the findings of previous studies carried out in the Punjab region [112]. The physicochemical properties, such as TDS, pH, and EC, also widely vary across the region. High TDS levels generally correlate with elevated uranium concentrations, and research indicates that the region has high salinity. For example, Southwest Punjab also has relatively high uranium levels in groundwater [4]. According to the observations of Bajwa et al. [4], uranium mobilization in the water phase may be influenced by saltwater conditions. The pH of groundwater in Punjab is typically observed to range from slightly acidic to alkaline, with a range of 7.1–9.2 [113, 114]. Another report has revealed that the Punjab region is facing a crisis due to high uranium levels in groundwater [115]. On the basis of these studies, the possible region for the high occurrence of uranium in groundwater is related to geomorphology and/or industrial growth in the region [4]. Owing to this industrial waste and excessive use of fertilizers, contaminated waste moves deeper into the earth and becomes mixed with groundwater. This groundwater is the primary source of water for drinking and other purposes; thus, residents are exposed to high uranium concentrations [116]. Other studies conducted in different parts of Panjab (Malwa; Tarn Taran and Amritsar; Southwest Punjab) have shown that the region has high uranium concentrations in groundwater [107, 113, 114, 116].

Several studies conducted in Uttarakhand focused on drinking water quality and highlighted the variation in uranium concentrations and associated physicochemical parameters such as pH, TDS and alkalinity (Table 1). pH and TDS levels are usually 6.5–8.5 and less than 500 mg/l, respectively, which are under the AERB standard [9].

5.2 *Comparative Analysis of Water Quality in Various Water Sources*

Many comparative analyses of water quality in different water systems between studies carried out at the national and international levels have revealed significant differences, and such results lead to better water resource management [117, 118]. In the Indian context, studies carried out by Singh et al. [113, 114] focused on contamination due to the presence of heavy metals in rivers caused by industrial waste disposal and agricultural runoff. This leads to the degradation of water quality and may pose health risks. Similar challenges have also been faced and caused by urbanization and chemical waste at international levels, such as in Europe, Canada, and the United States [119]. Over 97 million residents in the USA have been exposed to toxic, unregulated drinking water that significantly threatens their health. The Environmental Protection Agency (EPA) has established guidelines for more than 90 contaminants, yet recent research reveals that 27% of the 4800 public water systems analyzed show detectable levels of at least four harmful chemicals that the EPA does not regulate [120]. Even in the Indian context, supplying contaminated free potable water is quite challenging [121].

Table 1 Uranium concentration and Physicochemical properties in drinking water of Northern India

Locations		TDS (mg/L)	EC	pH	Mean Uranium Concentration (μg/L)	References
Punjab	Mansa District	1498.6	–	–	211.2	[107]
	Amritsar	427	934	7.5	8.6	[107]
	Gurdaspur	323	647	7.5	4.3	[107]
	Pathankot	297	592	7.5	3.0	[107]
	SW Punjab	599	1207	7.47	73.5	[4]
Uttarakhand	Pithoragarh	204.65	309.26	7.39	0.10–9.96	[16]
	Rudraprayag	550.9	822.2	7	0.71	[108]
	Tehri	578.1	859	7	0.89	[108]
	Uttarkashi	807.2	1201	7.1	2.86	[108]
	Chakrata	–	–	–	1.32	[109]
	Kedar valley	131	172	7.1	7	[111]
	Bageshwar	89.4	164.8	–	1.6	[110]

Various studies have examined physicochemical parameters (pH, TDS, BOD, COD, etc.) to assess water quality in samples collected from Ganges and Yamuna, which are the major sources of potable water for humans. In addition to physicochemical parameters, several heavy metals (Pb, Cu, Zn, Cr, Cd, and Ni) have also been assessed [122–124]. According to this research, the primary cause of this contamination is municipal pollutants, including untreated domestic and sewage waste, whereas industrial pollutants consist of heavy metals discharged into the river, which increases the metal load in the water. Hence, such unregulated factors increase the deterioration of the water quality of rivers due to the discharge of municipal and industrial waste materials and further pollute the groundwater of the area.

Various studies have also been conducted in the Indian–Himalayan region, viz. In the Jammu Kashmir, Himachal, Uttarakhand, and Punjab regions, the leaching of these rocks/soils is the main contributing factor for the slightly higher TDS and pH [111, 113, 114, 125]. In the Punjab region, industrial waste and excessive use of fertilizers are the main sources of pollutants in water [4]. A large-scale, nationwide study was conducted in India by the Bhabha Atomic Research Centre (BARC), Department of Atomic Energy (DAE), to assess uranium concentrations in drinking water sources and to examine the impact of various water quality parameters on uranium chemistry. As part of this National Uranium Project, a total of 55,554 water samples comprising both surface water and groundwater were collected, covering an area of approximately 1.2×10^6 km^2 [126]. The geometric mean uranium concentration was 0.8 μg/L in surface water (range: ≤0.2–22 μg/L) and 2.1 μg/L in groundwater (range: ≤0.2–4918 μg/L). The analysis included 12 key water quality parameters, namely, pH, temperature, TDS, ORP, DO, fluoride, chloride, nitrate, sulfate, phosphate, hardness, and alkalinity, which were measured following national and international protocols. Many of these parameters are reported to be higher than the acceptable limits prescribed by the Bureau of Indian Standards

(BIS) for drinking water. Specifically, exceedances were observed for sulfate (4.2%), chloride (12.9%), nitrate (14%), fluoride (20.5%), TDS (34.3%), alkalinity (45%), and hardness (51.6%). Notably, only 2% of the groundwater samples contained more uranium than the national radiological safety limit established by the Atomic Energy Regulatory Board (AERB) (60 μg/L). This extensive dataset highlights the complex interplay between uranium and multiple geochemical parameters across diverse hydrogeological settings. Some studies under this project have attempted correlation studies [127, 128]. Patni et al. [129] studied the correlation between uranium and other water quality indicators before and after the monsoon season and reported that establishing strong and consistent relationships between uranium and specific water quality indicators remains scientifically challenging because of the intricate nature of uranium mobility. Factors such as redox conditions, pH, ionic competition, and isotopic variation play significant roles in determining uranium transport behavior, particularly in groundwater systems. This study thus serves as a critical foundation for understanding how multiparameter interactions influence uranium occurrence and mobility in Indian water resources. Figure 4 shows the variation in the mean, median, and mode values of the selected water quality parameters [126].

6 Conclusion

Understanding water quality parameters is crucial for predicting the combined effects of these parameters on uranium solubility. Water quality parameters, including pH, dissolved organic matter, carbonate ions, turbidity, heavy metal ions and biological parameters in aquatic systems, are significantly related to each other. These factors collectively determine the occurrence and mobility of uranium in water sources. This chapter highlights the complex interplay between various water quality parameters and the presence of competing ions, which influence mainly the solubility and mobility of uranium in different water systems. These parameters significantly alter uranium speciation, thereby affecting its bioavailability and potential toxicity. On the basis of these studies, a high amount of dissolved organic matter leads to a greater number of complex ions in the water; thus, high uranium levels in the water were obtained in the northern part of India. Correlation studies between water quality parameters and uranium could be good indicators of uranium contamination in water sources. Despite several studies conducted around the world, there is a considerable knowledge gap in understanding the behavior of uranium in water aquifers. The present study shows that the interpretation of these factors in water sources is vital for designing effective remediation strategies to mitigate uranium and other heavy metal ion contamination in drinking water. A more comprehensive compiled hydrochemical dataset is necessary to understand the key factors controlling the mobility and enrichment of uranium in water sources, especially groundwater. The interaction of various parameters is essential for developing effective remediation strategies and a regulatory framework. Future research should

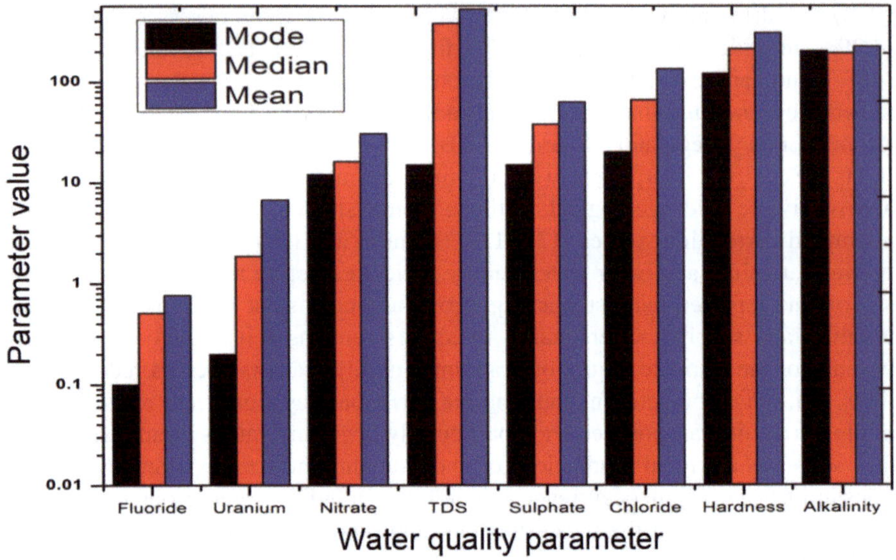

Fig. 4 Variation in uranium with other water quality parameters [126]

focus on long-term monitoring and advanced modeling approaches to predict uranium behavior under varying environmental conditions. Furthermore, integrating innovative treatment technologies with water quality management practices will be crucial in mitigating uranium contamination and ensuring safe drinking water supplies.

References

1. Morin-Crini N, Lichtfouse E, Liu G, Balaram V, Ribeiro ARL, Lu Z, Stock F, Carmona E, Teixeira MR, Picos-Corrales LA, Moreno-Piraján JC, Giraldo L, Li C, Pandey A, Hocquet D, Torri G, Crini G (2021) Emerging contaminants: analysis, aquatic compartments and water pollution. In: Morin-Crini N, Lichtfouse E, Crini G (eds) emerging contaminants Vol. 1. Environmental chemistry for a sustainable world, vol 65. Springer, Cham. https://doi.org/10.1007/978-3-030-69079-3_1
2. Zamora MLL, Zielinski JM, Moodie GB, Falcomer RA, Hunt WC, Capello K (2009) Uranium in drinking water: renal effects of long-term ingestion by an aboriginal community. Arch Environ Occup Health 64(4):228–241. https://doi.org/10.1080/19338240903241267
3. Jha PK, Tripathi P (2021) Arsenic and fluoride contamination in groundwater: a review of global scenarios with special reference to India. Groundw Sustain Dev 13:100576. https://doi.org/10.1016/j.gsd.2021.100576
4. Bajwa BS, Kumar S, Singh S, Sahoo SK, Tripathi RM (2017) Uranium and other heavy toxic elements distribution in the drinking water samples of SW-Punjab, India. J Radiat Res Appl Sci 10(1):13–19
5. Fierro P, Valdovinos C, Vargas-Chacoff L, Bertrán C, Arismendi I (2017) Macroinvertebrates and fishes as bioindicators of stream water pollution. In: Water quality, vol 2, pp 23–38

6. Forio MAE, Goethals PL (2020) An integrated approach of multi-community monitoring and assessment of aquatic ecosystems to support sustainable development. Sustainability 12(14):5603
7. Siddiqua A, Sadef Y, Ahmad SS, Farid M, Ali S (2021) Evaluation of water quality and its potential threats along river Chenab using geo statistical techniques. Pol J Environ Stud 30(6):5239–5254
8. Palamuleni L, Akoth M (2015) Physico-chemical and microbial analysis of selected borehole water in Mahikeng, South Africa. Int J Environ Res Public Health 12(8):8619–8630. https://doi.org/10.3390/ijerph120808619
9. BIS (Bureau of Indian Standard) (2012) Indian standards specifications for drinking water, IS: 10500, Second revision, Bureau of Indian Standards, India
10. World Health Organization (2017) Guidelines for drinking water quality, 4th edn. WHO
11. Adimalla N (2018) Groundwater quality for drinking and irrigation purposes and potential health risks assessment: a case study from semi-arid region of South India. Expo Health 11(2):109–123
12. Pal M, Ayele Y, Hadush M, Panigrahi S, Jadhav VJ (2018) Public health hazards due to unsafe drinking water. Air Water Borne Dis 7(1000138):2
13. Balaram V, Rani A, Rathore DPS (2022) Uranium in groundwater in parts of India and world: a comprehensive review of sources impact to the environment and human health analytical techniques and mitigation technologies. Geosyst Geoenviron 1(2):100043. https://doi.org/10.1016/j.geogeo.2022.100043
14. Vishwakarma CA, Sen R, Singh N, Singh P, Rena V, Kumari R, Mukherjee S (2018) Geochemical characterization and controlling factors of chemical composition of spring water in a part of eastern Himalaya. J Geol Soc India 92:753–763
15. Coyte RM, Singh A, Furst KE, Mitch WA, Vengosh A (2019) Co-occurrence of geogenic and anthropogenic contaminants in groundwater from Rajasthan, India. Sci Total Environ 688:1216–1227
16. Patni K, Pande C, Pande AP, Tewari G, Joshi T (2020) Distribution of naturally occurring uranium and other heavy toxic elements in selected spring water samples of Pithoragarh District, Uttarakhand, India. SN Appl Sci 2:1–9
17. Jia Y, Xi B, Jiang Y, Guo H, Yang Y, Lian X, Han S (2018) Distribution, formation and human-induced evolution of geogenic contaminated groundwater in China: a review. Sci Total Environ 643:967–993
18. Mazhar SN, Ahmad S (2020) Assessment of water quality pollution indices and distribution of heavy metals in drinking water in Ramganga aquifer, Bareilly District Uttar Pradesh, India. Groundw Sustain Dev 10:100304
19. Shrestha S, Kazama F (2007) Assessment of surface water quality using multivariate statistical techniques: A case study of the Fuji river basin, Japan. Environ Model Softw 22(4):464–475
20. Pesce SF, Wunderlin DA (2000) Use of water quality indices to verify the impact of Córdoba City (Argentina) on Suquía River. Water Res 34(11):2915–2926
21. Delpla I, Jung AV, Baures E, Clement M, Thomas O (2009) Impacts of climate change on surface water quality in relation to drinking water production. Environ Int 35(8):1225–1233
22. Kazi TG, Arain MB, Jamali MK, Jalbani N, Afridi HI, Sarfraz RA, Baig JA, Shah AQ (2009) Assessment of water quality of polluted lake using multivariate statistical techniques: A case study. Ecotoxicol Environ Saf 72(2):301–309
23. Craswell E (2021) Fertilizers and nitrate pollution of surface and ground water: an increasingly pervasive global problem. SN Appl Sci 3(4):518
24. Del Bello L (2020) Fluorosis: an ongoing challenge for India. Lancet Planet Heal 4(3):e94–e95
1. 25. Choubisa SL (2018) A brief and critical review of endemic hydrofluorosis in Rajasthan, India. Fluoride 51(1):13–33
26. Zainurin SN, Wan Ismail WZ, Mahamud SNI, Ismail I, Jamaludin J, Ariffin KNZ, Wan Ahmad Kamil WM (2022) Advancements in monitoring water quality based on various sensing methods: A systematic review. Int J Environ Res Public Health 19(21):14080

27. Latif M, Nasir N, Nawaz R, Nasim I, Sultan K, Irshad MA, Irfan A, Dawoud TM, Younous YA, Ahmed Z, Bourhia M (2024) Assessment of drinking water quality using water quality index and synthetic pollution index in urban areas of mega city Lahore: a GIS-based approach. Sci Rep 14(1):13416. https://doi.org/10.1038/s41598-024-63296-1
28. Stanley DM, Wilkin RT (2019) Solution equilibria of uranyl minerals: role of the common groundwater ions calcium and carbonate. J Hazard Mater 377:315–320
29. Singh S, Mosley LM (2003) Trace metal levels in drinking water on Viti Levu, Fiji Islands. South Pac J Nat Appl Sci 21(1):31–34
30. Orloff KG, Mistry K, Charp P, Metcalf S, Marino R, Shelly T, Melaro E, Donohoe AM, Jones RL (2004) Human exposure to uranium in groundwater. Environ Res 94(3):319–326
31. Sajwan RS, Joshi V, Ahamad T, Kumar N, Parmar P, Jindal MK (2024) Assessment of radon transportation and uranium content in the tectonically active zone of Himalaya, India. Sci Total Environ 926:171823
32. Brugge D, deLemos JL, Oldmixon B (2005) Exposure pathways and health effects associated with chemical and radiological toxicity of natural uranium: a review. Rev Environ Health 20(3):177–194
33. Gueniot B, Munier-Lamy C, Berthelin J (1988) Geochemical behavior of uranium in soils, part I. Influence of pedogenetic processes on the distribution of uranium in aerated soils. J Geochem Explor 31(1):21–37
34. Guthrie VA, Kleeman JD (1986) Changing uranium distributions during weathering of granite. Chem Geol 54(1–2):113–126
35. Fiedor JN, Bostick WD, Jarabek RJ, Farrell J (1998) Understanding the mechanism of uranium removal from groundwater by zero-valent iron using X-ray photoelectron spectroscopy. Environ Sci Technol 32(10):1466–1473
36. Chandrasekar T, Sabarathinam C, Viswanathan PM, Rajendiran T, Mathivanan M, Natesan D, Samayamanthula DR (2021) Potential interplay of uranium with geochemical variables and mineral saturation states in groundwater. Appl Water Sci 11:1–18. https://doi.org/10.1007/s13201-021-01396-3
37. Paradis CJ, Johnson RH, Tigar AD, Sauer KB, Marina OC, Reimus PW (2020) Field experiments of surface water to groundwater recharge to characterize the mobility of uranium and vanadium at a former mill tailing site. J Contam Hydrol 229:103581
38. Rosen MR, Burow KR, Fram MS (2019) Anthropogenic and geologic causes of anomalously high uranium concentrations in groundwater used for drinking water supply in the southeastern San Joaquin Valley. CA J Hydrol 577:124009. https://doi.org/10.1016/j.jhydrol.2019.124009
39. Prasanna MV, Chidambaram S, Gireesh TV, Jabir Ali TV (2011) A study on hydrochemical characteristics of surface and sub-surface water in and around Perumal Lake, Cuddalore district, Tamil Nadu, South India. Environ Earth Sci 63:31–47
40. Yidana S, Ophori D, Banoeng-Yakubo B (2008) Groundwater availability in the shallow aquifers of the southern voltaian system: a simulation and chemical analysis. Environ Geol 55:1647–1657
41. Thivya C, Chidambaram S, Thilagavathi R, Keesari T, Rao MS, Prasanna MV, Nepolian M (2015) Occurrence of high uranium and radon in hard rock aquifers of South India–Evaluating the temporal and spatial trends. Groundw Sustain Dev 1(1–2):68–77
42. Boulos PF, Jacobsen LB, Heath JE, Kamojjala SRI (2014) Real-time modeling of water distribution systems: a case study. J Am Water Works Assoc 106(9):E391–E401
43. Vasavi S, Rajeswari VNAS, Kalyan CV (2024) Enhancing water resource management: a GUI-based approach for water body classification and change detection in VHRS images. In: 2024 Third international conference on intelligent techniques in control, optimization and signal processing (INCOS). IEEE, pp 1–6. https://doi.org/10.1109/INCOS59338.2024.10527683
44. Ritter K, Solomon P, Sibley K, Hall P, Keen G, Mattu B, Linton L (2002) Sources, pathways, and relative risks of contaminants in surface water and groundwater: a perspective prepared for the Walkerton inquiry. J Toxicol Environ Health A 65(1):1–142

45. Sophocleous M (2002) Interactions between groundwater and surface water: the state of the science. Hydrogeol J 10:52–67
46. Mishra RK (2023) The effect of eutrophication on drinking water. Br J Multidiscip Adv Stud 4(1):7–20
47. Griffiths C, Klemick H, Massey M, Moore C, Newbold S, Simpson D, Walsh P, Wheeler W (2012) US environmental protection agency valuation of surface water quality improvements. Rev Environ Econ Policy
48. Hutter K, Visher D (1987) Lake hydraulics. Dev Hydraul Eng 4:1
49. Martin JL, Schottman RW, McCutcheon SC (2018) Hydrodynamics and transport for water quality modeling. CRC press
50. Chakraborty SK, Chakraborty SK (2021) Physiography of rivers: relevant hypothesis and theories. In: Riverine ecology volume 1: eco-functionality of the physical environment of rivers. Springer, Cham, pp 235–374
51. Xu YP, Zhang X, Ran Q, Tian Y (2013) Impact of climate change on hydrology of upper reaches of Qiantang River basin, East China. J Hydrol 483:51–60
2. 52. Wohl E (2017) The significance of small streams. Front Earth Sci 11:447–456
53. Lancaster B (2019) Rainwater harvesting for drylands and beyond: Volume 1, guiding principles to welcome rain into your life and landscape. Rainsource Press
54. Olsson G (2012) Water and energy Nexus. Life cycle assessment and water management-related issues-(Quaderns de medi ambient; 4), 137–164
55. Redvers N, Chischilly AM, Warne D, Pino M, Lyon-Colbert A (2021) Uranium exposure in American Indian communities: health, policy, and the way forward. Environ Health Perspect 129(3):035002
56. Campbell PGC (1995) Interactions between trace metals and aquatic organisms: a critique of the free-ion activity model. In: Metal speciation and bioavailability in aquatic systems. Wiley, London, pp 45–102
57. Li Z, Sun P, Zhang C, Zhu N, Xu N, Li D et al (2024) Translocation and transformation of uranium along the aquatic food chain: new insights into uranium risks to the environment. J Hazard Mater 478:135499
58. Sasaki T, Koukami T, Kobayashi T, Kirishima A, Murakami H, Amano Y, Mizuno T, Iwatsuki T, Sasamoto H, Miyakawa K (2017) Determination of dissolved natural thorium and uranium in Horonobe and Mizunami underground research laboratory groundwater and its thermodynamic analysis. J Nucl Sci Technol 54(3):373–381
59. Iqbal MM, Shoaib M, Farid HU, Lee JL (2018) Assessment of water quality profile using numerical modeling approach in major climate classes of Asia. Int J Environ Res Public Health 15(10):2258
60. Crawford SE, Lofts S, Liber K (2017) The role of sediment properties and solution pH in the adsorption of uranium (VI) to freshwater sediments. Environ Pollut 220:873–881
61. Ubuoh EA, Nwogu FU, Ossai-Abeh EQ, Ikwuemesi JC, Oke AO, Umoh JD (2024) Evaluation of hydro-chemical facies and surface water quality dynamics using multivariate statistical approaches in southern Nigeria. Sci Rep 14(1):31600
62. Akiya N, Savage PE (2002) Roles of water for chemical reactions in high-temperature water. Chem Rev 102(8):2725–2750
63. Swift MJ, Andren O, Brussaard L, Briones M, Couteaux MM, Ekschmitt K, Kjoller A, Loiseau P, Smith P (1998) Global change, soil biodiversity, and nitrogen cycling in terrestrial ecosystems: three case studies. Glob Chang Biol 4(7):729–743
64. Shankland TJ, Ander ME (1983) Electrical conductivity, temperatures, and fluids in the lower crust. J Geophys Res Solid Earth 88(B11):9475–9484
65. Akhtar N, Syakir Ishak MI, Bhawani SA, Umar K (2021) Various natural and anthropogenic factors responsible for water quality degradation: A review. Water 13(19):2660
66. Shen X, Sun T, Su M, Dang Z, Yang Z (2019) Short-term response of aquatic ecosystem metabolism to turbidity disturbance in experimental estuarine wetlands. Ecol Eng 136:55–61

67. Whipple GC, Hazen A, Soper GA, Fuller GW, Maignen JPA, Chester JN et al (1901) The decolorization of water. Trans Am Soc Civ Eng 46(2):141–181
68. Germolus NP, Brezonik PL, Hozalski RM, Finlay JC (2021) Long-term water color and flow trends in the Mississippi River headwaters, 1944–2010. Limnol Oceanogr 66(9):3552–3567
69. Chapman DV (2021) Water quality assessments: a guide to the use of biota, sediments and water in environmental monitoring. CRC Press
70. Khatri N, Tyagi S (2015) Influences of natural and anthropogenic factors on surface and groundwater quality in rural and urban areas. Front Life Sci 8(1):23–39
71. Aiken G, Cotsaris E (1995) Soil and hydrology: their effect on NOM. J Am Water Works Ass 87(1):36–45
72. Stumm W, Morgan JJ (2013) Aquatic chemistry: chemical equilibria and rates in natural waters. John Wiley & Sons
73. Caporale AG, Violante A (2016) Chemical processes affecting the mobility of heavy metals and metalloids in soil environments. Curr Pollut Rep 2(1):15–27. https://doi.org/10.1007/s40726-015-0024-y
74. Sonone SS, Jadhav S, Sankhla MS, Kumar R (2020) Water contamination by heavy metals and their toxic effect on aquaculture and human health through food chain. Lett Appl NanoBiosci 10(2):2148–2166
75. Bulbul Ali A, Mishra A (2022) Effects of dissolved oxygen concentration on freshwater fish: a review. Int J Fish Aquat Stud 10(4):113–127
76. Boyd CE, Tucker CS, Boyd CE, Tucker CS (1998) Water quality and aquaculture: preliminary considerations. In: Pond Aquaculture Water Quality Management, pp 1–7
77. Ramachandra TV (2016) Recurring fish mortality episodes in Bangalore lakes: sign of irresponsible and fragmented governance. Energy
78. Metcalf L, Eddy HP, Tchobanoglous G (1991) Wastewater engineering: treatment, disposal, and reuse, vol 4. McGraw-Hill, New York
79. Edberg SCL, Rice EW, Karlin RJ, Allen MJ (2000) Escherichia coli: the best biological drinking water indicator for public health protection. J Appl Microbiol 88(S1):106S–116S
80. Yates MV, Gerba CP, Kelley LM (1985) Virus persistence in groundwater. Appl Environ Microbiol 49(4):778–781
81. Abbas Z, Khan MK, Abbas RZ, Sindhu ZUD, Sajid MS, Munir A, Wahid A, Zafar A, Mughal MAS, Imran M, Nadeem M (2022) Molecular epidemiology of Cryptosporidium parvum and Giardia lamblia in different water bodies, soil, and vegetables in Pakistan. Health Secur 20(4):308–320
82. Efstratiou A, Ongerth JE, Karanis P (2017) Waterborne transmission of protozoan parasites: review of worldwide outbreaks-an update 2011–2016. Water Res 114:14–22
83. Walker HW (2014) Harmful algae blooms in drinking water: removal of cyanobacterial cells and toxins. CRC Press
84. Paerl HW, Otten TG (2013) Harmful cyanobacterial blooms: causes, consequences, and controls. Microb Ecol 65:995–1010
85. Lopes MBS (2017) The 2017 World Health Organization classification of tumors of the pituitary gland: a summary. Acta Neuropathol 134:521–535
86. Gündoğdu A (2023) Oxidation-reduction potential and heavy metals in aquatic systems, 233
87. Waseem A, Ullah H, Rauf MK, Ahmad I (2015) Distribution of natural uranium in surface and groundwater resources: a review. Crit Rev Environ Sci Technol 45(22):2391–2423
88. Ragnarsdottir KV, Charlet L (2000) Uranium behaviour in natural environments
89. Santschi PH (2023) Watershed controls on uranium concentrations tied into natural organic matter and iron interactions in streambeds and wetlands (No. TAMUG-2023-1). Texas A&M University, College Station, TX
90. Markich SJ (2002) Uranium speciation and bioavailability in aquatic systems: an overview. Sci World J 2(1):707–729
91. Malik DS, Sharma AK, Sharma AK, Thakur R, Sharma M (2020) A review on impact of water pollution on freshwater fish species and their aquatic environment. In: Advances in

environmental pollution management: wastewater impacts and treatment technologies, vol 1, pp 10–28
92. Jayaswal K, Sahu V, Gurjar BR (2018) Water pollution, human health and remediation. In: Water remediation, pp 11–27
93. Rajkhowa S, Sarma J, Das AR (2021) Radiological contaminants in water: pollution, health risk, and treatment. In: Contamination of water. Academic, pp 217–236
94. Naja GM, Volesky B (2017) Toxicity and sources of Pb, Cd, Hg, Cr, As, and radionuclides in the environment. In: Handbook of advanced industrial and hazardous wastes management. CRC Press, pp 855–903
95. Bajwa BS, Kumar S, Singh S, Sahoo SK, Tripathi RM (2017) Uranium and other heavy toxic elements distribution in the drinking water samples of SW-Punjab India. J Radiat Res Appl Sci 10(1):13–19. https://doi.org/10.1016/j.jrras.2015.01.002
96. Ahamad T, Nautiyal OP, Singh P, Sajwan RS, Joshi M, Bourai AA (2025) Hydrogeochemical assessment of radon exposure with associated ingestion doses and risks in groundwater of Lesser Kumaon Himalayan region India. Environ Geochem Health 47(12). https://doi.org/10.1007/s10653-025-02843-9
97. Ogidi OI, Akpan UM (2022) Aquatic biodiversity loss: impacts of pollution and anthropogenic activities and strategies for conservation. In: Biodiversity in Africa: potentials, threats and conservation. Springer Nature, Singapore, pp 421–448
98. Rajak P, Ganguly A, Nanda S, Mandi M, Ghanty S, Das K et al (2024) Toxic contaminants and their impacts on aquatic ecology and habitats. In: Spatial modeling of environmental pollution and ecological risk. Woodhead Publishing, pp 255–273
99. Bird G, Macklin MG, Brewer PA, Zaharia S, Balteanu D, Driga B, Serban M (2009) Heavy metals in potable groundwater of mining-affected river catchments, northwestern Romania. Environ Geochem Health 31:741–758. https://doi.org/10.1007/s10653-009-9259-0
100. Fan C, Liu Y, Liu C, Zhao W, Hao N, Guo W, Yuan J, Zhao J (2022) Water quality characteristics, sources, and assessment of surface water in an industrial mining city, southwest of China. Environ Monit Assess 194(4):259. https://doi.org/10.1007/s10661-022-09908-7
101. Narangarvuu D, Enkhdul T, Erdenetsetseg E, Enkhrii-Ujin E, Irmuunzaya K, Batbayar G, Oyundelger K, Yam RSW, Pfeiffer M (2024) Mining and urbanization affect river chemical water quality and macroinvertebrate communities in the upper Selenga River basin, Mongolia (revised version). Environ Monit Assess 196(11):1–17. https://doi.org/10.1007/s10661-024-13225-6
102. Winde F, Sandham LA (2004) Uranium pollution of south African streams–an overview of the situation in gold mining areas of the Witwatersrand. GeoJournal 61:131–149. https://doi.org/10.1007/s10708-004-2867-4
103. Amis J, Duthe-Taussig D, Morton K, Sibilski U (2001) A strategic plan for the efficient monitoring of trends in ground water quality affected by gold mining in South Africa. In: Conference on environmentally responsible mining in southern Africa, pp 25–28
104. Simonic M (1996) Geohydrological investigation of the groundwater pollution next to slime dam #1, Kloof Mine: Phase 2. Confident. Rep. Klo/97/072 to Kloof GM (Pty) Ltd.
105. Atibu EK, Oliveira JM, Malta M, Santos M, Mulaji CK, Mpiana PT, Carvalho FP (2021) Assessment of natural radioactivity in rivers sediment and soil from the Copper Belt Artisanal Mining Region, Democratic Republic of the Congo. J Geosci Environ Protect 9(7):1–20. https://doi.org/10.4236/gep.2021.97001
106. Stammler L, Uhl A, Mayer B, Keller F (2016) Renal effects and carcinogenicity of occupational exposure to uranium: a meta-analysis. Nephron Extra 6(1):1–11. https://doi.org/10.1159/000442827
107. Sharma N, Singh J (2016) Radiological and chemical risk assessment due to high uranium contents observed in the ground waters of Mansa District (Malwa region) of Punjab state, India: an area of high cancer incidence. Expo Health 8:513–525

108. Panwar P, Joshi A, Singh KP, Prasad M, Mehra R, Sahoo SK, Ramola RC (2024) Distribution of uranium and selected toxic heavy metals in drinking water of Garhwal Himalaya India. J Radioanal Nucl Chem 333(6):2927–2935. https://doi.org/10.1007/s10967-023-08998-y
109. Sharma S, Singh KP, Joshi A, Mehra R, Ahamad T, Thakur V, Bourai AA, Ramola RC (2026) Levels and risk due to uranium content in drinking water of the Chakrata region of Garhwal Himalaya India. Ann Nucl Energy 225:111778. https://doi.org/10.1016/j.anucene.2025.111778
110. Kumar R, Jha VN, Jha SK, Sahoo SK, Verma GP, Kulkarni MS (2024) Ambient radiological condition around an operating uranium mill tailings disposal Facility at Turamdih, India. MAPAN 39:1–7
111. Prasad M, Kumar GA, Sahoo SK, Ramola RC (2019) Health risks associated with the exposure to uranium and heavy metals through potable groundwater in Uttarakhand state of India. J Radioanal Nucl Chem 319:13–21
112. Singh H, Singh J, Singh S, Bajwa BS (2009) Uranium concentration in drinking water samples using the SSNTDs. Indian J Phys 83:1039–1044
113. Singh KP, Chandra S, Prasad M, Joshi A, Prasad G, Ramola RC (2024a) Estimation of radiation dose due to ingestion of radon in water samples of Garhwal Himalaya, India. J Radioanal Nucl Chem 333(6):2867–2879
114. Singh P, Kaur B, Bajwa BS (2024b) Analysis of uranium and radon concentration in water of north-west Punjab, India. J Radioanal Nucl Chem:1–10
115. Virk HS (2017) A crisis situation due to uranium and heavy metal contamination of ground waters in Punjab state, India: a preliminary report. Res Rev J Toxicol 7(2):6–11
116. Kumar V, Maity A, Kumar A, Saha S, Kay P, Singh B, Mukherjee T (2023) Critical review on uranium and arsenic content and their chemical mobilization in groundwater: A case study of the Malwa region Punjab, India. Sci Total Environ 885:163885
117. Ali S, Verma S, Agarwal MB, Islam R, Mehrotra M, Deolia RK, Kumar J, Singh S, Mohammadi AA, Raj D, Gupta MK (2024) Groundwater quality assessment using water quality index and principal component analysis in the Achnera block, Agra district, Uttar Pradesh, Northern India. Sci Rep 14(1):5381. https://doi.org/10.1038/s41598-024-56056-8
118. Frisbie SH, Mitchell EJ, Sarkar B (2013) World health organization increases its drinking-water guideline for uranium. Environ Sci Process Impacts 15(10):1817–1823. https://doi.org/10.1039/C3EM00381G
119. Saucier D, Registe PPW, Belanger M, O'Connell C (2023) Urbanization, air pollution, and water pollution: identification of potential environmental risk factors associated with amyotrophic lateral sclerosis using systematic reviews. Front Neurol 14:1108383
120. Maruzzo AJ, Hernandez AB, Swartz CH, Liddie JM, Schaider LA (2025) Socioeconomic disparities in exposures to PFAS and other unregulated industrial drinking water contaminants in US public water systems. Environ Health Perspect 133(1):017002. https://doi.org/10.1289/EHP14721
121. Khurana I, Sen R, India W (2008) Drinking water quality in rural India: issues and approaches
122. Kumar D, Malik DS, Kumar N, Gupta N, Gupta V (2020) Spatial changes in water and heavy metal contamination in water and sediment of river Ganga in the river belt Haridwar to Kanpur. Environ Geochem Health 42(7):2059–2079. https://doi.org/10.1007/s10653-019-00471-8
123. Sharma D, Kansal A (2011) Water quality analysis of River Yamuna using water quality index in the national capital territory India (2000–2009). Appl Water Sci 1(3-4):147–157. https://doi.org/10.1007/s13201-011-0011-4
124. Jaiswal M, Gupta SK, Chabukdhara M, Nasr M, Nema AK, Hussain J, Malik T (2022) Heavy metal contamination in the complete stretch of Yamuna river: a fuzzy logic approach for comprehensive health risk assessment. PLoS One 17(8):e0272562. https://doi.org/10.1371/journal.pone.0272562
125. Pandith TA, Gewali JP, Simnani S, Nazir S, Singh KP, Chakan MR (2024) Exploring tectonic sites with radon from groundwater sources and dose evaluation in various age groups in Baramulla, J&K, India. Groundw Sustain Dev 26:101232

126. Sahoo SK, Jha SK, Jha VN, Patra AC, Kulkarni MS (2021) Survey of uranium in drinking water sources in India. Curr Sci 120(9):1482–1490
127. Sar SK, Sahu M, Singh S, Diwan V, Jindal M, Arora A (2017) Assessment of uranium in ground water from Durg District of Chhattisgarh state and its correlation with other quality parameters. J Radioanal Nucl Chem 314:2339–2348. https://doi.org/10.1007/s10967-017-5587-1
128. Sar SK, Diwan V, Biswas S, Singh S, Sahu M, Jindal MK, Arora A (2018) Study of uranium level in groundwater of Balod district of Chhattisgarh state, India and assessment of health risk. Hum Ecol Risk Assess Int J 24(3):691–698. https://doi.org/10.1080/10807039.2017.1397498
129. Patni K, Pande C, Pande AP, Tiwari G, Joshi T (2021) Seasonal variation of uranium and physico-chemical parameters in spring water sources of Pithoragarh city, Uttarakhand, India. J Radioanal Nucl Chem 329(2):647–660

Natural and Anthropogenic Factors: Shapers of Soluble Uranium Levels

Kashyap Komal, Ayush Kumar, Yvon Ukwishaka, Sarthak Ranjan, Manish Kumar, Biswa Mohan Sahoo, Pardeep Singh, and Preetismita Borah

Abstract The complex interaction of natural and anthropogenic controlling factors controls the solubility of uranium within the environment. This chapter reviews the controlling factors related to uranium solubility and, more importantly, geological and environmental factors that affect uranium mobility and bioavailability. Among the natural factors controlling uranium solubility in different ecosystems, pH, redox conditions, mineral composition, and groundwater flow are crucial. On the other hand, mining activities, nuclear power generation, and agricultural runoff serve as strong human activities upon which uranium levels are influenced through changes in their chemical compositions with an increase in the risk of contamination. Recent studies on the factors affecting uranium solubility range from natural conditions to human-induced influences, as discussed in this chapter. It focuses on case studies of areas that have been affected by uranium mining and industrial operations. This chapter also underscores the measures for uranium contamination abatement in the environment by considering methods such as remediation techniques and policy approaches. This chapter also sheds light on how to manage the environmental impact of uranium by examining the interaction between natural processes and human interventions.

K. Komal · A. Kumar · S. Ranjan
Chandigarh University, Mohali, Punjab, India

Y. Ukwishaka
Nebraska Public Health Environmental Laboratory, Lincoln, NE, USA

M. Kumar · P. Borah (✉)
CSIO-Central Scientific Instruments Organization, Chandigarh, India
e-mail: preetismita.csio@csir.res.in

B. M. Sahoo
School of Pharmacy and Life Sciences, Centurion University of Technology & Management, Bhubaneswar, Odisha, India

P. Singh
Department of Environmental Studies, PGDAV College, University of Delhi, New Delhi, India

M. K. Jindal, K. Patni (eds.), *Water Quality-Uranium Solubility Nexus: Unlocking the Secrets of Uranium Solubility*, Earth and Environmental Sciences Library, https://doi.org/10.1007/978-3-031-98909-4_17

Keywords Uranium solubility · Natural factors · Anthropogenic factors · Geochemistry · Redox conditions · Environmental contamination

1 Introduction

Uranium is a radioactive heavy metal that can exist in a variety of oxidation states and chemical forms and is naturally present in various environmental matrices. Therefore, its solubility, speciation, mobility and bioavailability can be altered by natural (biogeochemical) processes as well as anthropogenic activities. The behavior of uranium in the environment is, however, controlled to an overwhelming extent by (bio)-geochemical processes and natural factors such as the acidity of ecosystems, redox potential, type of associated minerals, biological indicators and hydrological properties [1]. On the other hand, human activities that utilize soluble forms of uranium, such as those associated with uranium mining, large-scale use of nuclear energy, and agricultural phosphate fertilizers, may considerably lead to increased environmental contamination [2]. With both natural and man-made sources contributing to its environmental mobility, uranium is a naturally occurring radioactive element that is extensively dispersed throughout the Earth's crust (Fig. 1).

The main influences on uranium solubility and behavior in ecosystems are biogeochemical processes and several external factors. In addition to human-driven activities such as mining, nuclear power generation, and inappropriate waste disposal, these factors also include natural factors such as pH (acidity), redox potential, mineralogy, hydrological features, and biological interactions [3]. To evaluate uranium contamination, predict its mobility, and lessen its effects on the environment and human health, it is essential to understand these factors. The two primary oxidation states of uranium found in natural systems are U(IV) and U(VI). Under reducing conditions, U(IV) is relatively insoluble and forms solid minerals such as uraninite; under oxidizing conditions, U(VI) is soluble and extremely mobile, frequently generating uranyl ions (UO_2^{2+}). Redox potential is an important element in

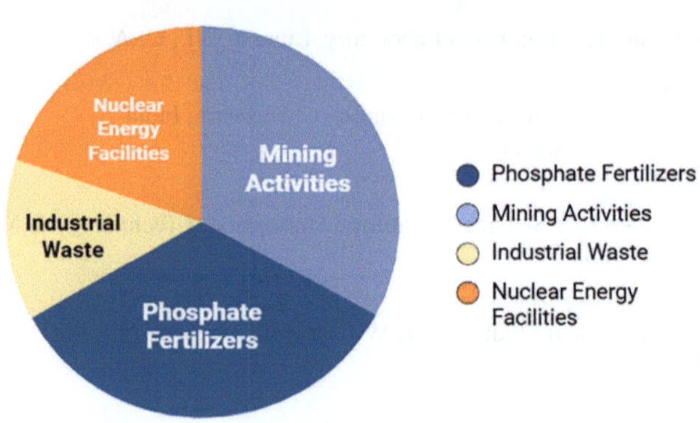

Fig. 1 Distribution of anthropogenic uranium sources

this transition because it encourages uranium solubilization, which mobilizes uranium at the surface and in groundwater by increasing the degree of oxidation. Similarly, the most important factor influencing uranium speciation and solubility is the pH of the surrounding environment. Because uranium-bearing materials dissolve in acidic conditions, uranium mobility is increased. [4] However, stable uranium–carbonate complexes, which improve uranium transport, are encouraged in alkaline environments. Groundwater flow, precipitation patterns, and aquifer characteristics are examples of hydrological elements that affect uranium transport and distribution in ecosystems. For example, when complexed with organic matter or bicarbonate, uranium can travel great distances in aquifers. The natural uranium cycle has been severely disrupted by human activity, which has also increased the amount of uranium in the environment. Since massive amounts of soluble uranium are released into soils and water systems during the extraction and processing of uranium ores, uranium mining and milling are two of the main anthropogenic causes of uranium contamination. If mining tailings are not adequately managed, they can cause uranium to leak into groundwater and other ecosystems over time [5]. Another source is the nuclear power sector, which releases soluble uranium into the environment through improper handling, mishaps, and nuclear waste disposal. Finally, uranium levels in soil and water, especially in agricultural areas, are influenced by the use of phosphatic fertilizers that contain trace amounts of uranium. Uranium mobility may be increased by the interplay of artificial and natural forces. For example, acid mine drainage (AMD) in mining operations reduces the pH of ecosystems, which increases the mobility and solubility of uranium-bearing minerals [6]. Similarly, changes in uranium speciation and increased risk of contamination might result from changes in redox conditions caused by industrial activities or the breakdown of organic matter [6, 7]. Given the toxicity and environmental persistence of uranium, comprehending how anthropogenic and natural factors affect solubility to forecast contamination risks and cleanup procedures is critical. To evaluate the behavior of uranium under various environmental situations, future research should concentrate on incorporating these aspects into prediction models [7].

2 Environmental Impact and Significance of Uranium

For a long time, naturally occurring radioactive uranium has been valued for its use in nuclear technologies and energy generation [8]. However, because of uranium's mobility, bioavailability, and toxicity, its occurrence in the environment, especially in soluble forms, is a significant human and environmental concern globally [9]. Many environmental factors, such as pH, redox potential, mineral composition, and groundwater hydrology, govern the complex geochemistry of uranium while regulating uranium solubility and mobility through interchangeability of its oxidation states (U^{4+} and U^{6+}). Among uranium oxidation states, in a reducing environment, the reduced form, U(IV), is less soluble and immobile, whereas under oxidizing conditions, the oxidized form, U(VI), is more soluble and mobile [9, 10].

In addition to these natural factors, human activities have had a significant impact on environmental uranium concentrations. Mining and milling operations expose uranium-containing ore to air, which can cause oxidation and make it soluble [10]. As a result, the ore becomes contaminated and is spread by wind and water. In the same way, the generation of nuclear energy pollutes the environment through radioactive waste disposal and the uranium enrichment process, which needs strict control to prevent environmental contamination [11]. Agricultural runoff is another important source of anthropogenic uranium; phosphate fertilizers, which are used to improve soil fertility, add trace amounts of uranium to soil and water, which changes the concentration of uranium in agricultural areas (Fig. 2).

For efficient management and cleanup, comprehending the anthropogenic and natural processes influencing uranium solubility is critical. [12] By incorporating elements such as pH and redox conditions, mineral compositions, and flow regimes, uranium solubility under a range of natural and man-made influences can be investigated. [12]

(a) Namibia: Mining for Uranium in Arid Land

Some of the largest open-pit uranium mines in the world, such as Husab, are located in Namibia [13]. The solubility and mobility of uranium are impacted by the arid climate, which aids in its dispersion. Uranium is generally found in low solubility states as uraninite (UO_2), but when it is oxidized to the hexavalent uranyl ion (UO_2^{2+}), it becomes mobile. Uranium ores are exposed to water and oxygen in the atmosphere during mining activities, which makes them more soluble [13, 14]. Sulfate and nitrate species predominate in uranium tailings wastewaters, which

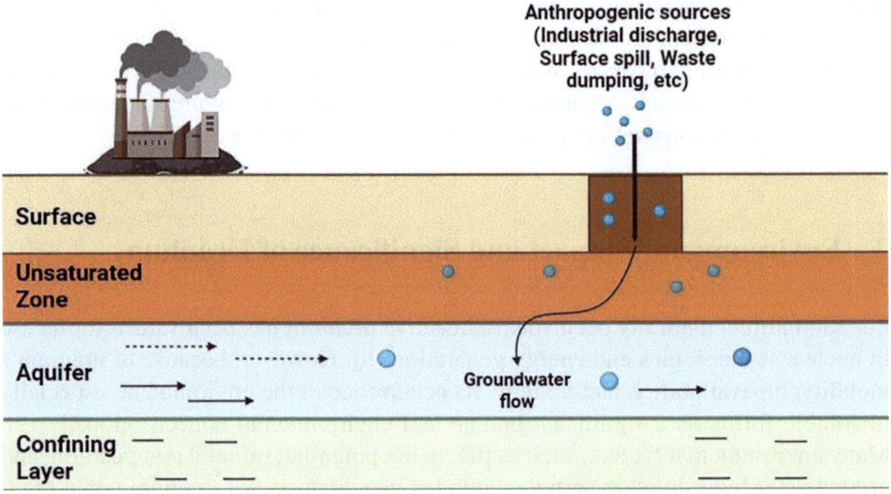

Fig. 2 Cross-sectional graphic depicting the influence of an anthropogenic uranium source on underground water systems. Through the unsaturated zone, contaminants move downward into the aquifer, where they are carried horizontally by groundwater flow. Below, a restricting layer prevents downward motion

further increase uranium solubility by creating highly soluble uranyl sulfate and carbonate complexes. Studies have revealed that aquifers that supply water to nearby settlements have high levels of uranium [14].

(b) Niger: Pollution of groundwater and artisanal mining

Other noteworthy uranium producers include those from Niger, whose history of artisanal mining has contaminated the surrounding environment [15]. Because of artisanal mines, uranium interacts with the water table in the semiarid Sahel region through oxidizing circumstances. Uranium is more soluble in acidic aquifers when exposed to mining waste sulfates [15, 16]. The health of nearby communities that depend on groundwater for irrigation and drinking is more vulnerable. Dust particles containing uranium that are produced at mining settlements can accumulate crops and lower fertility levels when they settle on soils [16].

(c) Fertilizer contamination in Indian agricultural belts

The primary cause of uranium mobilization in India's agricultural activities, particularly in Punjab and Andhra Pradesh, is the application of fertilizers based on phosphates [17]. Phosphate fertilizers occasionally include trace levels of uranium, which can seep into soils and then into water bodies. Uranyl carbonate complexes increase the solubility of uranium in India, where soil conditions are generally alkaline [18]. Owing to the use of water irrigation techniques, uranium concentrations in aquifers have increased to as high as 60 μg/L, which is higher than the WHO's recommended safety threshold of 30 μg/L. Research has indicated that uranium is absorbed by crops, potentially causing chronic exposure and entry into the food chain [17, 18].

2.1 *Uranium in Natural Ecosystems*

Certain rock types are closely associated with uranium under natural conditions (Table 1). Uranium is naturally found in rock types at higher concentrations than other rock varieties. Granites, sedimentary rocks, particularly black shales and phosphorites, contain high levels of uranium because they contain organic matter and can convert U(VI) to U(IV), which results in its mobilization within the rock matrix. The connection with reducing environments often hinders natural movement, but it also provides soluble uranium when rocks are disturbed or become oxidized through weathering processes.

2.1.1 Solubility and Mobility of Uranium in Soil and Water

Numerous parameters, such as pH, redox conditions, and the presence of complex minerals, influence the mobility of uranium in soil and water [1]. The production of carbonate ions, which can create soluble uranyl–carbonate complexes and allow

Table 1 Uranium associations with different rock types, reasons for their associations and the impacts of environmental changes

Rock Type	Uranium Association	Reason for Association	Impact of Environmental manipulation
Granitic	High concentrations of uranium	Naturally, contains higher uranium levels compared to other rock types	Disruption through weathering or oxidation can release soluble uranium
Volcanic	High concentrations of uranium	Similar to granitic rocks, volcanic rocks have natural uranium deposits	Can release soluble uranium when disturbed or oxidized
Sedimentary	Contains high uranium levels, especially in black shales and phosphorites	High organic matter content that can convert U(IV) to U(VI), leading to uranium mobilization	Disturbance or oxidation can release soluble uranium
Black shales	High levels of uranium	Rich in organic material that promotes the conversion of U(IV) to U(VI)	Increased mobility of uranium when disturbed or oxidized
Black shales	High levels of uranium	Rich in organic material that promotes the conversion of U(VI) to U(IV)	Increased mobility of uranium when disturbed or oxidized

uranium to flow through both surface water and groundwater, increases uranium mobility in alkaline environments. However, uranium tends to adsorb onto organic matter, clay minerals, and iron oxides under acidic conditions; this can decrease its mobility while increasing its concentration in soils [19].

The environment's redox potential largely dictates uranium behavior in aquatic systems. Under an oxidizing environment, U(VI), which is predominantly present as a uranyl ion (UO_2^{2+}), is readily transported by water. This form may result in bioavailability and uptake by aquatic organisms, causing biomagnification and bioaccumulation across food webs. However, the reduced uranium (U(IV)) precipitates and attaches to the material under reducing environments, such as a marsh or deeper anoxic sediments [1, 19].

2.1.2 Biological Uptake and Ecological Consequences of Uranium Exposure

Although uranium is typically not absorbed by plants or microorganisms in natural environments, its amount can vary on the basis of environmental circumstances, chemical form, and concentration [20]. In areas where soil uranium concentrations are relatively high, some plants can absorb uranium and add it to the local food chain. Because of its solubility, uranium is accessible to creatures such as fish, algae, and invertebrates in aqueous environments; however, this also presents a risk of bioaccumulation and biomagnification [19]. Ecological repercussions may result from the detrimental impacts of uranium exposure on a species's reproductive and physiological health [20].

2.1.3 Environmental Impact and Natural Sources of Uranium

When exposed to stable environments, the presence of natural uranium sources, such as weathered rocks and enriched soil, can produce baseline levels that are deemed low risk [21]. Because of their propensity to affect solubility and movement, natural occurrences such as wildfires, landslides, and flooding can alter uranium-bearing soils and rocks. Increased uranium levels in nearby water sources could result from these disturbances, which could harm the environment [21, 22]. To differentiate between a baseline uranium level and contamination caused by human activities, comprehending natural variables is essential. [22]

3 Geochemical Behavior of Uranium in the Environment

The different oxidation states of uranium, including U(IV) and U(VI), affect its behavior in the environment. Uranium is a naturally occurring radioactive element that can exist in a variety of oxidation states [23]. Of these, the most environmentally significant are U(IV), or tetravalent, and U(VI), or hexavalent. The redox potential of the environment in which uranium is found, which is an indicator of the oxidizing or reducing conditions, determines the oxidation states of the element [23, 24]. This ion can change into a uranyl ion, which significantly facilitates uranium transport. This movement is caused by the interaction of carbonate or phosphate ions with the substrate found in the soil and groundwater. As a result, it clarifies how elevated pH levels could cause uranium to seep into waterways, particularly in the vicinity of industrial or mining sites [24].

A highly mobile U(VI) transforms into a less soluble U(IV) under reducing and reducing redox conditions, allowing immobilization. Additionally, biological processes can impact uranium; certain microbes can convert U(VI) into U(IV) and vice versa; as a result, contaminated sites may be amenable to bioremediation. In light of uranium's spread and possible toxicity in both natural and man-made environments, these geochemical processes and related microbial interactions are significant [25]. The redox-sensitive nature of uranium, which is impacted by a variety of geochemical conditions, characterizes its geochemical behavior in the environment [25, 26]. It is therefore most stable in two oxidation states: U(IV), which is present as the highly soluble uranyl ion, UO_2^{2+}, and is insoluble and immobile under reducing circumstances. Its ability to form stable complexes with carbonate, sulfate, and phosphate ions, primarily under alkaline and acidic conditions, improves its mobility [26]. The solubility of uranium is significantly influenced by pH; complexation causes more mobility in acidic or alkaline environments, but adsorption onto minerals such as clays and iron oxides may impede uranium transport under favorable conditions [26, 27].

3.1 *pH and Redox Conditions*

The environmental conditions, particularly pH and redox state, significantly affect the behavior of uranium in natural systems (Table 2). These factors influence uranium solubility, speciation, and mobility, which are critical for understanding its transport in groundwater and potential risks to human health and ecosystems·

3.2 *Role of Complexing Agents and Ligands*

Certain ions, such as carbonate, phosphate, and sulfate ferrous ions, can act as complexing agents and significantly affect the solubility and mobility of uraninite [26]. U(VI) is converted to a stable complex under alkaline conditions with a high concentration of carbonate ions, increasing the solubility of uranium in these waters [26, 27]. On the other hand, phosphate ions combine with UO_2 ions to produce uranyl-phosphate complexes, which, under some circumstances, may precipitate to become more soluble and reduce uranium mobility. However, the mobility of uranium is dramatically increased in coastal and agricultural environments where these ligands are highly persistent [27].

Table 2 Impact of different environmental conditions on uranium speciation, solubility, and mobility

Condition	Effect on Uranium	Uranium Speciation	Mobility status
Alkaline (high pH)	Forms soluble complexes with carbonate ions	U(VI) in soluble complexes (e.g., $UO_2(CO_3)_3^{4-}$)	Increased mobility in groundwater systems
Acidic (low pH)	Adsorbs onto minerals like clays and iron oxides	U(IV) or U(VI) adsorbed on minerals	Decreased mobility, more stored in soil or sediments
Reducing conditions	Uranium becomes more stable and less mobile	U(IV) (less soluble)	Reduced mobility, more likely to be immobilized
Oxidizing conditions	Uranium becomes more soluble	U(VI) (more soluble)	Increased mobility in groundwater or surface water

3.3 *Sorption and Desorption Processes*

One important process regulating uranium mobility in soils and sediments is absorption into soil particles, minerals, and organic materials [1]. The mineral makeup, organic matter, and competing ions of the soil are the primary determinants of uranium sorption. Because uranium ions are firmly bonded by organic materials, clays, and iron oxides, there is less leaching of uranium ions into groundwater [4]. Most significantly, uranium may desorb and return to solution as a result of relatively minor changes in circumstances, such as a pH shift or the addition of competing ions, increasing its bioavailability [26, 27].

3.4 *Influence of Climatic Conditions*

Climate conditions have an indirect yet significant impact on uranium behavior. In general, uranium reacts more readily at higher temperatures, which affects the redox state of uranium as well as its complexation-related reactions [4]. The stability of uranium-containing minerals is also influenced by seasonal and climatic factors, such as drought or periods of rainfall, which can affect groundwater flow and variations in the soil moisture content. Owing to the high evaporation rate in arid climates, uranium may accumulate in soil and sediment [4, 23].

4 Natural Factors Influencing Uranium Solubility

Numerous geochemical and ecological factors that impact uranium mobility and availability determine its solubility in natural settings [24]. The way uranium interacts with soil and water systems depends on several elements, including the organic content, groundwater chemistry, microbial activity, and mineral composition [25]. These findings indicate that uranium speciation is mostly governed by natural factors, including pH, redox conditions, and the presence of complexing species. It progresses from stable, immobile forms to highly soluble, mobile forms. The latter knowledge is crucial for estimating the danger of contamination and for forecasting the behavior of uranium in ecosystems, particularly in areas with high natural uranium concentrations [25, 26].

4.1 Mineral Composition and Rock Types

The stability and mobility of uranium depend heavily on the minerals found in the geological matrix. Background uranium levels are typically determined by uranium-bearing minerals, such as carnotite ($K_2(UO_2)_2(VO_4)_2{\cdot}3H_2O$) and uraninite ($UO_2$) [26, 27]. Environmental factors such as pH and redox potential affect how soluble minerals are; uranium-bearing minerals dissolve more readily in oxidizing environments, which releases uranium into the soil and water. In some places, uranium availability is further influenced by sedimentary rocks, such as phosphorites, which naturally have relatively high uranium levels [28].

4.2 Role of Organic Matter

Uranium can become complex by organic matter in soils and sediments, which alters its mobility and solubility [29]. Organic substances can form soluble organo-uranium complexes with uranium ions, particularly fulvic and humic acids. This complexation makes uranium more mobile in soil and water, particularly in organic matter-rich environments. As a reducing agent that converts U(VI) to the less mobile U(IV) form, organic matter may aid in uranium precipitation under reducing conditions [28, 29].

4.3 Climate and Weathering Processes

The weathering process and climate have a major impact on uranium mobilization in natural settings [30]. Rainfall, temperature changes, and freeze–thaw cycles contribute to weathering, which breaks down any minerals that contain uranium. The resulting solute dissolves in water and soil when it is soluble. Because increased precipitation rates increase uranium solubility and mobility, uranium leaches from soils into groundwater more readily in these areas, especially if the water is somewhat acidic. Uranium mobility is also affected by temperature [31]. Chemical reactions, such as the dissolution of uranium minerals and the creation of uranium complexes, are accelerated by relatively warm temperatures. In temperate or monsoon climates, seasonal fluctuations may facilitate uranium transport through cyclical wet and dry cycles that improve the weathering and leaching of uranium[37.38].

4.4 Natural Groundwater Chemistry

Certain components, such as dissolved oxygen, sulfate, and chloride, affect the stability and migratory probability of uranium complexes [32]. For example, uranium becomes more soluble in oxygenated water since it promotes the oxidation of U(IV) to U(VI). Furthermore, under some circumstances, sulfate ions can interact with uranium to produce uranium-sulfate complexes, which can precipitate in some situations while increasing solubility in others. By creating uranium–chloride complexes, which improve uranium mobility, chloride, which is frequently present in coastal or saline groundwater systems, can also increase uranium solubility. Whether uranium remains immobile or becomes accessible depends on the groundwater's general chemical makeup, which includes pH, redox potential, and the presence of complexing agents [33].

4.5 Natural Events and Geological Disturbances

By upsetting geological formations or releasing minerals that contain uranium, natural calamities such as earthquakes, landslides, volcanic eruptions, and floods can have a substantial impact on the solubility and distribution of uranium [34]. Landslides and erosion can expose newly formed rock surfaces to oxidizing conditions, which can increase uranium solubility. Large tracts of sediments that have been severely contaminated by floods may carry uranium, which could contaminate habitats downstream [33, 34]. Additionally, by depositing volcanic debris and exchanging gases that promote its oxidation, volcanic activity can increase the amount of available soluble uranium. Although they are less frequent than changes caused by humans, natural disturbances have the potential to jeopardize ecosystems and water supplies and produce localized spikes in uranium concentrations [2].

4.6 Microbial Activity

Microbial activity has a major impact on the environmental dynamics of uranium, leading to important biochemical and redox changes [35]. Certain bacteria, especially those that thrive in anaerobic environments, can convert U(VI) to U(IV), which reduces uranium solubility and causes uranium to precipitate as insoluble uranium oxide minerals. In anoxic settings such as wetlands, where microbes play a major role in the uranium cycle, microorganisms such as Shewanella species and Geobacter are crucial for microbial reduction [2, 35]. As an alternative, certain microorganisms can increase uranium mobility by generating organic or acidic compounds that react with it and keep it in solution. The behavior of these microbial communities can either improve uranium flow through water systems or stabilize it in sediments [5].

4.7 Carbonate Content

Carbonate ions are significant complexing agents found in soil and groundwater that increase uranium solubility. Stable uranyl–carbonate complexes with U(VI) under alkaline conditions increase uranium mobility and solubility in groundwater

[36]. Areas containing limestone or dolomite tend to have relatively high levels of carbonate in their groundwater, which can lead to enhanced mobility of uranium. The effect is especially apparent under alkaline conditions (pH above 7), where uranium does not precipitate and remains stable in solution because of the stability of uranyl-carbonate complexes [37].

5 Anthropogenic Factors Impacting Soluble Uranium Levels

Numerous human activities significantly affect the amount of soluble uranium in the environment. Anthropogenic influences may increase uranium mobility and bio-availability, posing risks to human health and the environment [7].

5.1 Industrial Waste and Effluents

Uranium pollution in the environment is caused by a variety of industrial processes in addition to mining and agriculture [38]. Waste products containing uranium can be released by operations in sectors such as chemical manufacturing, metal refining, and phosphate synthesis. Long-term environmental consequences may arise from the loading of soluble uranium from industrial effluents into adjacent soils or water bodies. Notably, effluent from major phosphate processing facilities can introduce substantial quantities of uranium into nearby ecosystems. [38, 39]

5.2 Military Activity and Nuclear Weapons Testing

Depleted uranium (DU) has long been used in military applications, such as nuclear weapon testing and armor-piercing projectiles [40]. However, because of their high density and penetrating power, DU munitions are used in some combat situations [40, 41]. These explosives increase the solubility of munitions by releasing uranium into the soil, which may subsequently seep into groundwater [41].

5.3 Fossil Fuel Combustion

One important but frequently disregarded source of uranium in the environment is the burning of fossil fuels, especially in coal-fired power plants [42]. Traces of uranium are found naturally in coal and are emitted into the atmosphere after combustion. The uranium load in the surrounding surroundings may progressively rise as a result of uranium settling on adjacent soils and streams. Power plants located close to rivers may also release wastewater and uranium-bearing ash into aquatic systems, which can affect water quality and increase bioavailability in ecosystems downstream [43].

5.4 *Nuclear Waste Disposal and Storage*

One major concern for lowering contamination from this type of material is the disposal of nuclear waste, which contains high concentrations of uranium and other radioactive materials [44]. Deep geological repositories or specialized containment facilities are required for safe storage to stop the release of leftover nuclear fuel and waste. Uranium leakage from improper handling or insufficient containment can contaminate surrounding soils and groundwater. Despite strict rules, there are always risks associated with disposing of nuclear waste, especially at older storage sites where the containment system may deteriorate with time [2].

5.5 *Irrigation Practices in Agriculture*

Soil uranium levels are directly impacted by the availability of groundwater or river water in regions that rely on irrigation [45]. Over time, uranium can accumulate in agricultural soils as a result of irrigation with uranium-contaminated water, especially in arid areas where substantial quantities of natural ores are present [45, 46]. In addition to affecting the quality of the soil, this accumulation may lead to uranium absorption by crops, which could compromise the safety of food. Concern over the slow buildup of uranium in agricultural runoff from irrigation is growing in areas that mostly depend on agricultural soils and groundwater supplies, especially in areas where uranium levels are already high as a result of natural or man-made processes [47].

6 Human Health Risks Associated with Uranium Exposure

Soil uranium levels are directly impacted by the availability of groundwater or river water in regions that rely on irrigation. Over time, uranium can accumulate in agricultural soils as a result of irrigation with uranium-contaminated water, especially in arid areas where substantial quantities of natural ores are present [48]. In addition to affecting the quality of the soil, this accumulation may lead to uranium absorption by crops, which could compromise the safety of food. Concern over the slow buildup of uranium in irrigation runoff is growing in areas that mostly depend on agricultural soils and groundwater supplies, especially in areas where uranium levels are already high as a result of natural or man-made processes [49].

6.1 Routes of Exposure

One of the main routes of exposure by ingestion is through uranium in food and drinking water. Groundwater concentrations may be significant in areas where uranium is present or has been contaminated by mining and industry [7]. When consumed, uranium is mostly kept in the kidneys, where it can eventually have negative effects. People who engage in mining, grinding, or nuclear fuel manufacturing are most likely to be exposed to and inhale uranium particles during work. Aerosols and dust, which include uranium accumulation, can enter the bloodstream through the air. What does it mean? Chronic inhalation of uranium particles can result in respiratory diseases and lung cancer due to their radioactive disintegration [50]. Uranium can be absorbed via the skin, albeit this is less common, especially in jobs where employees handle enriched atomic elements [50, 51]. Although there is usually little chance of infection from cutaneous exposure, continuous exposure can nevertheless cause a substantial load on the body [51].

6.2 Health Impacts of Uranium Exposure: Effects on Renal Health and Carcinogenic Potential

The health implications of uranium exposure are broad and diverse, with particular emphasis on its negative effects on kidney health and carcinogenic potential [52]. Uranium, a heavy metal and radioactive element, may accumulate in many organs of the body, posing serious long-term health risks. Table 3 categorizes the key health hazards associated with uranium exposure, including nephrotoxicity, which directly impacts kidney function, and the possibility of cancer. The toxicity of uranium to the kidneys stems mostly from its ability to accumulate in the renal tubules, resulting in fibrosis, inflammation, and cellular destruction. Chronic exposure to even low quantities of uranium in drinking water can cause kidney damage, particularly in sensitive groups such as the elderly and children [52].

6.3 Community Health Implications and Public Health Responses

Health Disparities in Exposed Communities: Exposure hazards are frequently greater for populations residing close to uranium mining sites or in regions with natural uranium resources. Chronic uranium exposure often results in a greater prevalence of health problems, such as respiratory, reproductive, and renal difficulties, in these areas. Furthermore, these health inequities are exacerbated by the lack of access to healthcare in rural or indigenous populations close to mining sites.

Table 3 Summarized major health implications of excessive exposure to uranium

Category	Health Impact	Description
Renal health	Kidney toxicity	Uranium is easily absorbed by the kidneys, leading to nephrotoxicity. It accumulates in the renal tubules, causing fibrosis, inflammation, and cellular damage. Chronic exposure to low uranium levels in drinking water can result in proteinuria, decreased kidney function, and an increased risk of kidney disease.
	Kidney function and vulnerable populations	Vulnerable groups, such as elderly individuals, children, and individuals with preexisting renal conditions, are at higher risk of uranium-induced nephrotoxicity. Impaired renal function leads to higher uranium retention due to decreased clearance of toxins from the blood.
Carcinogenic potential	Lung cancer	Occupational exposure to high concentrations of uranium dust increases the risk of lung cancer. Uranium particles accumulate in the lungs over time, exposing lung tissue to alpha radiation, which can lead to cancer. Studies indicate uranium workers are more prone to lung cancer.
	Bone cancer and leukemia	Uranium accumulates in bones due to its affinity for phosphate groups, potentially leading to bone cancer and leukemia. Prolonged exposure to uranium may extend its half-life in human tissue, prolonging the risk of these cancers. Research suggests a possible link, although conclusive evidence is lacking.

Public Health Monitoring and Regulation: Public health organizations such as the EPA and WHO set permissible amounts of uranium in drinking water to reduce health concerns (EPA, 2000). However, efficient monitoring and control are still challenging, especially in isolated regions where uranium poisoning is pervasive. Public health initiatives must continue to protect susceptible groups from uranium exposure.

7 Conclusion

A complicated interaction between natural and human factors determines uranium's capacity to contaminate water, soil, and biota. Natural variables such as pH, redox potential, mineral composition, and groundwater dynamics have significant impacts on the chemical speciation and movement of uranium. Alkalinity reduces uranium mobility and solubility, making it a more dangerous substance for surface and groundwater systems in alkaline and oxidizing environments. Reducing conditions immobilize uranium and produce insoluble compounds that accumulate in soils and sediments, in contrast to greater amounts. Because of this twofold tendency, it is important to understand site-specific geochemical conditions for assessing the environmental impact of uranium.

Numerous activities, such as mining, nuclear energy production, agriculture, industrial processes, and military uses, contribute to uranium contamination problems. Through waste and runoff, these activities cause uranium-containing minerals to oxidize and introduce uranium into the surface and groundwater. Widely utilized in agriculture, phosphate fertilizers increase the spread of heavy metals and increase environmental pressure from military uses such as depleted uranium weapons, especially in areas plagued by conflict.

Because of its radiological and nephrotoxic characteristics, uranium exposure poses serious health hazards. Individuals who are exposed to uranium via contaminated drinking water, work-related exposure, or soil-grown food crops are more likely to develop lung cancer, kidney disease (Kidney Outcome Prediction and Health Network), developmental and reproductive problems, and possibly neurological impacts. In contrast, communities that are at risk, especially those that are close to mining sites or in regions with substantial uranium reserves, face long-term health implications as a result of inadequate resources. This is particularly troublesome. These health hazards can be addressed by implementing efficient public health programs, regularly monitoring health results, and granting access to medical care.

Uranium pollution must be managed and remediated via a combination of scientific research, contemporary remediation technology, regulatory frameworks, and policy enforcement. However, there could be serious dangers. Permeable barriers, chemical stabilization, and phytoremediation can all be used to prevent uranium mobility in contaminated environments. To ascertain their efficacy, alternative approaches must also be taken into account. For example, industrial emission management, safe limits for uranium exposure, and appropriate smelt disposal are all reliant on national, international, or local regulatory initiatives. Hazardous waste handling is also needed. More studies are needed to enhance preventive measures, even if the World Health Organization (WHO) and Environmental Protection Agency (EPA) have set criteria for uranium in drinking water. Therefore, collaborative, multidisciplinary studies on uranium behavior in many environments are essential. Important topics include analyzing the long-term repercussions of uranium on ecosystems, determining the cumulative effects of human activity, and developing creative cleanup strategies for the environment. An integrated management strategy that addresses environmental and public health issues is necessary to protect the environment, protect human health, and lessen the negative consequences of uranium pollution.

References

1. Abdelouas A (2006) Uranium mill tailings: geochemistry, mineralogy, and environmental impact. Elements 2(6):335–341. https://doi.org/10.2113/gselements.2.6.335
2. Brugge D, deLemos JL, Oldmixon B (2005) Exposure pathways and health effects associated with chemical and radiological toxicity of natural uranium: a review. Rev Environ Health 20(3):177–193. https://doi.org/10.1515/REVEH.2005.20.3.177

3. Choppin GR, Liljenzin JO, Rydberg J, Ekberg C (2013) Radiochemistry and nuclear chemistry, 4th edn. Elsevier
4. Langmuir D (1997) Aqueous environmental geochemistry. Prentice Hall
5. Morrison SJ, Metzler DR, Carpenter CE (2001) Uranium precipitation in a permeable reactive barrier by progressive irreversible dissolution of zero-valent iron. Environ Sci Technol 35(2):385–390. https://doi.org/10.1021/es0011876
6. Selvakumar S, Chandrasekar N, Kumar G (2017) Hydrogeochemical processes and uranium contamination in groundwater: a review of key mechanisms, influencing factors, and geochemical models. J Environ Manag 199:89–108. https://doi.org/10.1016/j.jenvman.2017.05.042
7. Zielinski RA, Meier AL (1988) The chemical and isotopic record of environmental uranium contamination. Appl Geochem 3(1):1–14. https://doi.org/10.1016/0883-2927(88)90002-4
8. Smith A, Jones B (2019) Environmental chemistry of uranium: mobility and bioavailability. Environmental Science Press. https://doi.org/10.1016/B978-0-12-815974-4.00012-4
9. Taylor R, Evans G (2018) Redox conditions and pH influence on uranium speciation in aquatic systems. Geochem J 45(3):210–225. https://doi.org/10.1016/j.gca.2018.06.015
10. Barker D (2020) Groundwater contamination and uranium mobility. Academic Publishing. https://doi.org/10.1016/B978-0-12-819460-8.00010-3
11. Lee C (2021) Impact of agricultural runoff on uranium levels in surface waters. Environmental Agriculture 15(4):453–467. https://doi.org/10.1007/s00128-021-03225-y
12. Johnson T (2017) Uranium mining in Namibia: environmental and socioeconomic implications. J Min Environ 23(2):115–128. https://doi.org/10.1080/12345678.2017.1105892
13. Patel S, Singh P (2022) Comparative uranium contamination studies in different global contexts. Glob Environm Stud. https://doi.org/10.1016/B978-0-12-821985-4.00013-8
14. Kumar P (2020) Regulatory frameworks for uranium waste management in India. J Policy Environ Res 34(1):67–80. https://doi.org/10.1016/j.jenvman.2020.110267
15. Hanson T, Leonard J (2018) Military use of depleted uranium and long-term soil contamination. Environ Monit Assess 190(2):112. https://doi.org/10.1007/s10661-017-6448-5
16. Chen Y et al (2021) Fossil fuel combustion as a source of uranium contamination: the case of coal-fired power plants. Energy Environ Sci 14(4):1056–1064. https://doi.org/10.1039/D1EE00652A
17. Rani D et al (2021) Urban development and uranium mobilization in groundwater. Environ Earth Sci 80(1):65. https://doi.org/10.1007/s12665-021-09314-7
18. Gauthier R, Thompson M (2020) Industrial sources of uranium contamination: phosphate production and environmental impacts. J Ind Ecol 24(6):1237–1250. https://doi.org/10.1111/jiec.13020
19. Brugge D, Goble R (2002) The history and health effects of uranium mining and milling among Navajo people. Am J Public Health 92(9):1410–1419. https://doi.org/10.2105/AJPH.92.9.1410
20. Campbell KM, Cramer SP (2009) Factors controlling uranium contamination of groundwater at legacy sites in the western United States. Appl Geochem 24(5):1004–1015. https://doi.org/10.1016/j.apgeochem.2009.02.023
21. Altair T, Sartori LM, Rodrigues F, de Avellar MGB, Galante D (2020) Natural radioactive environments as sources of local disequilibrium for the emergence of life. Astrobiology. arXiv preprint arXiv:2011.14839. https://arxiv.org/abs/2011.14839
22. International Atomic Energy Agency (IAEA) (2021) The environmental behaviour of uranium. IAEA. https://www-pub.iaea.org/MTCD/Publications/PDF/PUB_DOC_488_web.pdf
23. Santschi PH, Li Y-H, Bell J (1979) Natural radionuclides in the water of Narragansett Bay. Earth Planet Sci Lett 42(1):77–86. https://doi.org/10.1016/0012-821X(79)90121-9
24. Santschi PH, Honeyman BD (1989) A Brownian-pumping model for oceanic trace metal scavenging: evidence from Th isotopes. J Mar Res 47(4):869–892. https://doi.org/10.1357/002224089785076489
25. Santschi PH, Xu C (2020) Can the protein/carbohydrate (P/C) ratio of exopolymeric substances (EPS) be used as a proxy for their 'stickiness' and aggregation propensity? Mar Chem 218:103734. https://doi.org/10.1016/j.marchem.2020.103734

26. U.S. Geological Survey (2022) Living organisms take up uranium from their environment. U.S. Geological Survey. https://www.usgs.gov/news/state-news-release/living-organisms-take-uranium-their-environment
27. World Nuclear Association (2022) Uranium mining overview. World Nuclear Association. https://world-nuclear.org/information-library/nuclear-fuel-cycle/mining-of-uranium/uranium-mining-overview
28. Arnold T, Zorn T, Bernhard G, Nitsche H, Rickert P, Zanker H (2001) Sorption of uranium(VI) onto phyllite: experiments and modeling. Radiochim Acta 89(7):467–474. https://doi.org/10.1524/ract.2001.89.7.467
29. Murphy WM, Shock EL (1999) Environmental aqueous geochemistry of actinides. Rev Mineral Geochem 38(1):221–254. https://doi.org/10.2138/rmg.1999.38.8
30. Abdelouas A, Lutze W, Nuttall HE (1999) Uranium contamination in the subsurface: characterization and remediation. Radiochim Acta 79(3):159–167. https://doi.org/10.1524/ract.1999.79.3.159
31. Gascoyne M (1992) Geochemistry of the actinides and their daughters. Treatise Geochem 7:33–50. https://doi.org/10.1016/B978-0-08-095975-7.00602-3
32. Clark DL, Hecker SS, Jarvinen GD, Neu MP (2006) Actinide chemistry and nuclear waste management. Springer
33. Catalano JG, Brown GE (2005) Uranyl adsorption onto montmorillonite: evaluation of binding sites and the effects of cation exchange capacity. Geochim Cosmochim Acta 69(12):2995–3005. https://doi.org/10.1016/j.gca.2004.12.013
34. Langmuir D (1978) Uranium solution-mineral equilibria at low temperatures with applications to sedimentary ore deposits. Geochim Cosmochim Acta 42(6):547–569. https://doi.org/10.1016/0016-7037(78)90001-7
35. International Atomic Energy Agency (IAEA) (2021) The environmental behaviour of uranium and its mobilization in the biosphere. International Atomic Energy Agency. https://www-pub.iaea.org/MTCD/Publications/PDF/PUB_DOC_488_web.pdf
36. McDonald M, Taylor D (2016) Anthropogenic sources of uranium contamination and its impact on groundwater in the mining regions of Africa. Sci Total Environ 566:343–351. https://doi.org/10.1016/j.scitotenv.2016.05.032
37. Schmidt K, Kreutz E (2018) Uranium mobilization in groundwater due to mining and land use changes in uranium-rich regions. Environ Earth Sci 77(3):97–108. https://doi.org/10.1007/s12665-018-7631-1
38. Abed JM, Al-Omari A (2019) The role of agricultural runoff in the mobilization of uranium in arid regions. J Environ Radioact 204:77–86. https://doi.org/10.1016/j.jenvrad.2019.04.004
39. Albrecht MA, Williams SA (2017) Mining contamination and the release of uranium into surface water: a review of environmental impacts and remediation techniques. Environ Pollut 228:324–332. https://doi.org/10.1016/j.envpol.2017.05.034
40. Bonotto DM, Souza AP (2020) Uranium mobilization from tailings in mining areas: an assessment of anthropogenic impacts on groundwater contamination. Environ Geochem Health 42(5):1485–1497. https://doi.org/10.1007/s10653-019-00416-w
41. Cummings JA, Richter AA (2018) Agricultural runoff and uranium contamination in soils: the long-term impact of fertilization practices. Sci Total Environ 618:437–445. https://doi.org/10.1016/j.scitotenv.2017.10.195
42. Gauthier G, Glover SE (2021) Geochemical factors influencing uranium mobility in mining-impacted environments: a review of anthropogenic contributions. Environ Toxicol Chem 40(5):1512–1527. https://doi.org/10.1002/etc.4991
43. Krabbenhoft DP, Webster JC (2014) The influence of mining activities on uranium solubility in contaminated groundwater. J Contam Hydrol 166:36–49. https://doi.org/10.1016/j.jconhyd.2014.05.003
44. Salazar J, Molinar J (2016) Uranium and radium contamination in the vicinity of nuclear power plants: impact on aquatic ecosystems. J Environ Radioact 155:64–71. https://doi.org/10.1016/j.jenvrad.2015.10.015

45. Cocca S, Papa S (2019) Health effects of uranium exposure in contaminated environments: a case study from mining regions. Environ Health Perspect 127(5):570–576. https://doi.org/10.1289/EHP4359
46. De Oliveira PR, Siqueira LA (2020) Human health risks from uranium exposure through drinking water: a review of studies in contaminated regions. Sci Total Environ 726:138722. https://doi.org/10.1016/j.scitotenv.2020.138722
47. Fabrikant JI, Kelsey WA (2018) Uranium exposure and its impact on kidney function: a comprehensive review. Int J Environ Res Public Health 15(11):2435. https://doi.org/10.3390/ijerph15112435
48. Miller MD, Van Hoy MC (2017) Health risks from uranium exposure: a focus on kidney function and cancer. J Toxicol Environ Health 80(3):181–193. https://doi.org/10.1080/15287394.2017.1296981
49. Roberge JJ, Beaudoin G (2019) Uranium and radon exposure from drinking water: human health risks in affected regions. J Water Health 17(3):442–453. https://doi.org/10.2166/wh.2019.136
50. Gauthier G, Deneer HL (2021) Health impacts of uranium exposure in contaminated communities: a systematic review of epidemiological studies. Environ Health Perspect 129(2):025002. https://doi.org/10.1289/EHP7371
51. Lindell T, Eklund B (2020) The effects of chronic low-dose uranium exposure on human health: an overview of recent findings. Environ Toxicol 35(5):583–594. https://doi.org/10.1002/tox.22922
52. Santos DF, Almeida AM (2018) Uranium exposure and its potential impacts on human health: a review of the neurotoxic and carcinogenic effects. Environ Toxicol Pharmacol 63:76–85. https://doi.org/10.1016/j.etap.2018.01.006

Implication for Uranium Transport and Remediation Strategies

Manjulata Yadav and Vijita Diwan

Abstract One of the common radioactive elements found in drinking water is uranium. Drinking water containing trace amounts of uranium can cause major health problems. Consequently, the detection of uranium in drinking water has raised concerns among public regulators and policy makers. In addition to its radioactivity, excessive exposure to U in drinking water can harm biological structures through chemical toxicity, which is thought to be the main health risk. Uranium concentrations in drinking water below 30 µg/l are considered safe by the World Health Organization (WHO). Uranium transport and remediation strategies are critical components of managing uranium mining, nuclear energy, and environmental contamination. The transport of uranium poses significant safety risks due to its radioactive nature, requiring stringent safety measures, regulatory compliance, and secure transportation routes. The potential for accidents or spills necessitates specialized containers and monitoring systems to prevent environmental contamination. Effective remediation of uranium contamination involves both active and passive techniques for soil and water restoration, including in situ chemical reduction, bioremediation, and the removal of contaminated soil and water. Currently, it is important to use successful remediation techniques and environmental monitoring to prevent ecosystem damage.

Keywords Uranium transport · Remediation · Nanotechnology · Health impact

M. Yadav (✉)
Department of Physics, S.S.S.T.S.R. Govt. Degree College Nainidanda, Pauri Garhwal, India

V. Diwan
Department of Chemistry, SSPU Bhilai, Durg, Chhattisgarh, India

M. K. Jindal, K. Patni (eds.), *Water Quality-Uranium Solubility Nexus: Unlocking the Secrets of Uranium Solubility*, Earth and Environmental Sciences Library, https://doi.org/10.1007/978-3-031-98909-4_18

1 Overview of Uranium Transport and Remediation

Uranium is a radioactive element commonly found in soil, rock, water, and uranium mining areas [1]. The U symbol represents uranium, which is found on Earth in three different isotope forms. 92 electrons and an equivalent amount of protons make up the U. It belongs to the periodic table's actinide series. It is most commonly found in the hexavalent state in the form of a uranyl ion (UO_2^{2+}). All uranium isotopes have radioactive properties. Weakly radioactive, uranium decays slowly and releases an alpha particle. The uranium isotopes have half-lives that range from 4.5 billion years to 159,200 years. The most prevalent and stable isotope of uranium is U^{238}. On Earth, 99% of uranium is naturally present, while two isotopes, U^{235} and U^{234}, constitute less than 1% of uranium [2]. Despite being weakly radioactive, it causes substantial health problems for living beings. Uranium is also chemically toxic to humans. When consumed, uranium primarily impacts the kidneys and can cause kidney damage at high concentrations. The renal tubules are particularly affected [3]. To safeguard public health, national and international advisory organizations such as the International Commission on Radiological Protection (ICRP), the U.S. Environmental Protection Agency (EPA), the World Health Organization (WHO), and the Codex Alimentarius Commission (Codex) have established guidelines and ensured their implementation through regulatory bodies. The transport and remediation of uranium are essential for the responsible management of uranium, especially in the contexts of nuclear energy production, mining, and environmental protection. Being radioactive, uranium requires careful handling throughout its lifecycle—from extraction to transport and, ultimately, to its disposal or remediation in cases of contamination. Various treatment technologies, ranging from conventional methods to advanced methods, have been developed to provide uranium-free drinking water. These technologies include adsorption, ion exchange, membrane filtration, photocatalysis, electrocoagulation, and biological methods [4–15].

2 Origin Occurrences and Transport Mechanism

Uranium minerals such as Fe–Mn (oxy) hydroxides, silicates, and phosphates can be found in different types of rocks and soils. Felsic rocks (4.8 mg/kg) typically contain higher concentrations of uranium than do mafic rocks (0.6 mg/kg) [16]. This difference arises from the incompatibility of uranium in silicate magmas, combined with its large ionic radius and high valence. During partial melting and crystal fractionation, uranium is preferentially incorporated into high-temperature metaluminous melts and subsequently accumulates in granitic and pegmatitic rocks [17]. In sedimentary deposits, uranium concentrations are usually approximately 3–4 mg/kg [18]. In carbonate fluorapatite rocks, high uranium concentrations—up to 1070 mg/kg—have been found in the Keuper aquifer region of Germany. Its host rocks and sediments transfer U to groundwater by mineral breakdown and/or U desorption from the mineral surface [19–21]. These mineals attributed the uranium to leach out from the host rocks under oxic to suboxic conditions [22]. Some studies [23, 24] have also shown that calcium and carbonates in groundwater react with uranium to form U-carbonate complexes and Ca-U-carbonate complexes. These complexes are highly soluble in water, which facilitates their rapid movement in subsurface water. Other factors influencing uranium release into groundwater include overexploitation of water resources, water–rock interactions, pH, redox potential (Eh), organic matter, bicarbonates, and other ions. Alkaline pH, organic matter, and bicarbonates increase the release of uranium from host minerals, whereas lower Eh conditions tend to reduce uranium release [19]. The uranium contents of various rock, coal and fly ash samples are presented in Fig. 1.

The elevated uranium levels are primarily due to the presence of structurally substituted carbonate in the crystal structure, as well as the alteration of apatite during weathering [16, 25]. Certain regions, such as the Hetao basin in Inner Mongolia, also present high uranium concentrations (0.44–323 μg/L), particularly in alluvial soils, unlike flat plains [26]. Additionally, elevated uranium concentrations have been observed in India (Punjab), mainly as a result of weathering of the Siwalik Mountains, with minor contributions from the coal industry (fly ash) and agricultural fertilizers [27–29]. High uranium concentrations have also been reported in groundwater in countries such as the India, United States, Pakistan, Switzerland, South Korea, Australia, Brazil, Egypt, and Finland [30, 31]. Uranium concentrations in surface waters vary widely, depending on factors such as local geology, land use, pH, and mineralization sources. Uranium concentrations in river waters are generally low, typically approximately 4 μg/L or less, with a worldwide average of 0.3 μg/L [32]. In France, river concentrations range from 0.08 to 0.35 μg/L [33]. In Switzerland, the Dischmar River has a concentration of 2.4 μg/L [34]. Indian rivers concentrations ranging from 0.92 to 5.37 μg/L [35] and also reported uranium in river sediment samples [36]. In the River Clyde, Scotland, concentrations ranged

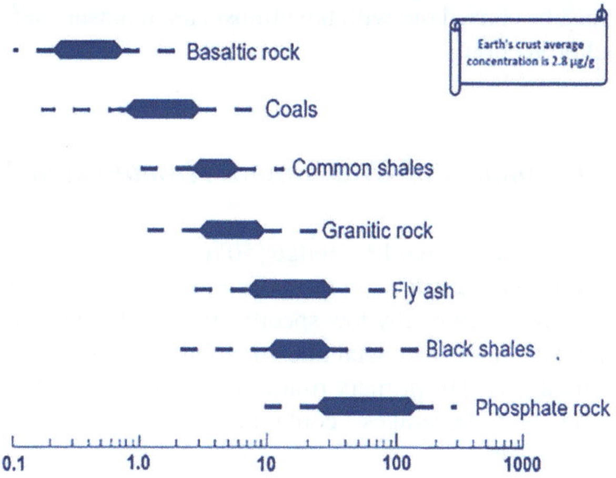

Fig. 1 The amount of uranium in different rocks, coal, and fly ash (Source: [1])

from <0.02 to 1.24 μg/L [37]. The Yellow River, China, contributes approximately 1% of the global dissolved uranium flux to the sea, with concentrations in 2010 ranging from 3.85 to 7.57 μg/L [38]. The application of phosphate fertilizers has been identified as a source of uranium in surface waters. For example, in northern Everglades, USA, drainage from phosphate-fertilized peat resulted in increased uranium concentrations of 0.3–2.4 μg/L compared with those in unfertilized peat (<0.1 μg/L) [39]. Agricultural areas also present elevated concentrations of uranium, with values >2 μg/L being prevalent [40]. The uranium concentration was found to be relatively high in alkaline and saline lakes. In lakes from the Rift Valley of Ethiopia, uranium concentrations range from 0.2 to 73.5 μg/L, with one sample reaching 100 μg/L at pH 10 [41]. Mono Lake, California, had a significantly higher concentration of 554 μg/L [42]. Eastern Mongolia's soda lakes had concentrations between 57 and 1490 μg/L, with a typical pH of 10 [43]. In areas with acid mine drainage (e.g., the Hejiacun U mine, China), uranium concentrations can be very high. For example, at the Poços de Caldas U Mine, Brazil, concentrations ranged from 6000 to 14,000 μg/L, and the water had a low pH of 2.7 [44]. The Curilo U mine, Bulgaria, also has elevated concentrations (100–2750 μg/L), with low pH and high metal concentrations [45]. Uranium concentrations can fluctuate due to changes in pH, redox conditions, rainfall, and temperature, which affect the solubility of uranium in water [46, 47]. In open seawater, uranium concentrations are generally approximately 3.3 μg/L. Uranium behaves conservatively, mainly existing as uranyl (U(VI)) complexed with carbonate and calcium/magnesium. The residence time of uranium in seawater is long, on the order of hundreds of thousands of years [48, 49].

3 Uranium Transport Implications and Challenges

The transport of uranium, a radioactive element commonly used in nuclear energy generation, has significant implications and challenges because of its hazardous nature. While uranium is essential for nuclear power production, its transportation must be carried out with the utmost care to ensure safety, security, and environmental protection.

3.1 Impact of Uranium on Humans and Animals

A study by Maynard & Hodge [50] on uranium ingestion and reproduction reported adverse health effects in rats. Uranium, the heaviest naturally occurring alpha emitter, has exceptionally low specific activity. Despite being radioactive, uranium does not have a known metabolic function in mammals; thus, it is classified as nonessential [51]. The primary routes of exposure to uranium are through the ingestion of food and beverages containing uranium isotopes and the inhalation of uranium-contaminated dust. When assessing uranium toxicity, both its chemical and radiological properties must be considered. Unlike other radioactive elements, uranium concentrations in water are often specified in the same way as chemical toxins are, using units such as parts per billion (ppb or μg/l).

Most of the uranium that enters the body is derived from gastrointestinal absorption after the ingestion of food or water. Once ingested, uranium typically does not enter the bloodstream and is mostly eliminated within a few days. The kidneys and bones are the primary sites for uranium deposition, with approximately 12% of absorbed uranium accumulating in the kidneys and 22% in bones. The remaining uranium is distributed throughout the body and then eliminated [52]. While most of the uranium deposited in the kidneys is excreted within a few days, the portion deposited in bones can remain for many years. On average, the human body contains approximately 90 μg of uranium from normal intake via water, food, and air. Given the limited direct effects of uranium exposure in humans, much of the knowledge about uranium toxicity is derived from animal studies. Many experiments on rats and mice were conducted in the 1940s and 1950s, which highlighted various health effects, particularly in the proximal tubules, when exposed to uranium through inhalation, ingestion, or dermal absorption [53]. The physiological processes of uranium exposure and their effects on the human body are presented in Fig. 2.

In one 2-week study by Ortega et al. [55], 40 male Sprague–Dawley rats were administered uranyl ethanoate dehydrate in their drinking water at doses ranging from 0 to 16 mg/kg body weight. A variety of metabolic effects were observed in the

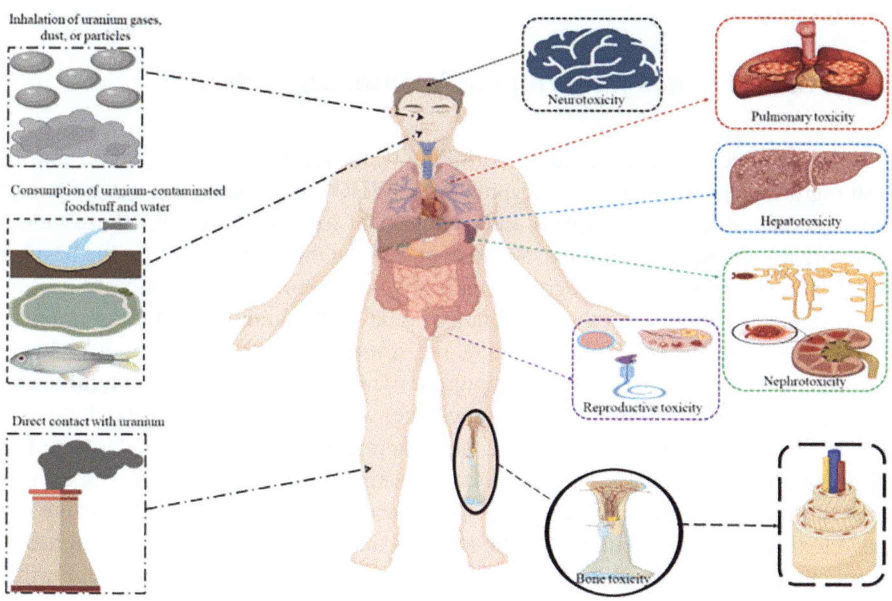

Fig. 2 outlines the mechanisms of uranium exposure and its impact on the human body (Source: [54])

rats. A separate study by Gilman et al. [56] involved 15 male and 15 female Sprague–Dawley rats exposed to water containing 0.001 to 600 mg/l uranyl nitrate hexahydrate for 91 days. Histopathological changes were most notable with 2% uranyl nitrate hexahydrate. In another experiment, Domingo et al. [57] administered doses of uranyl acetate dihydrate ranging from 0 to 50 mg/kg to pregnant Swiss mice, revealing fetal toxicity and growth abnormalities, including external and internal malformations, at uranium doses above 2.8 mg/kg body weight per day.

High doses of soluble uranium compounds have been shown to induce bone cancer in experimental animals via injection or inhalation. However, no evidence of cancer has been reported in animals that consume either soluble or insoluble uranium compounds [58]. Several clinical studies have explored the effects of uranium exposure in humans. A study of 324 individuals in Nova Scotia, Canada, who were exposed to varying levels of natural uranium in drinking water (up to 0.7 mg/l) revealed no direct correlation between uranium exposure and overt kidney disease or other health complaints. However, elevated urinary β-2-microglobulin excretion was noted, which suggested potential kidney stress. Research in Finland, which involved individuals exposed to uranium-contaminated well water with a median concentration of 28 μg/l, revealed a correlation between uranium exposure and changes in kidney function, such as increased excretion of calcium, phosphate, glucose, and calcium in urine, indicating potential mild effects on proximal tubule function [3]. However, no impact on glomerular function was observed, and the study's small sample size and variation in the unexposed population made it difficult to establish clear conclusions.

3.2 Global Limits for Uranium in Drinking Water

The permissible limit for uranium in drinking water varies across different countries and regions. The World Health Organization (WHO) periodically revises its guidelines, considering both chemical and radiological toxicity. However, these guidelines are not universally adopted. Some countries set their own limits on the basis of local factors, including geological conditions, socioeconomic considerations, and the presence of uranium in the environment. The WHO guideline value for uranium in drinking water is set at 30 μg/l [59], but some countries adopt higher or lower limits, depending on their specific regulatory frameworks and the potential health risks to their populations. Regulatory agencies assess the risk of establishing safe levels of uranium in drinking water, yet discrepancies in these standards reflect ongoing challenges in achieving a global consensus on what constitutes an acceptable limit for uranium in drinking water. In conclusion, while uranium exposure can have toxic effects, particularly on the kidneys and bones, the risk to human health from environmental levels of uranium in drinking water remains uncertain. Further research is needed to better understand the health implications and to harmonize international standards regarding acceptable uranium concentrations in drinking water.

4 Uranium Remediation Technologies

Because of their chemical and radioactive characteristics, uranium and materials containing uranium are considered serious threats to the environment and living things [4]. The long-lived radioactive nature of uranium and its chemical toxicity require careful and effective remediation techniques to protect human health, the environment, and ecosystems. Below are several key soil and water remediation techniques used for addressing uranium contamination.

4.1 Electrocoagulation

Electrocoagulation (EC) is an advanced coagulation process in which coagulants are generated in situ by applying an electric charge between the cathode and anode. Metal hydroxide flocs are created when the hydroxide ions produced on the cathode during water splitting interact with the metal ions produced by the oxidation of the anode. When attached to gas bubbles produced in the electrochemical cell, these flocs subsequently absorb dissolved uranium and precipitate by sedimentation or move toward the liquid surface via froth flotation [60–62]. EC offers several advantages over traditional coagulation methods, such as in situ coagulant generation, low secondary sludge formation, high removal efficiency, and minimal impact from coions. Factors such as pH, initial U(VI) concentration, electrode type, applied current, and operating time influence U(VI) removal efficiency. For example, adding organic ligands, such as amino, hydroxyl, and carboxyl groups, enhances U(VI) removal efficiency. Li et al. [8] conducted experiments with Fe as an anode and organic ligands (OGL) for U(VI) removal, achieving optimal results with a 3:1 OGL to U(VI) mole ratio, resulting in 99.65% removal efficiency under certain conditions (pH = 6, current density = 0.6 mA/cm^2, operating time = 24 min). The removal mechanism involves U(VI)-OGL chelation, which results in a negative charge and increases the colloidal size, facilitating floc formation and settling. Anode materials such as Mg and Al showed lower efficiencies because of the lower generation of Mg flocs and the coverage of Al anode surfaces by Al oxides. In contrast, Fe anodes produce $Fe(OH)_3$, which precipitates, avoiding secondary pollution [8].

4.2 Membrane Filtration

Membrane filtration is a pressure-driven process in which membranes serve as barriers to contaminants such as suspended solids, bacteria, viruses, and mono- and multivalent ions. There are four types of membrane processes: microfiltration (MF), ultrafiltration (UF), nanofiltration (NF), and reverse osmosis (RO). MF and UF are low-pressure membranes with large pores, making them ineffective for removing mono- and multivalent ions. On the other hand, RO membranes are nonporous, dense membranes (<1 nm) that use chemical diffusion to remove ions, whereas NF membranes have intermediate pore sizes, effectively removing multivalent ions but only partially removing monovalentions [63]. NF membranes have gained attention for their high removal efficiency, low space requirement, and selectivity, making them effective options for U(VI) removal.

4.3 Ion Exchange Method

In addition to recovering uranium resources, ion exchange (IE) is a promising method for removing uranium from wastewater while minimizing the adverse environmental effects of uranium mining. Quaternary amine groups [$-N(CH_3)_3OH$], a broad pH range of operation, and good selectivity for uranium are characteristics of strong base anion (S_{BA}) exchange resins. Therefore, the most often utilized IE resin for treating wastewater containing uranium is S_{BA}. The ability of the resin to remove uranium was attributed to the creation of binary anionic uranyl sulfate complexes and their subsequent bonding with the resin [64, 65]. This removal mechanism is shown below:

$$[S_{BA}\ \text{Polymer-anion}] \xrightarrow{UO_2(SO_4)n^{m-}} [S_{BA}\ \text{Polymer-}\ UO_2(SO_4)n^{m-}] + \text{anion}$$

Rather than the saturation of functional groups, deposits on the resin surface may be the primary cause of the deactivation of the resins [64].

4.4 Photocatalytic Technology

This method breaks down and extracts uranium from wastewater by using photocatalysts, which are usually semiconductor materials, to capture solar energy and initiate chemical processes. Because photocatalysts are readily regenerable and reusable, the procedure is more economically feasible and results in less waste. The semiconductor photocatalyst is photoexcited to produce electron–hole pairs when exposed to light. The electrons produced reduce U(VI) into U(IV), which is presented in Fig. 3. Consequently, U(VI) is reduced and immobilized [66, 67]. The recovery of uranium from radioactive wastewater or seawater is, however, limited by a number of factors, including the presence of organic matter, low uranium content, a large number of competing ions, and significant acidity or alkalinity in particular environments throughout the practical application process.

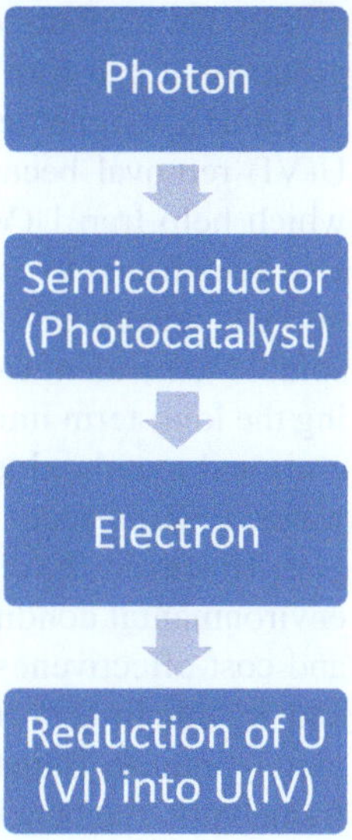

Fig. 3 Mechanism of removal of U(VI) by photocatalysis

4.5 Adsorption Technique

Adsorption is a widely used physicochemical process in water treatment, where an adsorbate from the feed solution is attracted to and retained on the surface of the adsorbent. Its popularity in U(VI) removal is attributed to its low initial cost, ease of operation, and minimal energy requirements. The effectiveness of adsorption largely depends on the type and properties of the adsorbents employed. Recent studies (after 2017) have explored various materials, including clay minerals, metal oxides, layered double hydroxides (LDHs), graphene oxides (GOs), and metal–organic frameworks (MOFs), for U(VI) removal from water [15]. Different types of clay minerals have attracted significant attention for U(VI) adsorption because of their high specific surface area, high availability, low cost, and ion-exchange capacity [68]. The surfaces of clay minerals possess both positive and negative charges due to isomorphous substitution (such as $Al^{+3} \rightarrow Si^{+4}$; $Mg^{+2} \rightarrow Al^{+3}$) and edge defects, enabling them to adsorb both cations and anions by neutralizing charges [69]. Several clay minerals, such as bentonite (montmorillonite), kaolinite, and palygorskite, have demonstrated efficient U(VI) adsorption in both their natural and modified forms.

The modification of these clay minerals can further enhance their adsorption capacity. Modifications, particularly the introduction of amine groups or other functional groups, significantly increase the surface area and enhance the ability of clay minerals to bind U(VI). This modification often leads to increased ion exchange capacity and better adsorption kinetics, making modified clay minerals promising materials for large-scale U(VI) removal in contaminated water sources. Metal oxides, particularly those containing iron (Fe) and manganese (Mn), are also widely studied for their ability to adsorb U(VI) from water. These materials possess a high affinity for uranium due to their surface reactivity and the availability of hydroxyl groups that can form complexes with uranium ions. Iron oxides, such as goethite (FeOOH), magnetite (Fe_3O_4), and hematite (Fe_2O_3), are particularly effective in U(VI) removal because of their large surface area and variable oxidation states, which help trap U(VI) ions. Manganese oxides, such as birnessite (MnO_2), have also been found to exhibit high adsorption capacities for U(VI). These metal oxides not only adsorb U(VI) but also reduce it from its soluble hexavalent form to the less soluble tetravalent form [U(IV)], thereby decreasing uranium mobility and enhancing the long-term immobilization of uranium in contaminated water. Both clay minerals and metal oxides offer distinct advantages in uranium removal from drinking water, making them attractive materials for adsorption-based treatment technologies. However, the choice of adsorbent largely depends on factors such as the local environmental conditions, the concentration of uranium, and the desired efficiency and cost-effectiveness of the treatment process. The coordination of metal centers and organic linkers creates a family of inorganic–organic hybrid materials known as metal–organic frameworks (MOFs). By adding functional groups to the MOF framework, the benefits of MOFs—like their high porosity—are combined with the potent attraction of functional groups to uranium ions [70].

4.6 Biological Methods (Bioremediation and Phytoremediation)

Bioremediation is an effective technique that leverages microorganisms (MOs) and their enzymes to transform pollutants into less toxic, less mobile forms. In microbial bioremediation, the use of mixed cultures has been found to increase uranium (U(VI)) removal compared with that of pure cultures. For example, Bender et al. [71] employed a microbial mat containing a group of MOs immobilized in silica gel particles (SMPs) to achieve significant U(VI) removal, reducing the concentration from 2.5 mg/L to 3 μg/L. The potential of SMP for field applications was evaluated in a 200 L barrel containing 2 mg/L U(VI) under low Eh conditions. The process involved the sorption of U(VI) ions onto SMP, followed by the reduction of U(VI) to U(IV) by extracellular reductases secreted by microorganisms such as Rhodopseudomonas and sulfate-reducing bacteria (SRB). The reoxidation of U(IV) was prevented by hydrogen (H_2) and hydrogen sulfide (H_2S) produced by cyanobacteria and SRB.

In another study, Deinococcus radiodurans, known for its resistance to radiation and extreme conditions, was used for U(VI) removal. Despite its robustness, large volumes of culture are needed because of poor biomass yield, making the process inefficient and costly [72]. To improve biomass yield, Manobala et al. [73] genetically modified D. radiodurans and incorporated green fluorescent protein to increase biofilm formation. This approach resulted in a removal efficiency of 75% with a U(VI) concentration of 1000 mg/L. The biofilm biomass, which was immobilized in the extracellular polymeric matrix, proved to be both economical and feasible, as there was no need to separate the biomass from the bulk liquid waste. Functional groups such as carboxyl, phosphate, and hydroxyl groups on biofilm biomass facilitate efficient U(VI) removal [74].

Similarly, the extracellular polysaccharide (EPS) secreted by *Pseudomonas aeruginosa* exhibited a high adsorption capacity (985 mg/g) and a reaction rate constant of 0.013 g/mg/min at pH 5. FT-IR analysis confirmed that U(VI) strongly bonded to carboxyl groups at pH 5, whereas at pH < 5, bonding occurred through amino and hydroxyl groups [75]. Certain cations and anions, such as Pb^{2+}, Cd^{2+}, Co^{2+}, K^{+}, Ca^{2+}, Ni^{2+}, Al^{3+}, Na^{+}, Cl^{-}, SO_4^{2-}, NO_3^{-}, and PO_4^{3-}, had minimal effects on U(VI) removal, whereas the presence of Th^{4+}, Fe^{2+}, Fe^{3+}, and CO_3^{2-} significantly decreased U(VI) removal efficiency due to competition for binding sites and the formation of soluble U-carbonate complexes [76].

Phytoremediation is a low-cost, eco-friendly technique suitable for treating pollutants at low concentrations [77]. This technique uses plants to absorb, accumulate, or degrade uranium from contaminated environments. Certain plant species, such as Pteris vittata (Chinese brake fern), have been identified for their ability to accumulate uranium in their tissues. Phytoremediation is particularly useful in large, spread-out contaminated areas and offers a more environmentally friendly and less disruptive method than excavation does. Selecting appropriate plant species for U(VI) removal is challenging because of factors such as tolerance, accumulation, kinetics, biomass production, and biomagnification. Among various phytoremediation techniques, phytofiltration using aquatic macrophytes has shown significant promise. A study by Pratas et al. [78] used Callitriche stagnalis, a native plant, for U(VI) accumulation from polluted water. The laboratory experiment revealed an 85.5% removal efficiency at an initial U(VI) concentration of 500 μg/L. The functional groups (amino, carboxyl, sulfhydryl, phosphoryl, and hydroxyl) present in the plant cell wall interact with U(VI), increasing uptake into the plant. However, concentrations above 1000 μg/L led to plant decay, and concentrations exceeding 2000 μg/L caused severe chlorosis. Despite these limitations, phytoremediation is considered effective for treating large volumes of water with low U(VI) concentrations.

4.7 *Nanotechnology for Uranium Removal from Water*

Nanotechnology has shown great promise in addressing the challenge of removing uranium from water, particularly in contaminated groundwater and wastewater systems. The unique properties of nanoparticles, such as their large surface area-to-volume ratio and enhanced reactivity, make them highly effective at adsorbing and immobilizing uranium. Nanomaterials (NMs), including metal oxide nanoparticles (such as Fe_3O_4, TiO_2, and CeO_2), carbon-based materials (e.g., graphene and carbon nanotubes), and composite nanomaterials, have been used extensively for uranium adsorption. These materials possess high surface areas and reactive sites that facilitate the binding of uranium ions. NMs can generally be classified into three categories for elimination purposes, i.e., based on their action as nanoadsorbents, nanomembranes and nanocatalysts. Magnetic nanoparticles, especially those made of iron oxide (Fe_3O_4), are widely used for uranium removal because of their ease of separation from water via an external magnetic field after the removal process. These nanoparticles can be functionalized with various groups to increase their selective affinity for uranium. Nanofiltration (NF) membranes are another promising application of nanotechnology for uranium removal. These membranes, with pore sizes between 1 and 10 nm, are capable of rejecting uranium ions on the basis of size exclusion and charge effects. Nanofiltration membranes can selectively filter out uranium ions from water while allowing other smaller ions and molecules to pass through. The development of novel, high-performance NF membranes with improved selectivity for uranium and resistance to fouling is an ongoing area of research.

Different nanoparticles combine with other materials (e.g., polymers, clays, or carbon) from a nanocomposite to enhance the stability and efficiency of uranium removal. These materials enhance the performance and support the removal of uranium from nanoparticles under diverse environmental conditions. For example, polymeric materials can be combined with metal nanoparticles to improve uranium ion integrity over a long period of use.

4.8 In Situ Remediation Technologies

In situ chemical reduction (ISCR): This technique involves injecting reducing agents into contaminated groundwater, which chemically converts soluble uranium (U(VI)) into an insoluble form (U(IV)), causing the uranium to precipitate and be immobilized in the soil or aquifer.

5 Advances in Transport Containment and Safety

Efficient and safe transportation of uranium involves advanced containment and safety technologies that ensure minimal risk to the environment and human health. The development of advanced materials, such as high-density polyethylene and corrosion-resistant alloys, has increased the structural integrity of transport containers. To minimize radiation exposure, uranium is transported in containers equipped with radiation shielding, often using lead or other dense materials to prevent harmful exposure to surrounding environments.

Modern containers are often fitted with sensors that can detect temperature, pressure, and radiation levels. These sensors provide real-time data that help track the safety of uranium during transit and alert authorities in the case of anomalies.

To monitor uranium shipments and ensure compliance with safety regulations, automated tracking systems such as GPS and RFID tags are now widely used. These systems provide real-time location updates, improving the ability to manage logistics and ensuring that uranium is transported securely.

Advanced monitoring systems, including satellite and drone surveillance, are employed to track uranium shipments and ensure that any deviations from the designated route are flagged immediately, enhancing security measures.

Advanced technologies are now capable of detecting leaks or spills during uranium transport. Automated systems can alert authorities immediately and activate emergency containment protocols, preventing the spread of contamination.

Spill containment systems, such as secondary containment basins and mobile decontamination units, have been developed to rapidly contain and clean up potential leaks or spills during uranium transport.

International bodies such as the International Atomic Energy Agency (IAEA) have established strict safety standards and protocols for the transport of uranium. Advances in compliance technologies help ensure adherence to these standards, including tamper-proof seals, incident reporting systems, and the certification of transport routes.

6 Emerging Trends in Sustainable Practices

Currently, some of the emerging trends and techniques used to remove uranium from water focus on sustainability and green practices. For example:

1. Green Remediation Approach

This approach is highly environmentally friendly, such as the use of biodegradable surfactants or plant-based materials for phytoremediation, and is gaining popularity. These approaches reduce the reliance on toxic or hazardous substances, leading to more sustainable and eco-friendly practices.

2. Circular Economy Approaches:

Uranium waste management and remediation are moving toward circular economy models, where waste products such as uranium mine tailings are repurposed or recycled. For example, certain tailings are being explored for their potential as secondary sources of uranium or other valuable metals, reducing waste and supporting resource recovery.

3. Renewable Energy Powered Remediation:

As part of a push toward sustainability, there is growing interest in using renewable energy sources (such as solar or wind power) to power remediation systems. For example, solar-powered electrokinetic systems or wind energy-driven water pumps for groundwater treatment reduce the carbon footprint of uranium remediation efforts.

4. Sustainable Water Treatment Technologies:

The integration of low-energy, sustainable water treatment methods is an ongoing trend in uranium remediation. Techniques such as passive treatment systems, which rely on natural processes (e.g., bacterial activity and plant absorption), are becoming more common in areas where energy costs or environmental concerns limit the use of more traditional methods.

5. Enhanced Monitoring and Data Analytics:

The use of data analytics and remote sensing technologies to monitor uranium contamination in real time is improving the sustainability of remediation efforts. Drones, sensors, and GIS mapping are being utilized to track contamination levels, the effectiveness of remediation strategies, and potential risks, all of which enable more targeted and efficient interventions.

6. Public and Stakeholder Engagement:

Sustainable practices are also driven by increased public and stakeholder engagement. The communities affected by uranium contamination are becoming more involved in decision-making processes, helping ensure that remediation and transport practices are transparent, socially responsible, and aligned with local environmental priorities.

7. Environmental Remediation:

After mining activities cease, the site undergoes remediation to remove or contain any remaining contaminants. This may involve revegetation of the area, groundwater treatment, and additional stabilization measures to prevent the spread of contamination.

7 Conclusion

Uranium concentrations in ground and surface waters can vary significantly depending on local geological conditions, mining activities, land use (such as phosphate fertilizer application), and hydrological factors. While concentrations in river and stream waters are typically low (approximately 0.3–4 μg/L), elevated levels are often found in areas with uranium mineralization or human activities, such as mining and agricultural runoff. In certain lakes and mine drainage areas, concentrations can be much higher, often exceeding hundreds to thousands of micrograms per liter. Technological innovations in uranium transport and remediation are evolving rapidly, with a strong emphasis on safety, efficiency, and sustainability. Advances in containment technologies, along with new approaches in remediation (such as nanotechnology, bioremediation, and hybrid systems), are significantly improving the effectiveness of uranium cleanup and transport. Additionally, emerging trends in sustainable practices, such as green chemistry, renewable energy use, and circular economy models, are shaping the future of uranium remediation, offering promising solutions for environmental restoration and resource management. These innovations will be key to mitigating the long-term environmental and health risks associated with uranium contamination. In conclusion, nanotechnology offers a wide range of innovative solutions for the removal of uranium from water. As research and development in this field progresses, more efficient, cost-effective, and environmentally friendly methods are expected to emerge for addressing uranium contamination in water sources.

References

1. Balaram V, Rani A, Rathore DPS (2022) Uranium in groundwater in parts of India and world: a comprehensive review of sources, impact to the environment and human health, analytical techniques, and mitigation technologies. Geosyst Geoenviron 1(2):100043
2. Babu MS, Somashekar RK, Kumar SA, Shivanna K, Krishnamurthy V, Eappen KP (2008) Concentration of uranium levels in groundwater. Int J Environ Sci Technol 5:263–266
3. Kurttio P, Harmoinen A, Saha H, Salonen L, Karpas Z, Komulainen H, Auvinen A (2006) Kidney toxicity of ingested uranium from drinking water. Am J Kidney Dis 47(6):972–982
4. Akash S, Sivaprakash B, Raja VV, Rajamohan N, Muthusamy G (2022) Remediation techniques for uranium removal from polluted environment–review on methods, mechanism and toxicology. Environ Pollut 302:119068
5. Amphlett JTM, Sharrad CA, Ogden MD (2018) Extraction of uranium from non-saline and hypersaline conditions using iminodiacetic acid chelating resin Purolite S930+. Chem Eng J 342:133–141
6. Gupta DK, Chatterjee S, Mitra A, Voronina A, Walther C (2019) Uranium and plants: elemental translocation and phytoremediation approaches. In: Uranium in Plants and the Environment, pp 149–161
7. Helal AS, Hemadi M, Lomas JS, Ammar S, Abdelhafiz A, El-Sheikh SM et al (2025) Uranium removal from environmental water and nuclear waste: nanomaterial solutions and their environmental sustainability. Chem Eng J 507:160298
8. Li P, Wang J, Wang Y, Liang J, Pan D, Qiang S, Fan Q (2019) An overview and recent progress in the heterogeneous photocatalytic reduction of U (VI). J Photochem Photobiol C: Photochem Rev 41:100320
9. Muzzarelli RA (2011) Potential of chitin/chitosan-bearing materials for uranium recovery: an interdisciplinary review. Carbohydr Polym 84(1):54–63
10. Newsome L, Morris K, Trivedi D, Atherton N, Lloyd JR (2014) Microbial reduction of uranium (VI) in sediments of different lithologies collected from Sellafield. Appl Geochem 51:55–64
11. Shen J, Schäfer A (2014) Removal of fluoride and uranium by nanofiltration and reverse osmosis: a review. Chemosphere 117:679–691
12. Tatarchuk T, Shyichuk A, Mironyuk I, Naushad M (2019) A review on removal of uranium (VI) ions using titanium dioxide based sorbents. J Mol Liq 293:111563
13. Wang LK, Hung YT, Shammas NK (2009) Handbook of advanced industrial and hazardous wastes treatment. CRC Press
14. Wang T, Zheng X, Wang X, Lu X, Shen Y (2017) Different biosorption mechanisms of Uranium (VI) by live and heat-killed Saccharomyces cerevisiae under environmentally relevant conditions. J Environ Radioact 167:92–99

15. Xie Y, Chen C, Ren X, Wang X, Wang H, Wang X (2019) Emerging natural and tailored materials for uranium-contaminated water treatment and environmental remediation. Prog Mater Sci 103:180–234
16. Banning A (2021) Geogenic arsenic and uranium in Germany: large-scale distribution control in sediments and groundwater. J Hazard Mater 405:124186
17. Cuney M (2010) Evolution of uranium fractionation processes through time: driving the secular variation of uranium deposit types. Econ Geol 105(3):553–569
18. Pękala A (2017) Thorium and uranium in the rock raw materials used for the production of building materials. In: IOP conference series: materials science and engineering, vol 245, 2nd edn. IOP Publishing, p 022033
19. Alam MS, Cheng T (2014) Uranium release from sediment to groundwater: influence of water chemistry and insights into release mechanisms. J Contam Hydrol 164:72–87
20. Chen SB, Zhu YG, Hu QH (2005) Soil to plant transfer of 238U, 226Ra and 232Th on a uranium mining-impacted soil from southeastern China. J Environ Radioact 82(2):223–236
21. Jerden JL Jr, Sinha AK (2006) Geochemical coupling of uranium and phosphorous in soils overlying an unmined uranium deposit: Coles Hill, Virginia. *J Geochem Explor* 91(1–3):56–70
22. Ulrich S, Gillow J, Roberts S, Byer G, Sueker J, Farris K (2019) Hydrogeochemical and mineralogical factors influencing uranium in background area groundwater wells: Grants, New Mexico. J Hydrol Reg Stud 26:100636
23. Langmuir D (1978) Uranium solution-mineral equilibria at low temperatures with applications to sedimentary ore deposits. Geochim Cosmochim Acta 42(6):547–569
24. Stewart BD, Mayes MA, Fendorf S (2010) Impact of uranyl– calcium– carbonato complexes on uranium (VI) adsorption to synthetic and natural sediments. Environ Sci Technol 44(3):928–934
25. Banning A, Rüde TR (2015) Apatite weathering as a geological driver of high uranium concentrations in groundwater. Appl Geochem 59:139–146
26. Guo H, Jia Y, Wanty RB, Jiang Y, Zhao W, Xiu W et al (2016) Contrasting distributions of groundwater arsenic and uranium in the western Hetao basin, Inner Mongolia: implication for origins and fate controls. Sci Total Environ 541:1172–1190
27. Kumar A, Karpe RK, Rout S, Gautam YP, Mishra MK, Ravi PM, Tripathi RM (2016) Activity ratios of 234U/238U and 226Ra/228Ra for transport mechanisms of elevated uranium in alluvial aquifers of groundwater in south-western (SW) Punjab, India. J Environ Radioact 151:311–320
28. Lapworth DJ, Brauns B, Chattopadhyay S, Gooddy DC, Loveless SE, MacDonald AM et al (2021) Elevated uranium in drinking water sources in basement aquifers of southern India. Appl Geochem 133:105092
29. Patnaik R, Lahiri S, Chahar V, Naskar N, Sharma PK, Avhad DK et al (2016) Study of uranium mobilization from Himalayan Siwaliks to the Malwa region of Punjab state in India. J Radioanal Nucl Chem 308:913–918
30. Jindal MK, Pandit SA, Karunakara N et al (2024a) High uranium dose from the groundwater in a granitic terrain in the eastern part of Karnataka, India. J Radioanal Nucl Chem 333:2905–2913. https://doi.org/10.1007/s10967-023-09053-6
31. UNSCEAR (United Nation Scientific Committee on the effect of Atomic Radiations) (2016) Sources, effects and risks of ionizing radiation. United Nations, New York, p 512. https://www.unscear.org/docs/publications/2016/UNSCEAR_2016_Report-CORR.pdf
32. Amrane M, Oufni L (2017) Determination for levels of uranium and thorium in water along Oum Er-Rabia river using alpha track detectors. J Radiat Res Appl Sci 10(3):246–251
33. Smedley PL, Kinniburgh DG (2023) Uranium in natural waters and the environment: distribution, speciation and impact. Appl Geochem 148:105534
34. Regenspurg S, Margot-Roquier C, Harfouche M, Froidevaux P, Steinmann P, Junier P, Bernier-Latmani R (2010) Speciation of naturally-accumulated uranium in an organic-rich soil of an alpine region (Switzerland). Geochim Cosmochim Acta 74:2082–2098

35. Singh P, Rana NPS, Azam A, Naqvi AH, Srivastava DS (1996) Levels of uranium in waters from some Indian cities determined by fission track analysis. Radiat Meas 26(5):683–687
36. Jindal MK, Patni K, Joshi T, Pande AP (2024b) Study of radionuclides in river basins of India: current status, risk assessment, and possible solutions. In: River Basin Ecohydrology in the Indian sub-continent. Elsevier, pp 241–257
37. Smedley PL, Bearcock JM, Fordyce FM, Everett PA, Chenery S, Ellen R (2017) Stream-water geochemical atlas of the Clyde Basin
38. Juanjuan S, Zhigang Y, Bochao X, Wenhua D, Dong X, Xueyan J (2014) Concentrations and fluxes of dissolved uranium in the Yellow River estuary: seasonal variation and anthropogenic (Water-Sediment Regulation Scheme) impact. J Environ Radioact 128:38–46
39. Zielinski RA, Simmons KR, Orem WH (2000) Use of 234U and 238U isotopes to identify fertilizer-derived uranium in the Florida Everglades. Appl Geochem 15(3):369–383
40. Birke M, Rauch U, Lorenz H (2009) Uranium in stream and mineral water of the Federal Republic of Germany. Environ Geochem Health 31:693–706
41. Rango T, Bianchini G, Beccaluva L, Tassinari R (2010) Geochemistry and water quality assessment of Central Main Ethiopian rift natural waters with emphasis on source and occurrence of fluoride and arsenic. J Afr Earth Sci 57:479–491
42. Simpson HJ, Trier RM, Toggweiler JR, Mathieu G, Deck BL, Olsen CR, Hammond DE, Fuller C, Ku TL (1982) Radionuclides in mono lake, California. Science 216:512–514
43. Linhoff BS, Bennett PC, Puntsag T, Gerel O (2011) Geochemical evolution of uraniferous soda lakes in eastern Mongolia. Environ Earth Sci 62:171–183
44. Ladeira ACQ, Gonçalves CR (2007) Influence of anionic species on uranium separation from acid mine water using strong base resins. J Hazard Mater 148:499–504
45. Groudev S, Spasova I, Nicolova M, Georgiev P (2007) Acid mine drainage cleanup in a uranium deposit by means of a passive treatment system. Physicochem Probl Miner Process 41:265–274
46. Ma R, Zheng C, Prommer H, Greskowiak J (2011) Modelling field-scale uranium mass transfer at the Hanford IFRC site. In: 7th international conference on calibration and reliability in groundwater modelling: managing groundwater and the environment, ModelCARE 2009, pp 141–146
47. Mochizuki A, Hosoda K, Sugiyama M (2016) Characteristic seasonal variation in dissolved uranium concentration induced by the change of lake water pH in Lake Biwa, Japan. Limnology 17:127–142
48. Barnes CE, Cochran JK (1990) Uranium removal in oceanic sediments and the oceanic U balance. Earth Planet Sci Lett 97:94–101
49. Henderson GM, Anderson RF (2003) The U-series toolbox for paleoceanography. Rev Mineral Geochem 52:493–531
50. Maynard EA, Hodge HC (1949) Studies of the toxicity of various uranium compounds when fed to experimental animals. In: Voegtlin C, Hodge HC (eds) Pharmacology and toxicology of uranium compounds with a section on the pharmacology and toxicology of fluorine and hydrogen fluoride. McGraw-Hill Book Co., Inc, New York, pp 309–376
51. Berlin M, Rudell B (1986) Uranium. In: Handbook on the toxicology of metals, vol 2. Academic Press, p 2
52. WHO (2011) Uranium in drinking water. WHO Geneva
53. ATSDR (2013) Toxicological profile for uranium. Atlanta, Georgia
54. Faqir Y, Li Z, Gul T, Jiang Z, Yu L, Tan C et al (2025) Uranium's hazardous effects on humans and recent developments in treatment. Ecotoxicol Environ Saf 293:118043
55. Ortega A, Domingo JL, Llobet JM, Tomas JM, Paternain JL (1989) Evaluation of the oral toxicity of uranium in a 4-week drinking-water study in rats. Bull Environ Contam Toxicol 42(6):935–941
56. Gilman AP, Villeuve DC, Secours VE, Yagminas AP, Tracy BL, Quinn JM et al (1998) Uranyl nitrate: 28-day and 91-day toxicity studies in the Sprague-Dawley rat. Toxicol Sci 41(1):117–128

57. Domingo JL, Paternain JL, Llobet JM, Corbella J (1989) The developmental toxicity of uranium in mice. Toxicology 55(1–2):143–152
58. Wrenn ME, Durbin PW, Howard B, Lipsztein J, Rundo J, Still ET, Willis DL (1985) Metabolism of ingested U and Ra. Health Phys 48(5):601–633
59. WHO (2005) Uranium in drinking-water. WHO/SDE/WSH/03.04/118
60. Gandhi TP, Sampath PV, Maliyekkal SM (2022) A critical review of uranium contamination in groundwater: treatment and sludge disposal. Sci Total Environ 825:153947
61. Mollah MYA, Schennach R, Parga JR, Cocke DL (2001) Electrocoagulation (EC)—science and applications. J Hazard Mater 84(1):29–41
62. Raj SK, Carrier AJ, Youden BC, Servos MR, Oakes KD, Zhang X (2024) Electrochemical techniques for uranium extraction from water. Chem Eng J 492:152341
63. Shih MC (2005) An overview of arsenic removal by pressure-drivenmembrane processes. Desalination 172(1):85–97
64. Chen F, Lv M, Ye Y, Miao S, Tang X, Liu Y et al (2022a) Insights on uranium removal by ion exchange columns: the deactivation mechanisms, and an overlooked biological pathway. Chem Eng J 434:134708
65. Kosari MR, Sepehrian H, Salamat AR (2016) Uranium removal from aqueous solution using ion-exchange resin DOWEX® 2x8 in the presence of sulfate anions. Int J Eng 29(12):1677–1683
66. Chen T, Yu K, Dong C, Yuan X, Gong X, Lian J et al (2022b) Advanced photocatalysts for uranium extraction: elaborate design and future perspectives. Coord Chem Rev 467:214615
67. Zhu W, Li X, Wang D, Fu F, Liang Y (2023) Advanced photocatalytic uranium extraction strategies: Progress, challenges, and prospects. Nanomaterials 13(13):2005
68. Wu W, Chen D, Li J, Su M, Chen N (2018) Enhanced adsorption of uranium by modified red muds: adsorption behavior study. Environ Sci Pollut Res 25:18096–18108
69. Bachmaf S, Planer-Friedrich B, Merkel BJ (2008) Effect of sulfate, carbonate, and phosphate on the uranium (VI) sorption behavior onto bentonite. Radiochim Acta 96(6):359–366
70. Henry M, Rinsant D, Maynadié J, Carboni M (2024) Uranium extraction by metal-organic frameworks: advanced materials for new sorption possibilities. Dalton Trans 53:17110–17122
71. Bender J, Lee RF, Phillips P (1995) Uptake and transformation of metals and metalloids by microbial mats and their use in bioremediation. J Ind Microbiol 14:113–118
72. Battista JR (1997) Against all odds: the survival strategies of Deinococcus radiodurans. Ann Rev Microbiol 51(1):203–224
73. Manobala T, Shukla SK, Rao TS, Kumar MD (2019) Uranium sequestration by biofilm-forming bacteria isolated from marine sediment collected from southern coastal region of India. Int Biodeterior Biodegradation 145:104809
74. Manobala T, Shukla SK, Rao TS, Kumar MD (2021) Kinetic modelling of the uranium biosorption by Deinococcus radiodurans biofilm. Chemosphere 269:128722
75. Kazy SK, Sar P, D'Souza SF (2008) Studies on uranium removal by the extracellular polysaccharide of a Pseudomonas aeruginosa strain. Bioremediat J 12(2):47–57
76. Sar P, D'Souza SF (2001) Biosorptive uranium uptake by a pseudomonas strain: characterization and equilibrium studies. J Chem Technol Biotechnol 76(12):1286–1294
77. Naga Jyothi MSV, Gayathri S, Pushparaj Gandhi T, Maliyekkal SM (2021) Dissolved arsenic in groundwater bodies: a short review of remediation technologies. In: Pollution control technologies: current status and future prospects. Springer, Singapore, pp 75–111
78. Pratas J, Paulo C, Favas PJ, Venkatachalam P (2014) Potential of aquatic plants for phytofiltration of uranium-contaminated waters in laboratory conditions. Ecol Eng 69:170–176

Conclusion: Summary and Recommendations for Future Research

Emphasis on Real-World Application: Uranium Solubility in Real-World Samples: A Global Report

Abhishek Joshi, Krishna Pal Singh, R. S. Aswal, and R. C. Ramola

Abstract Uranium, a radionuclide, is highly toxic in nature. Its occurrence at elevated concentrations in different matrices is a worldwide topic of concern. Therefore, the present study is designed to provide a picture of uranium solubility in various environmental matrices. In this context, we highlighted its uses in different fields and sources in aqueous systems. In addition, it also highlights the fluctuations in uranium content at different geological locations, thereby focusing on regions with high-level prevalence. We discuss analytical techniques that are efficiently applied for precise measurement. Emphasis is given to the human health issues posed by the consumption of uranium-contaminated water. We also briefly discuss remedial strategies adopted for uranium removal and future perspectives of this study with advancements toward sustainable practices for uranium-affected zones.

Keywords Uranium solubility · Redox potential · Uranium contamination · Analytical techniques · Remedial strategies

1 Introduction

Uranium is a silvery white heavy metal with a slightly paramagnetic nature. Uranium is generally found in three isotopic states, ^{238}U, ^{235}U, and ^{234}U, in fractional amounts of 99.27, 0.72, and 0.01%, respectively. Owing to the pyrophoric nature of uranium,

A. Joshi · R. C. Ramola
Department of Physics, H.N.B. Garhwal University, Badshahi Thaul Campus, Tehri Garhwal, India

K. P. Singh (✉)
Department of Physics, P.N.G. Govt. P.G. College Ramnagar, Nainital, India

R. S. Aswal
Department of Environmental Sciences, H.N.B. Garhwal University, Badshahi Thaul Campus, Tehri Garhwal, India

M. K. Jindal, K. Patni (eds.), *Water Quality-Uranium Solubility Nexus: Unlocking the Secrets of Uranium Solubility*, Earth and Environmental Sciences Library, https://doi.org/10.1007/978-3-031-98909-4_19

it reacts with cold water. It can be found in water, soil, and air in varying amounts. It easily becomes oxidized in contact with air and occurs mainly in oxidized form in nature. It can be present in almost all soil, rock, water (underground and surface), and plants. This leads to its availability in trace amounts in food and drinking water. Almost all types of radionuclides found in drinking water that are naturally occur. Their concentration in water depends on the nature of the source. Solubility plays a vital role in the toxicology of uranium, affecting the targeted body organs, the type of toxic effects observed, and the methods of excretion. When soluble uranium compounds (such as UO_2) are inhaled and settle in the lungs, they dissolve and quickly absorbed into the bloodstream, allowing for rapid distribution to other body organs or elimination via the urine. Kidneys and bones are particularly vulnerable to the effects of dissolved uranium. Like other heavy metals, uranium can harm kidney tissues, categorizing it as a chemical toxicant. Additionally, dissolved uranium can accumulate in bone and persist for extended periods, potentially delivering a significant radiation dose if sufficiently enriched, thus posing a radiological risk to bone [1, 2]. There are established toxicity levels for insoluble and soluble compounds. Highly soluble uranium salts are known to exhibit significant toxic properties. Typically, hexavalent compounds of uranium are more soluble than their tetravalent compounds [3]. The quick movement of soluble uranium compounds within the body often results in a significant amount being absorbed. These soluble compounds may also penetrate the skin. The severe toxic effects associated with insoluble compounds are caused primarily by lung irradiation from inhaled particles. This material is gradually excreted from the animal's lungs [4].

This chapter explores the complex characteristics of uranium in groundwater, including its origin, geochemical behavior, movement, and health hazards. A comprehensive overview of its geological origins, mobility, and remediation effectiveness is provided. This study investigated various analytical techniques used to detect and quantify uranium. This chapter explores future perspectives and mitigation strategies to address this critical issue. Understanding the intricacies of uranium behavior in groundwater is crucial for developing effective management and remediation strategies to safeguard water resources and public health.

2 Application of Uranium in Different Fields

Uranium is crucial in several applications because of its radioactive characteristics. Although uranium is important in many sectors, it must be treated cautiously because of its radioactive nature and potential environmental hazards. Its future is closely associated with developments in nuclear energy-based technology and efforts for sustainable energy solutions. Some of the key areas for potential applications of uranium are briefly depicted in Table 1.

Table 1 Application of uranium and its isotopes in different sectors

S.N.	Type/Category of Uranium	Beneficiary sector	Application of uranium
1.	^{235}Uranium	Energy	In nuclear reactors as fuel to produce electricity.
2.	Uranium isotopes	Health	In cancer treatment, particularly in radiation therapy, and some compounds for targeted medical therapies.
3.	Enriched uranium	Defense	In nuclear weapon production.
4.	Depleted uranium	Defense	In armour-piercing ammunition.
5.	Uranium isotopes	R&D	In dating rocks and fossils.
6.	Uranium	Defense	In submarines and aircraft carriers.
7.	Uranium compounds	Commercial	In production of fluorescent lights and various chemical processes.

R & D = Research and Development

3 Sources of Uranium in Water and Its Guideline Value

Uranium is a lithophilic element found in almost every type of rock on the Earth's crust, with an average amount of 2.7 ppm [5]. The primary uranium ores are uraninite (UO_2), pitchblende (U_3O_8) and davidite. Water with a low redox potential (anoxia) forms U_4^+ or UO_2^+. U(IV) generally precipitates and is immobile in nature, while UO_2^+ forms soluble complexes [6]. However, in toxic water, UO_2^{2+} [U(VI)] forms stable radially soluble ionic and/or natural complexes with high mobility. It is crucial in uranium transport during the weathering process. The presence of complex ions, such as dissolved inorganic carbon, can increase the mobility and solubility of U(VI) [7–9].

A study conducted in the central district of South Greenland by Armour-Brown et al. [10] reported high pH values in analyzed water samples ascribed to carbonate occurrence and much higher uranium values in such samples. However, the high organic level in stream sediment in this area also suggested that secondary environmental effects were responsible for at least a few very high uranium values. Limited follow-up prospects in this district have shown many uraniferous fractures and faults in granite, eight of which contain primary pitchblende. In addition to the Iddefjord granite, several Precambrian granites may have high uranium contents [11, 12]. Haakonde et al. [13] reported that water infiltrated from mine wastewater dams probably contaminates the potable groundwater of boreholes and shallow wells, resulting in elevated uranium values.

In addition to the mining and processing of uranium, another possible influencing factor for high uranium content in surface water may be surface runoff from broad agricultural farms via the use of phosphate fertilizers around dams and streams [14, 15]. Owing to the ability of bicarbonate to leach uranium from soil and sediment, its formation during water seepage through soil might also be responsible for uranium contamination in groundwater. Annual and seasonal rainfall affects the moisture content [16], redox potential, and potential of hydrogen (pH) value of tailings piles. These characteristics further influence uranium dissolution, migration, and precipitation through water–rock interactions along the sampling profile. In addition, the composition of rock and soil might act as a uranium source in water via dissolution and/or the process of chemical weathering, which is strongly pH dependent [17]. The positive covariability shown by electrical conductivity (EC) and total dissolved solids (TDS) also reveals their influences on uranium levels. Consequently, high EC and TDS indicate increased solubility of uranium in groundwater [18]. Several factors, such as variable degrees of evaporation, mineral weathering, and uranium contents in the host rock, might explain the unevenness in the quantity of uranium in groundwater [19].

Reports have shown that an elevated uranium concentrations (> 5 ppm) have often found in the granites, syenites, pegmatites, acid volconic rocks, and gneiss [20]. The suggested value of uranium in potable water is less than 30 μg/l according to the World Health Organization (WHO). Another guidance level for determining the drinking water radionuclide content is the total indicator dose (TID). It involves the doses received from all radionuclides except radon, tritium, and K-40 [21]. The limit value of TID is 0.1 mSv/y. A daily dose of 100 μg/l uranium can account for 0.1 mSv/y.

4 Environmental Factors Influencing Uranium Prevalence

The primary influences of environmental and climatic elements on uranium occurrence are as follows:

(i) **Geology and rock formations**: Uranium naturally occurs in various geological formations, with granite and sedimentary rocks typically present at relatively high concentrations. For example, the elevated uranium concentrations observed in groundwater samples from the southwestern region of Punjab (India) might be due to leaching from adjoining/basement granite-rich rock formations [22].

(ii) **Sediment transport**: Erosion can transport uranium-containing sediments to downstream areas, which can impact aquatic ecosystems. Previous study [23] have reported that high uranium (> 30 μg/l) contents are linked mainly to alluvial aquifers in Punjab and Rajasthan states, India.

(iii) **Water chemistry**: Acidic conditions (low pH values) generally increase uranium solubility, aiding uranium transport in water. Variations in redox conditions can affect its oxidation state and impact its movement as well as its availability. Post et al. [19] revealed that a lower uranium concentration in a studied water sample might be most likely characterized by a low redox potential and high dissolved ferrous on concentration, indicating that local variability can be high.

(iv) **Climate and precipitation**: Intense rainfall can mobilize uranium by increasing water flow and dissolving additional uranium from landscapes. Elevated temperatures can accelerate chemical processes, potentially enhancing uranium mobility. Stalder et al. [24] noted that the uranium content in water might fluctuate over time because of fluctuations in precipitation, snow melt, or other factors.

(v) **Groundwater contamination**: Uranium can seep into groundwater, particularly in regions with fractured rock, potentially affecting drinking water supplies. For example, the presence of high uranium in phosphate fertilizers might have played a crucial role in uranium contamination of groundwater in the Mansa district of Punjab, India [25].

(vi) **Uranium release and processing**: Mining operations release considerable amounts of uranium into the environment, including radioactive dust, radon gas, and contaminated water, which can greatly affect local ecosystems. The Mexican state of Chihuahua is notable for the presence of natural radionuclides, and the depth of the wells, combined with the activities of a uranium factory, results in increased levels of uranium in the groundwater [26].

(vii) **Disposal of tailings**: Waste products from uranium milling, known as tailings, may contain high levels of uranium and other radioactive substances, pose contamination risks, if not manage appropriately. In Kyrgyzstan, water sourced from artesian and dug wells displays increased concentrations of uranium, which is affected by drainage water originating from mine tailings that have uranium levels exceeding 200 times its standard limit, i.e., 30 μg/l [27, 28].

Table 2 Levels of uranium concentrations in potable water from different countries around the world

Investigated location	Range of U in water (Ppb)	References	Investigated location	Range of U in water (Ppb)	References
Afghanistan	0.5–515.1	[29]	India	<8649	[30]
Argentina	0.04–11.0	[31]	Iraq	200–3500	[32]
Burundi	<700	[19]	Ireland	>300	[33]
Canada	0.05–4.21	[34]	Myanmar	<1–45	[35]
China	550–3360	[17]	Norway	>2.5–750	[36]
Finland	<0.01–25	[37]	Switzerland	0.05–92.02	[24]
Greece	0.015–1.4	[38]	Turkey	0.24–17.65	[39]
Germany	75.1–890	[40]	United States	≤5400	[41]
Hungary	<1–10.60	[42]	Zambia	110.03–135.30	[13]

5 Disparity in Uranium Content in Different Geological Regions

Variations in uranium concentrations can be obtained on the basis of their geological setting. Table 2 contains the details of the variation in uranium levels in groundwater for different regions of the world and including India. Studies conducted in different parts of India have suggested that the variation levels of uranium in Rajasthan (northern part), Punjab (southwestern part), and Haryana (western part) are very high in groundwater samples because of acid volcanic rocks and granites with high heat production. However, some other studies reported that a few samples had values that surpassed the suggested level of the WHO. Moreover, studies in various parts of India reported uranium concentrations within the safe limits.

Many studies from various parts of the world have reported different uranium levels in various sources of water (Table 2). Higher levels of uranium concentrations in groundwater (tubewells, borewells, and handpumps) have been found throughout the world, as recommended by the WHO. Studies carried out in Argentina, Canada, Finland, Greece, Hungary, and Turkey reported uranium levels within the safer limit of the WHO. Very high uranium concentrations were detected in water samples from Afghanistan, Burundi, China, Germany, India, Iraq, Ireland, Norway, Switzerland, the United States, and Zambia.

6 Instrumental Techniques Used for Uranium Measurement

New developments in instrumental analytical methods are making it possible to analyze smaller samples more precisely and accurately via multielement and isotope analysis [43]. Some of the most extensively used analytical methods for measuring the amounts of uranium and its isotopes in water samples are as follows:

(i) **Inductively coupled plasma–optical emission spectrometry (ICP–OES)**: ICP–OES is one of the most extensively used analytical methods for measuring uranium and its isotopic concentration in water samples. The studies performed previously via this method are referenced here [17, 44, 45].
(ii) **Inductively coupled plasma–mass spectrometry (ICP–MS):** The identification of uranium in collected water via ICP–MS (quadrupole) is a highly efficient and precise technique. There are various models of this technique (ICP-MS-7500cx, ICP-MS with a CETAC 260 autosampler, ICP-MS, Agilent 7700 Series, etc.) used by different researchers worldwide [11, 22, 37, 46–48].
(iii) **High-resolution inductively coupled plasma–mass spectrometry (HR-ICP-MS)**: HR-ICP-MS employs a magnetic sector mass spectrometer to differentiate ions according to their m/e (mass-to-charge) ratio, which enhances discrimination and minimizes potential interference at designated resolution settings. This method has been used by various researchers [49–51].

(iv) **Fluorimetry** (laser & light emitting diode): Laser-induced fluorimetry [14, 39, 40] and light-emitting diode fluorimetry [52–55] have been used in various studies in India as well as by researchers in other parts of the world.

(v) **Raman spectroscopy**: Raman spectroscopy is the most effective analytical approach for identifying the composition (chemical) of numerous substances (environmental), such as natural water. This approach is highly effective for swiftly detecting uranium in collected water samples and is therefore used by various researchers [56, 57].

(vi) **Other analytical methods:**

Gamma-ray spectrometry is primarily used for detecting the gamma rays emitted by uranium decay products, allowing for the determination of total uranium activity in a sample.

X-ray fluorescence (XRF) can be utilized for elemental analysis of uranium in solid samples, providing information about the concentration of uranium without the need for isotopic identification.

7 Health Implications Associated with the Consumption of Uranium-Contaminated Water

Uranium accounts for chemical and radiological toxicity. The radiological risk generally occurs because of prolonged exposure to insoluble uranium compounds or enriched uranium. However, natural or depleted uranium commonly accounts for chemical toxicity [58]. The uranium ingested through food or water is absorbed in the gastrointestinal tract in a small fraction (0.3–7.8%) [59]. The amount of uranium absorbed can be deposited in different organs, such as bones (primarily), with fewer deposits in the liver, kidney, and other soft tissues [60, 61].

Once uranium is introduced into the human body, elimination from the system involves a complex deposition and transfer process. An experimental study suggested that 90% of the uranium deposited in bones is left there after 1.5 years [62]. Research (population: 50; uranium value: <1–781 μg/l) conducted in Nova Scotia, Canada, revealed a statistically important association between uranium content in water and glucose excretion for both males and females [63]. Similarly, another study (population: 325; uranium concentration: 0.001–1920 μg/l) carried out in southern Finland revealed a correlation between exposure to uranium and increased fractional excretion of calcium [64] as well as with increased tubular function. However, researchers have noted that increased leakage of calcium into urine could increase vulnerability to osteoporosis. Seldén et al. [65] carried out research in Sweden and reported a strong correlation between urine uranium and well-known uranium levels. However, there was no clear evidence that uranium in potable water was nephrotoxic at its measured (<0.2–470 μg/l) values. Moreover, there are signs of nephrotoxicity when the urine uranium level is used as an exposure marker.

8 Remedial Technologies for Uranium Removal

Several techniques have been developed for the remediation and mitigation of uranium contamination in groundwater. The choice of a particular treatment or remediation technique depends on the properties of the contaminant as well as the environmental conditions. The available remediation techniques can be categorized broadly into three groups: physical, chemical, and bioremediation, as shown in Fig. 1.

Apart from the abovementioned techniques, a few mitigation techniques can also be used for risk management due to uranium exposure through potable water. It involves practices such as source control (identification, isolation, or treatment of potential sources), water treatment at the point of use (use of an RO system), blending of fresh water with low-level uranium water, and opting for an alternate water source. Remedial measures largely rely on its speciation, the occurrence of other pollutants, and the composition of aqueous solutions. Among these methods, the use of membrane separation technologies, defluorination, and biochar are a few commonly used strategies. In addition, phytoremediation can also be an efficient removal technique for uranium-affected zones. An inclusive approach involving experts working in different fields is needed.

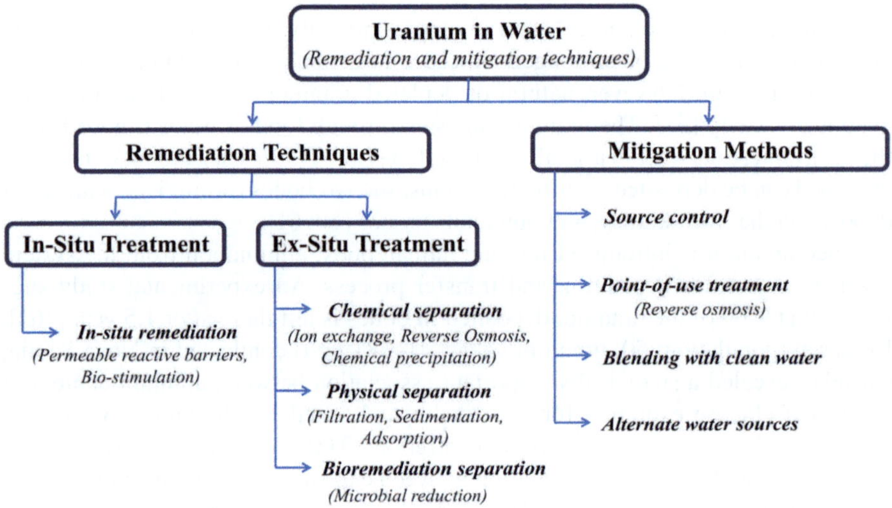

Fig. 1 Classification of remediation and mitigation techniques for uranium in potable water

9 Future Perspectives

The successful management of uranium contamination and its health-related implications basically depend on frequent monitoring of potable surface and groundwater sources through advanced and sophisticated instruments. This monitoring should be a seasonal and spatial basis for a longer period, i.e., decades, to obtain a more comprehensive picture of the water quality data trends. Compared with short-term exposure, long-term exposure to uranium can toxicologically affect inhabitants. Therefore, fluctuations in precipitation, snow melt, or other factors should be considered in uranium-based studies [24]. This type of study could be helpful in the long-term monitoring of uranium in aqueous media.

In addition, the application of real-time water quality monitoring systems may provide continuous data on different physicochemical parameters of water bodies of interest. Some of these physicochemical variables might influence the uranium concentration. A comprehensive geochemical investigation of uranium-affected regions can provide a broad picture of uranium distribution considering both aspects, i.e., site and depth. Artificial intelligence (AI) and machine learning (ML) techniques can be helpful in identifying water quality trends and associations established between different pairs of analyzed water quality characteristics. Moreover, collaboration among biologists, engineers, geologists, hydrologists, and technologists is essential to determine the root cause and possible solution. Finally, the active participation of stakeholders would also be helpful in tackling this issue.

10 Concluding Remarks

This chapter provides a comprehensive overview of uranium, focusing on its presence in groundwater. This chapter highlights the diverse applications of uranium, from nuclear energy and medical treatments to military uses and scientific research, while emphasizing the crucial need for careful handling due to its radioactive nature and potential environmental hazards. The fundamental properties of uranium, including its isotopic form, chemical reactivity, and natural occurrence in various geological formations, have been explored. Different analytical techniques used for uranium measurement are reviewed in this chapter. The potential health implications of consuming uranium-contaminated water, ranging from kidney damage to potential links with cancer, were addressed. Finally, various remedial technologies for uranium removal, encompassing physical, chemical, and bioremediation approaches, are outlined. The future perspectives are explored, emphasizing the need for advanced monitoring systems, real-time data collection, and the application of artificial intelligence (AI) and machine learning (ML) for predictive modeling and effective management of uranium in groundwater. Understanding the complex dynamics of uranium in groundwater is crucial for developing effective strategies to protect water resources and ensure public health.

References

1. Stannard, J; Baalman Jr, RW (1988), Radioactivity and health: a history. https://www.osti.gov/biblio/6608787
2. Eidson AF (1994) The effect of solubility on inhaled uranium compound clearance: a review. Health Phys 1:1–14. https://journals.lww.com/health-physics/abstract/1994/07000/The_Effect_of_Solubility_on_Inhaled_Uranium.1.aspx
3. Coetzee H, Rademeyer M (2006) Lead isotope ratios as a tracer for contaminated waters from uranium mining and milling. In: Uranium in the environment: mining impact and consequences. Springer, Berlin, Heidelberg, pp 663–670. https://doi.org/10.1007/3-540-28367-6_67
4. Sakaguchi T, Nakajima A (1987) Recovery of uranium from seawater by immobilized tannin. Sep Sci Technol 22:1609–1623. https://doi.org/10.1080/01496398708058421
5. Siegel MD, Bryan CR (2003) Environmental geochemistry of radioactive contamination. TrGeo 9:612. https://doi.org/10.1016/B0-08-043751-6/09049-6
6. Gascoyne M, Barber JH (1992) The mobility of U, Ra and Rn in a granitic batholith on the Canadian shield. Radiochim Acta 58–59:281–284. https://doi.org/10.1524/RACT.1992.5859.2.281/MACHINEREADABLECITATION/RIS
7. Langmuir D (1978) Uranium solution-mineral equilibria at low temperatures with applications to sedimentary ore deposits. Geochim Cosmochim Acta 42:547–569. https://doi.org/10.1016/0016-7037(78)90001-7
8. Ching-kuo Daniel Hsi D, Langmuir (1985) Adsorption of uranyl onto ferric oxyhydroxides: application of the surface complexation site-binding model. GeCoA 49:1931–1941. https://doi.org/10.1016/0016-7037(85)90088-2
9. Kohler M, Curtis GP, Meece DE, Davis JA (2004) Methods for estimating adsorbed uranium(VI) and distribution coefficients of contaminated sediments. Environ Sci Technol 38:240–247. https://doi.org/10.1021/ES0341236
10. Brown A, Steenfelt A, Kunzzenorf H (1983) Uranium districts defined by reconnaissance geochemistry in South Greenland. J Geochem Explor 19:127–145
11. Killeen PG, Heier KS (1975) A uranium and thorium enriched province of the Fennoscandian shield in southern Norway. Geochim Cosmochim Acta 39:1515–1524. https://doi.org/10.1016/0016-7037(75)90153-2
12. Banks JA (1995) Multicultural education and curriculum transformation. J Negro Educ 64:390. https://doi.org/10.2307/2967262
13. Haakonde T, Yabe J, Choongo K, Chongwe G, Islam MS (2020) Preliminary assessment of uranium contamination in drinking water sources near a uranium mine in the Siavonga District, Zambia, and associated health risks. Mine Water Environ 39:735–745. https://doi.org/10.1007/S10230-020-00731-5
14. Bhangare RC, Tiwari M, Ajmal PY, Sahu SK, Pandit GG (2013) Laser flourimetric analysis of uranium in water from Vishakhapatnam and estimation of health risk. Radiat Prot Environ 36:128–132. https://doi.org/10.4103/0972-0464.137478
15. Patni K, Pande C, Pande AP, Tewari G, Joshi T (2020) Distribution of naturally occurring uranium and other heavy toxic elements in selected spring water samples of Pithoragarh District, Uttarakhand, India. SN Appl Sci 2. https://doi.org/10.1007/s42452-020-03947-w
16. Gaskova OL, Boguslavsky AE (2013) Groundwater geochemistry near the storage sites of low-level radioactive waste: implications for uranium migration. Proc Earth Planet Sci 7:288–291. https://doi.org/10.1016/J.PROEPS.2013.03.130

17. Ma W, Gao B, Guo Y, Sun Z, Zhang Y, Chen G, Zhu X, Zhang C (2020) Occurrence and distribution of uranium in a hydrological cycle around a uranium mill tailings pond, Southern China. Int J Environ Res Publ Heal 17:773. https://doi.org/10.3390/IJERPH17030773
18. Yumnam G, Alam W (2023) Chemotoxicity and radiotoxicity risk of naturally occurring uranium in groundwater resources of Manipur Valley, India. Environ Eng Sci 40:136–152. https://doi.org/10.1089/EES.2022.0165;REQUESTEDJOURNAL:JOURNAL:EES;JOURNAL:JOURNAL:EES;WGROUP:STRING:PUBLICATION
19. Post VEA, Vassolo SI, Tiberghien C, Baranyikwa D, Miburo D (2017) High uranium concentrations in groundwater in Burundi. Proc Earth Planet Sci 17:524–527. https://doi.org/10.1016/J.PROEPS.2016.12.132
20. Aekerblom G, Lindgren J (1997) Mapping of groundwater radon potential. https://inis.iaea.org/search/search.aspx?orig_q=RN:29009838
21. Östergren I, Åkerblom G, Ek B (2005) Mätningar av naturlig radioaktivitet i och från filter vid några vat, Inis.Iaea.OrgI Östergren, G Åkerblom, BM Ekinis.Iaea.Org. https://inis.iaea.org/records/vsf05-nwz63/files/42022582.pdf
22. Bajwa BS, Kumar S, Singh S, Sahoo SK, Tripathi RM (2017) Uranium and other heavy toxic elements distribution in the drinking water samples of SW-Punjab, India. J Radiat Res Appl Sci 10:13–19. https://doi.org/10.1016/j.jrras.2015.01.002
23. Pant D, Keesari T, Sharma D, Rishi M, Singh G, Jaryal A, Sinha UK, Dash A, Tripathi RM (2017) Study on uranium contamination in groundwater of Faridkot and Muktsar districts of Punjab using stable isotopes of water. J Radioanal Nucl Chem 313:635–639. https://doi.org/10.1007/S10967-017-5284-0
24. Stalder E, Blanc A, Haldimann M, Dudler V (2012) Occurrence of uranium in Swiss drinking water. Chemosphere 86:672–679. https://doi.org/10.1016/j.chemosphere.2011.11.022
25. Sharma T, Litoria PK, Bajwa BS, Kaur I (2021) Appraisal of groundwater quality and associated risks in Mansa district (Punjab, India). Environ Monit Assess 193:1–21. https://doi.org/10.1007/S10661-021-08892-8
26. Colmenero F, Fernández AM, Cobos J, Timón V (2018) Thermodynamic properties of uranyl-containing materials based on density functional theory. J Phys Chem C 122:5254–5267. https://doi.org/10.1021/ACS.JPCC.7B12341/SUPPL_FILE/JP7B12341_SI_001.PDF
27. Corcho Alvarado JA, Balsiger B, Röllin S, Jakob A, Burger M (2014) Radioactive and chemical contamination of the water resources in the former uranium mining and milling sites of Mailuu Suu (Kyrgyzstan). J Environ Radioact 138:1–10. https://doi.org/10.1016/J.JENVRAD.2014.07.018
28. Bjørklund G, Semenova Y, Pivina L, Dadar M, Rahman MM, Aaseth J, Chirumbolo S (2020) Uranium in drinking water: a public health threat. Arch Toxicol 94:1551–1560. https://doi.org/10.1007/S00204-020-02676-8
29. Kato M, Azimi MD, Fayaz SH, Shah MD, Hoque MZ, Hamajima N, Ohnuma S, Ohtsuka T, Maeda M, Yoshinaga M (2016) Uranium in well drinking water of Kabul, Afghanistan and its effective, low-cost depuration using Mg-Fe based hydrotalcite-like compounds. Chemosphere 165:27–32. https://doi.org/10.1016/J.CHEMOSPHERE.2016.08.124
30. Jindal MK, Pandit SA, Karunakara N, Chandrashekara MS, Kumara S, Kumar V, Salim D, Srinivasan R (2024) High uranium dose from the groundwater in a granitic terrain in the eastern part of Karnataka, India. J Radioanal Nucl Chem 333:2905–2913. https://doi.org/10.1007/S10967-023-09053-6
31. Bomben AM, Equillor HE, Oliveira AA (1996) 226Ra and natural uranium in Argentinian bottled mineral waters. Radiat Prot Dosim 67:221–224. https://doi.org/10.1093/oxfordjournals.rpd.a031820
32. Ali AH, Mheemeed AK, Hassan HI (2015) Concentration measurements of uranium, thorium and their daughter products in water produced from and near oil fields in north of Iraq using SSNTRD's passive method. J Radioanal Nucl Chem 303:959–965. https://doi.org/10.1007/S10967-014-3573-4
33. Papageorgiou F, McDermott F, Van Acken D (2022) Uranium in groundwaters: insights from the Leinster granite, SE Ireland. Appl Geochem 139:105236. https://doi.org/10.1016/J.APGEOCHEM.2022.105236

34. OMEE (1996) Effluent monitoring and effluent limits—electric power generation sector, Ontario Regulation 2i5/95 made under the Environmental Protection Act, Ontario
35. Bacquart T, Frisbie S, Mitchell E, Grigg L, Cole C, Small C, Sarkar B (2015) Multiple inorganic toxic substances contaminating the groundwater of Myingyan Township, Myanmar: Arsenic, manganese, fluoride, iron, and uranium. Sci Total Environ 517:232–245. https://doi.org/10.1016/J.SCITOTENV.2015.02.038
36. Frengstad B, Midtgård Skrede AK, Banks D, Reidar Krog J, Siewers U (2000) The chemistry of Norwegian groundwaters: III. The distribution of trace elements in 476 crystalline bedrock groundwaters, as analysed by ICP-MS techniques. Sci Total Environ 246:21–40. https://doi.org/10.1016/S0048-9697(99)00413-1
37. Turtiainen T, Kostiainen E, Hallikainen A (2011) 210Pb and 210Po in Finnish cereals. J Environ Radioact 102:438–442. https://doi.org/10.1016/J.JENVRAD.2010.09.009
38. UNSCEAR (2008) Sources and effects of ionizing radiation: United Nations Scientific Committee on the Effects of Atomic Radiation (UNSCEAR), Report to the General Assembly, with Scientific Annexes: United Nations, Volume I, United Nations
39. Kumru MN (1997) Possible uranium rich areas in the Aegean region of Turkey. Appl Radiat Isot 48:295–299. https://doi.org/10.1016/S0969-8043(96)00047-4
40. Baumann N, Arnold T, Lonschinski M (2011) TRLFS study on the speciation of uranium in seepage water and pore water of heavy metal contaminated soil. J Radioanal Nucl Chem 291:673–679. https://doi.org/10.1007/S10967-011-1389-Z
41. Nolan J, Weber KA (2015) Natural uranium contamination in major U.S. aquifers linked to nitrate. Environ Sci Technol Lett 2:215–220. https://doi.org/10.1021/acs.estlett.5b00174
42. Csondor K, Baják P, Surbeck H, Izsák B, Horváth Á, Vargha M, Erőss A (2020) Transient nature of riverbank filtered drinking water supply systems - a new challenge of natural radioactivity assessment. J Environ Radioact 211:106072. https://doi.org/10.1016/J.JENVRAD.2019.106072
43. Balaram V (2021) New frontiers in analytical techniques—opportunities and challenges in geochemical research. J Geol Soc India 97(4):331–334. https://doi.org/10.1007/S12594-021-1690-6
44. Popov BB, Najman S, Karapetkovska Hristova V, Ahmad A (2020) Inductively coupled plasma-optical emission spectroscopy (icp-oes) approach for the determination of heavy metals, metalloid and trace element in soil and vegetables. Indian Horticult J 4(2):98–104. https://www.indianjournals.com/ijor.aspx?target=ijor:ihj1&volume=4&issue=2&article=005
45. Chandrasekaran K, Karunasagar D, Arunachalam J (2011) Dispersive liquid–liquid micro extraction of uranium(VI) from groundwater and seawater samples and determination by inductively coupled plasma–optical emission spectrometry and flow injection–inductively coupled plasma mass spectrometry. Anal Methods 3:2140–2147. https://doi.org/10.1039/C1AY05329A
46. Rani A, Singh S, Duggal V, Balaram V (2013) Uranium estimation in drinking water samples from some areas of Punjab and Himachal Pradesh, India using ICP-MS. Radiat Prot Dosim 157:146–151. https://doi.org/10.1093/RPD/NCT112
47. Babu MNS, Somashekar RK, Kumar SA, Shivanna K, Krishnamurthy V, Eappen KP (2008) Concentration of uranium levels in groundwater. Int J Environ Sci Technol 5:263–266. https://doi.org/10.1007/BF03326020
48. Saha S, Das K, Hidaka H, Kimura K, Chakraborty PP, Hayasaka Y (2016) Detrital zircon geochronology (U-Pb SHRIMP and LA-ICPMS) from the Ampani Basin, Central India: implication for provenance and mesoproterozoic tectonics at East Indian cratonic margin. Precambrian Res 281:363–383. https://doi.org/10.1016/j.precamres.2016.06.011
49. Cheng Z, Zheng Y, Mortlock R, Van Geen A (2004) Rapid multi-element analysis of groundwater by high-resolution inductively coupled plasma mass spectrometry. Anal Bioanal Chem 379:512–518. https://doi.org/10.1007/S00216-004-2618-X

50. Satyanarayanan M, Balaram V, Hussin MS, Jemaili MAR, Rao TG, Mathur R, Dasaram B, Ramesh SL (2007) Assessment of groundwater quality in a structurally deformed granitic terrain in Hyderabad, India. Environ Monit Assess 131:117–127. https://doi.org/10.1007/S10661-006-9461-9
51. Mittal S, Rani A, Mehra R, Balaram V, Satyanarayanan M, Sawant SS (2017) Assessment of uranium in correlation with physico-chemical properties of drinking water of Northern Rajasthan. J Geol Soc India 90:233–238. https://doi.org/10.1007/S12594-017-0704-X
52. Duggal V, Sharma S, Singh A (2021) Toxicological risk and age-dependent radiation dose assessment of uranium in drinking water in Southwest-Central Districts of Haryana State, India. Groundw Sustain Dev 13:100577. https://doi.org/10.1016/J.GSD.2021.100577
53. Prakash R, Bhartariya KG, Singh S, Singh K, Rajak M, Kaushik YB (2020) Uranium and its correlation with other geogenic contaminants in ground water of Ganga Yamuna Doab, Fatehpur District, Uttar Pradesh, India. J Geol Soc India 95:359–365. https://doi.org/10.1007/S12594-020-1444-X
54. Panwar P, Joshi A, Singh KP, Prasad M, Mehra R, Sahoo SK, Ramola RC (2024) Distribution of uranium and selected toxic heavy metals in drinking water of Garhwal Himalaya, India. J Radioanal Nucl Chem 333:2927–2935. https://doi.org/10.1007/S10967-023-08998-Y
55. Yadav M, Singh KP, Dutt S, Mehta V, Kaur J, Ramola RC (2024) Risk assessment due to ingestion of uranium and radon in the drinking water samples of the Budhakedar region in Garhwal Himalaya, India. J Radioanal Nucl Chem 333:3065–3074. https://doi.org/10.1007/s10967-023-09262-z
56. Ho DML et al (2015) Raman spectroscopy of uranium compounds and the use of multivariate analysis for visualization and classification. Forens Sci Int 251:61. https://www.sciencedirect.com/science/article/pii/S0379073815000900
57. Lin DHM et al (2014) The use of different dispersive Raman spectrometers for the analysis of uranium compounds. Vibrat Spectrosc 73:102–110. https://www.sciencedirect.com/science/article/pii/S0924203114000903
58. Missimer TM, Teaf C, Maliva RG, Danley-Thomson A, Covert D, Hegy M (2019) Natural radiation in the rocks, soils, and groundwater of Southern Florida with a discussion on potential health impacts. Int J Environ Res Publ Heal 16:1793. https://doi.org/10.3390/IJERPH16101793
59. Wrenn ME, Durbin PW, Howard B, Lipsztein J, Rundo J, Still ET, Willis DL (1985) Metabolism of ingested U and Ra. Health Phys 48:601–633. https://doi.org/10.1097/00004032-198505000-00004
60. ICRP (1994) Human respiratory tract model for radiological protection. ICRP Publication 66. https://journals.sagepub.com/doi/pdf/10.1177/ANIB_24_1-3
61. Leggett RW (1994) Basis for the ICRP's age-specific biokinetic model for uranium. Health Phys 67:589–610. https://doi.org/10.1097/00004032-199412000-00002
62. Brugge D, De Lemos JL, Oldmixon B (2005) Exposure pathways and health effects associated with chemical and radiological toxicity of natural uranium: A review. Rev Environ Health 20:177–193. https://doi.org/10.1515/REVEH.2005.20.3.177
63. Zamora M, Tracy B, Zielinski JM, Meyerhof DP, Moss MA (1998) Chronic ingestion of uranium in drinking water: a study of kidney bioeffects in humans. Toxicol Sci. https://academic.oup.com/toxsci/article-abstract/43/1/68/1644334
64. Kurttio P, Auvinen A, Salonen L, Saha H, Pekkanen J, Mäkeläinen I, Väisänen SB, Penttilä IM, Komulainen H (2002) Renal effects of uranium in drinking water. Environ Health Perspect 110:337–342. https://doi.org/10.1289/EHP.02110337
65. Seldén AI, Lundholm C, Edlund B, Högdahl C, Ek BM, Bergström BE, Ohlson CG (2009) Nephrotoxicity of uranium in drinking water from private drilled wells. Environ Res 109:486–494. https://doi.org/10.1016/J.ENVRES.2009.02.002

Water Quality–Uranium Solubility Nexus: Integrated Approaches, Challenges, and Future Directions

Manoj Kumar Jindal

Abstract The water quality–uranium solubility nexus represents a complex, integrated challenge requiring multidisciplinary investigation. This chapter explores the path to understanding this nexus and the related challenges, along with appropriate future directions. Uranium, a naturally occurring radioactive element, has highly variable solubility in water and is influenced by a wide range of chemical, geological, hydrological, and anthropogenic factors. Its mobility and complexity in terms of solubility are governed by its hexavalent uranyl form, which poses serious environmental and public health risks due to its presence in diverse water sources, including groundwater, surface water, and even treated effluents. The solubility of uranium is governed by multiple interdependent water quality parameters, surrounding environmental conditions, and dynamic geochemical interactions. This chapter highlights all such cases that influence uranium behavior in both natural and human-impacted settings. This work further discusses transformation mechanisms, competitive ion binding, time-dependent leaching, and the role of radioactive decay products in modifying uranium mobility in the environment. Recognizing the multifactorial complexity of uranium solubility, this chapter advocates for integrated and interdisciplinary strategies. A holistic approach is essential for ensuring long-term water security and safeguarding both ecological and human health in regions affected by uranium contamination.

Keywords Water quality–uranium solubility nexus · Uranium solubility · Water quality · Sustainable approaches for uranium management · Integrated approaches · Uranium mobility · Future directions for uranium studies in water

M. K. Jindal (✉)
Center for Sustainable Drinking Water, Indian Institute of Technology Jodhpur, Jodhpur, India

M. K. Jindal, K. Patni (eds.), *Water Quality-Uranium Solubility Nexus: Unlocking the Secrets of Uranium Solubility*, Earth and Environmental Sciences Library, https://doi.org/10.1007/978-3-031-98909-4_20

1 The Chemistry of Water and Uranium

The existence of water is well known, and its scientific properties are complex. Water is universally recognized for its fundamental role in sustaining life, supporting ecosystems, regulating climate, and maintaining environmental cycles and various environmental matrices. Its chemical nature makes it highly interactive with a vast range of natural and anthropogenic substances, including minerals, rocks, gases, and contaminants. Among the many elements that water interacts with, uranium has become a growing concern because of its radioactive nature and potential impact on human and environmental health. Uranium, a naturally occurring radioactive element, has been found in both surface water and groundwater across diverse geologies. Its presence in water is not limited to a specific source; it can occur in oceans, rivers, lakes, shallow aquifers, and deep boreholes. Numerous global studies have reported uranium contamination in drinking water sources [1], making uranium contamination a pressing issue for researchers, governments, and public health institutions.

The core concern lies in uranium's mobility and solubility, especially in its hexavalent oxidation state (U(VI)), where it forms uranyl ion complexes that are highly soluble in water. This mobility allows uranium to travel long distances, contaminate remote water sources, and remain bioavailable for uptake by plants, animals, and humans. Understanding what controls uranium solubility in water is, therefore, essential for both environmental monitoring and the development of effective mitigation strategies. It can be easily transported due to its mobility, raising questions about how it dissolves in water, what environmental or chemical conditions facilitate this process, and what other substances may act as catalysts or binding agents.

2 Geological Sources and Natural Availability

Uranium is present in almost all rock types, although in varying concentrations. Granitic rocks, sedimentary formations such as shale and phosphate-rich strata, and metamorphic terrains are known to host relatively high concentrations of uranium [2, 3]. These geological sources contribute to uranium release into the hydrosphere, especially through weathering, leaching, and mineral dissolution. However, the presence of uranium in rocks alone does not guarantee its solubility. The availability of soluble uranium in water depends heavily on environmental and chemical conditions, including the presence of complexing agents, redox conditions, pH levels, and contact time between water and mineral substrates.

3 Chemical Factors Influencing Uranium Solubility

The solubility of uranium in water is governed by its oxidation state and interactions with water quality parameters. In its oxidized U(VI) form (e.g., uranyl ions, UO_2^{2+}), uranium is highly soluble and mobile, whereas the reduced U(IV) state precipitates as insoluble oxides such as UO_2. Several water quality parameters influence the redox transition and mobility of uranium: pH: Water pH greatly affects uranium speciation. Acidic or highly basic environments may enhance the dissolution of uranium-bearing minerals, whereas a near-neutral pH may limit solubility unless other agents are present. Carbonate and bicarbonate ions: These ions can bind strongly with uranyl ions to form soluble uranyl-carbonate complexes (e.g., $UO_2(CO_3)_3^{4-}$), which greatly increase uranium mobility in groundwater systems [4]. Sulfates and Phosphates: Similar to carbonates, these anions can form complex structures with uranium and increase its solubility, similar to nitrate [5]. Redox Conditions: In oxidizing environments, uranium exists as U(VI) and remains soluble. In reducing environments, it is likely to precipitate as U(IV) and become immobile. Organic matter: Natural organic compounds can either bind uranium and promote its transport or reduce it to insoluble forms, depending on their nature and environmental conditions [6].

The uranium–water quality relationship is not isolated. Rather, uranium interacts with a suite of physicochemical parameters in complex ways. These include total dissolved solids (TDS), which are a general indicator of all ionic contents in water, including potentially uranium. High TDS is often correlated with relatively high uranium levels, especially in aquifers rich in carbonate or sulfate ions. Electrical Conductivity (EC): High EC values may indicate the presence of dissolved species conducive to uranium mobility. Hardness: Hard water often contains high calcium and magnesium levels, which may compete with uranium for carbonate complexation. Alkalinity: High bicarbonate alkalinity stabilizes soluble uranyl-carbonate complexes. The interplay of these parameters creates a dynamic matrix where uranium solubility depends on the net effect of competing influences. For example, high carbonate contents may override reducing conditions, maintaining U(VI) solubility even at neutral pH. Such complexity necessitates site-specific assessments to predict uranium behavior in aquatic systems. All of these parameters interact dynamically, creating a matrix of conditions that either promote or inhibit uranium solubility. Each water body must therefore be assessed individually to understand the net impact of these factors.

4 Competitive Binding and Complex Interactions

The solubility of uranium is dependent not only on its own chemical behavior but also on its ability to compete with other metals and ions for binding agents. For example, heavy metals, such as cadmium, lead, or iron, may outcompete uranium for available ligands, thereby suppressing uranium solubility. Some metals may outcompete uranium for ligand binding because of their relatively high reactivity, reducing uranium mobility. Some metals may outcompete uranium for ligand binding because of their relatively high reactivity, reducing uranium mobility. Radioactive Families: Uranium decay products (such as radium or thorium) may also interact with available ligands, altering the chemical environment in ways that affect uranium behavior. Catalysts and Inhibitors: Some dissolved compounds or particulate matter may facilitate uranium precipitation or interfere with complexation, thus influencing its solubility pathway. Others may form co-complexes that stabilize uranium in the aqueous phase, increasing its transport potential. Hence, the presence of heavy metals and other radionuclides can influence uranium solubility in multiple ways.

5 Hydrogeological Mobility and Dilution Effects

Even in areas with low geological uranium contents, soluble uranium may still be detected in water because of hydrological processes. As water flows from uranium-rich zones to uranium-poor zones, it can transport dissolved uranium over long distances. This mobility challenges the assumption that high uranium levels at a given location always reflect local geological conditions. Conversely, water movement also leads to dilution effects. In regions where uranium concentrations are high, the inflow of fresh water, particularly after rainfall or snowmelt, can reduce uranium levels through dilution. This dual nature of water as both a transporter and a diluter makes its role in uranium solubility highly complex. Uranium mobility in the environment is governed by both chemical speciation and hydrological pathways. In natural systems, flowing water (streams, rivers) can transport dissolved uranium across long distances from the source. Infiltration and recharge may carry uranium into deeper groundwater systems. Dilution occurs where low-uranium waters mix with contaminated sources, often lowering concentrations but not eliminating risk. Conversely, evaporation in arid zones may concentrate uranium in residual water, particularly in shallow aquifers or surface ponds. The distance from the source often determines whether uranium concentrations remain elevated or diminish. However, as several case studies have shown, transported uranium can create new hotspots far from the original geological or anthropogenic source.

6 Rainfall and Seasonal Variability

Rainfall events can either enhance or suppress uranium solubility, depending on local environmental conditions: dilution scenario: rainwater can dilute uranium concentrations in surface water or shallow groundwater. Mobilization scenario: Acid rain or runoff from contaminated surfaces can leach uranium from rocks and soils, increasing its concentration in water bodies. Post-monsoon studies in several regions have shown both trends, indicating that seasonal variation is a significant factor in uranium dynamics and must be incorporated into long-term monitoring programs [7].

7 Climate Influence: Seasonal and Long-Term Effects

Climate variability, particularly precipitation and temperature changes, strongly influences uranium solubility: Monsoon and post-monsoon periods often have dilution effects or may increase uranium leaching. In dry seasons or during drought, limited recharge and increased evaporation may lead to uranium concentrations in stagnant waters. Climate change also presents long-term concerns, as groundwater levels shift and water chemistry evolves, and uranium solubility may increase in regions previously considered safe. A rise in temperature can alter microbial activity, redox conditions, and even mineral solubility, all of which affect the chemical behavior of uranium.

8 Transformation Processes and Inhibitory Conditions

Uranium solubility is also strongly influenced by transformation reactions in the aqueous phase. For example, uranium may initially dissolve under favorable conditions but later precipitate due to the formation of insoluble complexes or changes in pH and redox conditions. This dynamic equilibrium between soluble and insoluble forms is heavily dependent on reaction inhibitors or competing chemical reactions occurring simultaneously in the water body. In certain scenarios, uranium may interact with dissolved oxygen, sulfate reducers, or other reductants, shifting its oxidation state from soluble U(VI) to insoluble U(IV). This reduction pathway, often observed in anaerobic or stagnant conditions, can significantly reduce uranium mobility. Similarly, some environmental catalysts may draw oxygen away from uranium oxidation processes, thereby limiting their ability to remain in solution. Moreover, the adsorption of uranium onto mineral surfaces such as iron oxides, clays, or organic matter particles is another key factor in reducing solubility. In such cases, uranium exists not in the dissolved phase but as a sorbed species on the surface of solids, potentially remobilized under changing environmental conditions.

9 Anthropogenic Contributions and Indirect Loadings

In addition to naturally occurring uranium, human activities contribute significantly to the presence of uranium in water bodies. For example, agricultural runoff often contains phosphate-based fertilizers that are known to host uranium as impurities

[8]. The extensive use of these fertilizers in certain regions can gradually increase uranium concentrations in groundwater through leaching. Industrial discharges and uranium mining waste can also contaminate surrounding water bodies, sometimes hundreds of kilometers away from the original source, due to long-range hydrological transport. Improper disposal of uranium-rich materials in landfills or open land areas allows for the infiltration of uranium-bearing leachates into aquifers. In some cases, uranium is not directly introduced but rather appears in water as a byproduct of treatment systems. For example, reverse osmosis (RO) units, while effective at removing uranium from drinking water, may concentrate it in reject water or saltwater streams, leading to local hotspots of uranium contamination if disposal is not properly managed [9].

10 Time-Dependent Solubility and Leaching Dynamics

Uranium solubility is not only a matter of chemistry but also of time and exposure. Certain environmental settings may lead to slow uranium release through prolonged water–rock interactions. This is particularly true in confined aquifers where the water residence time is high. Short-term exposure may not cause measurable uranium dissolution, but over months or years, minerals may gradually release uranium into water, especially under fluctuating pH and redox conditions. Conversely, in highly dynamic systems such as surface water bodies, dilution and flushing often prevent long-term uranium buildup unless there is a continuous source. Understanding these time-dependent processes is essential for designing monitoring protocols, as a single-time-point measurement may not capture the full picture of uranium mobility and risk.

11 Radioactive Decay and Solubility Impacts

As a radioactive element, uranium undergoes self-decay over time, transforming into various daughter isotopes, such as thorium, radium, and polonium. While uranium itself may be relatively stable under given conditions, these daughter products may react with different water constituents, contribute to overall radioactivity and alter local chemistry in ways that feedback into uranium's own solubility profile. In some scenarios, decay can enhance solubility by breaking down solid matrices or altering oxidation states. In addition, it may lead to secondary mineral formation that traps uranium in insoluble forms.

12 Holistic Approach and Sustainable Uranium Mitigation

To understand the real-world implications of uranium solubility in groundwater, Fig. 1 presents a case study from the Chikkaballapur and Kolar districts of Karnataka, where extremely high uranium concentrations have been reported. According to Jindal et al. [10], uranium levels exceeding 3000 μg/L were observed, with the highest concentration reaching 8649 μg/L, a value significantly above global and national safety guidelines [14]. This dataset has been compared with other major reports from India to highlight regional variability and geological influences. For instance, Sahoo et al. [13] reported 55,554 samples, including 54,618 from groundwater and 936 from surface water, with a maximum uranium concentration of 4918 μg/L in groundwater. Similarly, Coyte et al. [12] compiled data from 68 studies, encompassing 324 primary samples and 4769 secondary data points, and reported a maximum of 2074.8 μg/L. Additionally, the Central Groundwater Board [11, 15] analyzed 14,377 and 11,445 shallow aquifer groundwater samples, with the highest concentration recorded as 2876 μg/L. The significantly high value reported by Jindal et al. [10] is noteworthy not only because of its magnitude but also because it originates from a region geologically similar to those cited by CGWB [11] and Coyte et al. [12], suggesting a strong geological basis for uranium mobilization in these zones. This highlights the need for site-specific evaluations of uranium behavior, especially in uranium-prone regions.

The health risks posed by such elevated uranium concentrations are not only chemical but also radiological. Figure 2 (adapted from Ma et al. [16]) illustrates the pathways and effects of uranium in the human body, emphasizing the dual hazard from both chemical toxicity and ionizing radiation. In such high-concentration scenarios, conventional treatment technologies such as reverse osmosis may prove inadequate. Research by Srinivasan et al. [9] has shown that RO systems may fail to effectively reduce uranium levels when concentrations exceed 1000 μg/L. Moreover, RO systems tend to generate reject water with uranium concentrations two to three times higher than those of the input water, posing secondary contamination risks if not managed properly.

Addressing uranium contamination in groundwater, such as regions such as the Chikkaballapur district of Karnataka, requires a comprehensive and holistic approach grounded in natural, community-based solutions. This involves a detailed study of all available local water bodies, assessing both their quantity and quality to evaluate their potential to meet the annual water demands of local communities. An essential component of this approach is the integration of time series data to understand the historical patterns of water availability and quality and to ensure long-term sustainability. In addition to traditional sources, alternative strategies such as rainwater harvesting must be promoted. These methods, when implemented with community participation, can enhance local water security without introducing

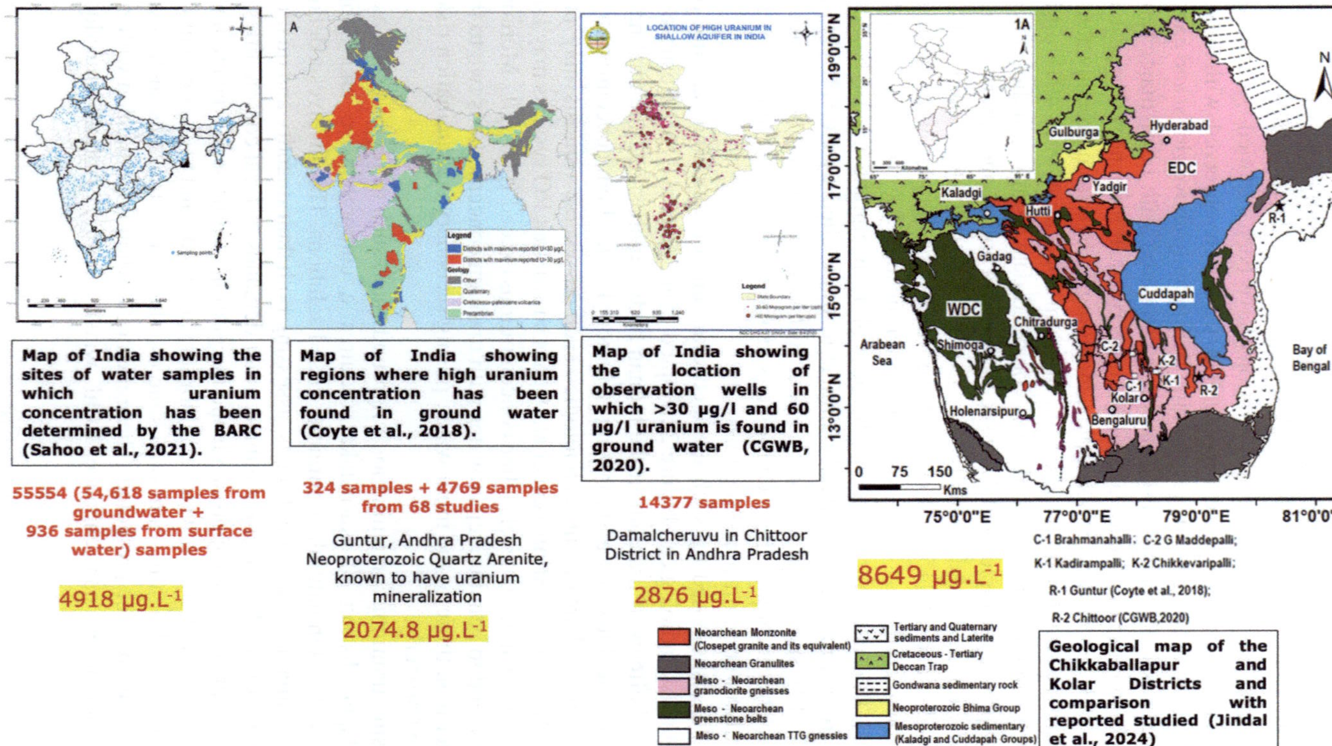

Fig. 1 Example illustrating high uranium concentrations with comparative analysis of reported studies to understand the impact of source and geological factors. Reproduced from Jindal et al. [10] with permission from Springer Nature; additional data from open sources [11–13]

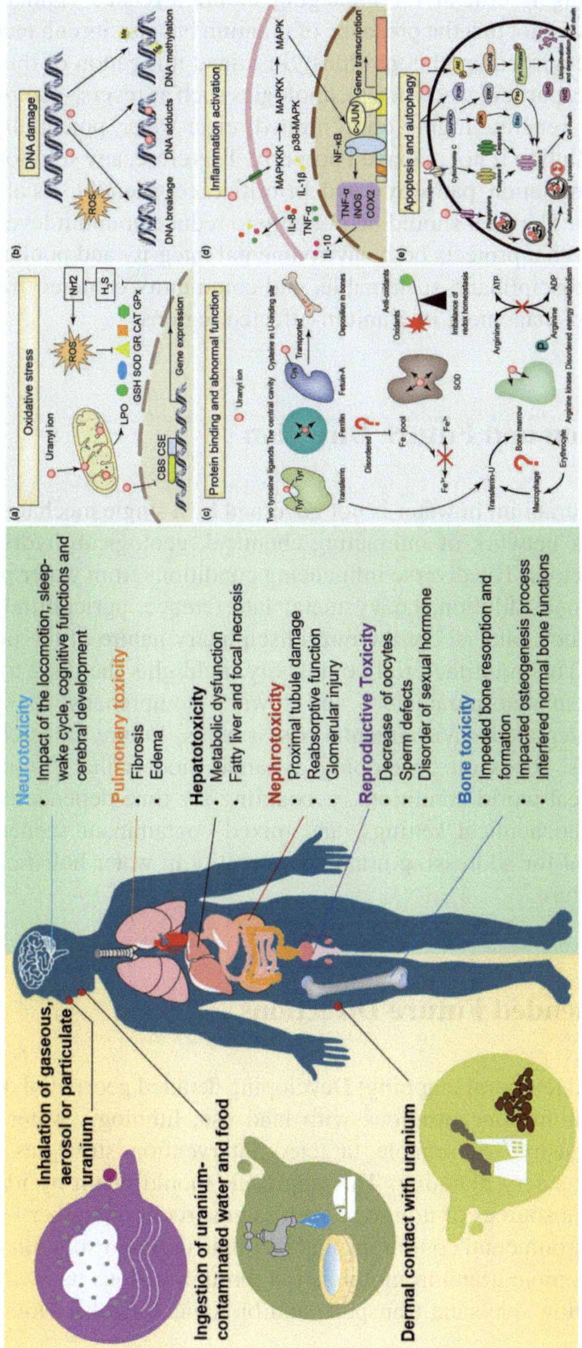

Fig. 2 Effects of uranium in the human body and associated functions. Reproduced from Ma et al. [16] with permission from Elsevier

secondary contamination risks. As emphasized by Srinivasan et al. [17], this type of integrated planning is crucial for managing uranium-related challenges in such regions. The issue is not just the presence of uranium but also its enhanced solubility under favorable environmental conditions. Sustained mitigation of this contamination is a challenge, particularly when technologies such as reverse osmosis, although widely used, can generate highly concentrated reject water, potentially worsening local uranium pollution if not managed properly. Therefore, any solution must avoid secondary contamination pathways and prioritize ecofriendly, locally adaptable treatment options. The goal should be not only to reduce uranium levels but also to do so in a manner that protects both environmental integrity and public health. This calls for an interdisciplinary, sustainable, and community-centered framework for long-term water management in uranium-affected regions.

13 Conclusion and Future Direction

The solubility of uranium in water is not governed by a single mechanism or source but by a complex network of interacting chemical, geological, hydrological, and anthropogenic factors. The diverse influencing conditions from water pH and redox potential to seasonal dilution, heavy metal interference, agricultural inputs, and industrial byproducts underscore the multidisciplinary nature of the uranium solubility challenge. This multifactorial complexity highlights the need to move away from studying isolated parameters and toward comprehensive, system-based approaches or integrated environmental assessments. Future research and policy development must focus on developing dynamic models that simulate uranium behavior under real-world conditions, accounting for time-dependent transformations, variable geochemical settings, and mixed contaminant scenarios. Various aspects are crucial for addressing uranium solubility in water holistically and sustainably in the future.

14 Recommended Future Directions

1. Geospatial and temporal mapping: Developing detailed geospatial databases that integrate uranium concentrations with land use, lithology, water quality, and population density will enable targeted intervention strategies and region-specific water safety planning. This approach should begin by identifying the primary natural sources of uranium, such as host rocks and other favorable geological or environmental conditions that facilitate uranium solubility. These conditions can promote uranium mobilization through various pathways, including hydrological flow, physical transport, and biological uptake. Notably, uranium

consumed by living organisms can be reintroduced into the environment through urine or other biological excretions, further contributing to its movement across environmental matrices. Mapping must also consider how uranium moves through hydrological systems, food chains, and human activities, such as through urine excretion, wastewater discharge, and agricultural reuse. Understanding this broader movement requires tracking both natural and secondary anthropogenic pathways. Establishing baseline maps will be useful for future interventions and risk prediction. Recognizing the host sources and favorable conditions at the outset is essential for understanding the uranium–water quality nexus.

2. Real-Time Monitoring and Smart Sensors: Advances in smart sensors and field-deployable devices have the potential to facilitate opportunities for real-time uranium detection in water. Such tools can monitor rapid fluctuations in uranium concentrations and their correlation with environmental parameters, enabling early warning systems for contamination. However, practical challenges remain. The cost, scalability, and adaptability, particularly in rural or resource-constrained regions, must be addressed to make real-time sensing a globally viable solution.
3. Predictive Modeling Using Machine Learning: Considering the complexity and number of influencing factors, machine learning and artificial intelligence (AI)-based models can play crucial roles in identifying patterns, predicting contamination hotspots, and optimizing monitoring networks. Artificial intelligence can greatly enhance our ability to understand and predict uranium behavior. Machine learning models can process large datasets encompassing historical uranium levels, seasonal trends, climatic variability, host rock profiles, and co-contaminant behavior. Future models should aim to integrate not only environmental conditions but also competitive ion interactions, transport mechanisms, and human-induced changes, creating dynamic simulations that guide targeted interventions. Integrating these multidimensional parameters enhances the accuracy and local relevance of predictions, making them more actionable for risk management and resource planning.
4. Holistic Policy Frameworks: Policies need to evolve from static thresholds to context-sensitive regulatory systems that reflect actual exposure risks. Land-use planning, groundwater extraction control, and contamination zoning must be integrated into water quality governance, especially in uranium-prone regions. Additionally, identifying and mapping high-risk areas is crucial for long-term monitoring and preparedness. Risk assessments must consider localized geochemical behavior, population vulnerability, and historical contamination trends.
5. Community Engagement and Awareness: As uranium contamination can gradually develop and often go unnoticed, public engagement is essential. Educating local communities about safe water practices, source selection, and early warning indicators can significantly reduce long-term exposure risks. A major challenge in addressing uranium solubility is the lack of accessible, integrated data, which hampers the scientific community's ability to take a historical and holistic approach. To overcome this, a collaborative framework involving industrial

stakeholders, commercial entities, government agencies, and academic researchers is crucial. All relevant datasets directly related to uranium levels and those indirectly affecting uranium mobility (e.g., land use, chemical discharge, water treatment data, and hydrological movement) should be consolidated into a centralized freely accessible platform. This would allow researchers and policymakers to better understand uranium solubility patterns, model its transport behavior, and design effective, long-term preventive strategies. Such a community-driven and interdisciplinary approach is essential to move beyond short-term fixes and toward sustainable environmental solutions.

6. Water Treatment Innovations: Research should prioritize the development of low-cost, locally adaptable treatment technologies that not only remove uranium efficiently but also ensure the safe disposal of uranium-enriched residues. Promising innovations such as biosorption, nanofiltration, and selective complexation should be further scaled and optimized for use in rural and semiurban areas, where access to centralized treatment is limited. While methods such as reverse osmosis are currently employed for uranium removal, they present long-term sustainability concerns. RO may not completely eliminate uranium from water and often generates a concentrated reject stream that contains 2–3 times higher uranium levels than the original input. This can ultimately increase the uranium burden on the environment if the waste is not properly managed. As such, RO may not be a viable long-term solution in regions with already elevated uranium concentrations. To address this issue sustainably, holistic treatment strategies that are effective, adaptable, and environmentally safe are needed. In particular, adsorbent-based technologies must include plans for the disposal or regeneration of used materials, as the improper handling of uranium-laden adsorbents could lead to secondary pollution risks. Future innovations in water treatment should therefore emphasize circular use, waste minimization, and ecological safety, making sustainable uranium removal a priority for both immediate and long-term water security.
7. Coupling with climate models: Climate change directly impacts water chemistry through altered recharge patterns, rainfall intensity, and evaporation rates. These changes influence uranium mobility and solubility. Future studies must integrate climate models with uranium risk assessments, particularly in arid and semiarid zones where groundwater resources are both limited and vulnerable. Predictive tools that combine hydrogeochemical simulations with climatic forecasts will be key to anticipating emerging hotspots and planning future mitigation efforts.
8. Interdisciplinary Collaboration: Uranium solubility is not solely a chemical issue; it is a multidisciplinary challenge requiring input from geologists, chemists, hydrologists, environmental scientists, public health professionals, engineers, and policymakers. Effective solutions demand collaborative frameworks across institutions and sectors both nationally and globally. Cross-disciplinary cooperation will be vital for translating science into policy, optimizing monitoring strategies, and developing sustainable treatment systems.

The uranium–water nexus is not just a scientific problem but also a societal responsibility. As nations invest in clean water access, nuclear energy, and climate resilience, understanding the solubility behavior of uranium becomes increasingly important not only for environmental integrity but also for the safety and well-being of millions. The insights presented in this chapter are a step toward that understanding, but much remains to be explored. Through sustained research, informed policymaking, and community participation, we can manage uranium solubility not only as a hazard but also as a solvable challenge in our shared environment in the future.

References

1. UNSCEAR (United Nation Scientific Committee on the effect of Atomic Radiations) (2016) Sources, effects and risks of ionizing radiation. New York, United Nations, p 512. https://www.unscear.org/docs/publications/2016/UNSCEAR_2016_Report-CORR.pdf
2. Nash JT (1979) Uranium and thorium in granitic rocks of northeastern Washington and northern Idaho, with comments on uranium resource potential. U.S. Geological Survey Open-File Report 79-233. https://pubs.usgs.gov/of/1979/0233/report.pdf
3. Sheldon RP (1959) Geochemistry of uranium in Phosphorites and black shales of the Phosphoria formation, Contributions to Geochemistry U.S. Geological Survey Bulletin 1084-D. https://pubs.usgs.gov/bul/1084d/report.pdf
4. Stanley DM, Wilkin RT (2019) Solution equilibria of uranyl minerals: role of the common groundwater ions calcium and carbonate. J Hazard Mater 377:315–320. https://doi.org/10.1016/j.jhazmat.2019.05.101
5. Bonotto DM, Wijesiri B, Goonetilleke A (2019) Nitrate-dependent uranium mobilisation in groundwater. Sci Total Environ 693:133655. https://doi.org/10.1016/j.scitotenv.2019.133655
6. Spirakis CS (1996) The roles of organic matter in the formation of uranium deposits in sedimentary rocks. Ore Geol Rev 11(1–3):53–69. https://doi.org/10.1016/0169-1368(95)00015-1
7. Devaraj N, Panda B, Chidambaram S, Prasanna MV, Singh DK, Ramanathan AL, Sahoo SK (2021) Spatio-temporal variations of uranium in groundwater: implication to the environment and human health. Sci Total Environ 775:145787. https://doi.org/10.1016/j.scitotenv.2021.145787
8. Singh G, Kainth HS, Singh G, Rani N, Duggal H, Upmanyu A et al (2022) Measurement of uranium in phosphate fertilizers for groundwater contamination employing X-ray and γ-ray spectroscopic techniques. J Radioanal Nucl Chem 331(4):1715–1722. https://doi.org/10.1007/s10967-022-08269-2
9. Srinivasan R, Pandit SA, Khatei G, Karunakara N, Kumara KS, Riotte J et al (2022) Reverse osmosis units in groundwater based public water supply system in rural Eastern Karnataka, India: an analysis. Curr Sci. https://doi.org/10.18520/cs/v123/i12/1493-1498
10. Jindal MK, Pandit SA, Karunakara N, Chandrashekara MS, Kumara S, Kumar V et al (2024) High uranium dose from the groundwater in a granitic terrain in the eastern part of Karnataka, India. J Radioanal Nucl Chem 333(6):2905–2913. https://doi.org/10.1007/s10967-023-09053-6
11. Central Ground Water Board (CGWB) report (2020) Uranium occurrence in shallow aquifers in India. http://cgwb.gov.in/WQ/URANIUM_REPORT_2019-20.pdf
12. Coyte RM, Jain RC, Srivastava SK, Sharma KC, Khalil A, Ma L, Vengosh A (2018) Large-scale uranium contamination of groundwater resources in India. Environ Sci Technol Lett 5(6):341–347. https://doi.org/10.1021/acs.estlett.8b00215

13. Sahoo SK, Jha SK, Jha VN, Patra AC, Kulkarni MS (2021) Survey of uranium in drinking water sources in India: interim observations. Curr Sci 120(9):1482–1490. https://doi.org/10.18520/cs/v120/i9/1482-1490
14. WHO (2012) Uranium in drinking water. Background document for development of WHO guidelines for drinking-water quality. WHO, Geneva, pp 1–21
15. Central Ground Water Board (CGWB) report (2024) Annual Ground Water Quality Report 2024. Ministry of Jal Shakti, Department of Water Resources, River Development and Ganga Rejuvenation, Government of India. https://cgwb.gov.in/cgwbpnm/public/uploads/documents/17363272771910393216file.pdf
16. Ma M, Wang R, Xu L, Xu M, Liu S (2020) Emerging health risks and underlying toxicological mechanisms of uranium contamination: lessons from the past two decades. Environ Int 145:106107. https://doi.org/10.1016/j.envint.2020.106107
17. Srinivasan R, Pandit SA, Jindal MK, Ganesan G, Ramkumar KD, Naregundi K et al (2024) Sustainable solution for drinking water supply in rural India affected by groundwater pollution. Curr Sci 127(6):719. https://doi.org/10.18520/cs/v127/i6/719-728

MIX
Papier aus verantwortungsvollen Quellen
Paper from responsible sources
FSC® C105338

If you have any concerns about our products,
you can contact us on
ProductSafety@springernature.com

In case Publisher is established outside the EU,
the EU authorized representative is:
Springer Nature Customer Service Center GmbH
Europaplatz 3, 69115 Heidelberg, Germany

Printed by Libri Plureos GmbH
in Hamburg, Germany